Reviews of Physiology, Biochemistry and Pharmacology 158

Reviews of

158 Physiology Biochemistry and Pharmacology

Editors
S.G. Amara, Pittsburgh • E. Bamberg, Frankfurt
B. Fleischmann, Bonn • T. Gudermann, Marburg
S.C. Hebert, New Haven • R. Jahn, Göttingen
W.J. Lederer, Baltimore • R. Lill, Marburg
A. Miyajima, Tokyo • S. Offermanns, Heidelberg
R. Zechner, Graz

With 14 Figures

Springer

ISBN 978-3-642-09089-9 e-ISBN 978-3-540-71791-1

ISSN 0303-4240

Springer is a part of Springer Science+Business Media

springer.com

Editor: Simon Rallison, London
Desk Editor: Anne Clauss, Heidelberg

Cover: WMXDesign GmbH, Heidelberg

Rev Physiol Biochem Pharmacol (2006)
DOI 10.1007/112_0604

E. I. Christensen · R. Nielsen

Role of megalin and cubilin in renal physiology and pathophysiology

Published online: 28 September 2006

Abstract Megalin and cubilin are endocytic receptors highly expressed in the endocytic apparatus of the renal proximal tubule. These receptors appear to be responsible for the tubular clearance of most proteins filtered in the glomeruli. Cubilin is a peripheral membrane protein, and therefore it does not have an endocytosis signaling sequence. It appears that megalin is responsible for internalization of cubilin and its ligands in addition to internalizing its own ligands. The proteinuria observed in megalin-deficient mice, in dogs lacking functional cubilin, and in patients with distinct mutations of the cubilin gene illustrates the importance of the receptors.

Introduction

The cells of the renal proximal tubule have a high capacity for uptake of proteins filtered in the glomeruli, reflected by an extensively developed endocytic apparatus and the normal very low urinary excretion of plasma proteins. The cutoff molecular mass for glomerular filtration of plasma proteins under physiological conditions is generally considered to be in the range of 60 kDa, corresponding approximately to the molecular mass of serum albumin. However, the diversity of proteins filtered is large and includes even larger proteins such as transferrin (81 kDa). Thus, the ability to recognize a large number of different proteins and at the same time provide for an efficient uptake must be a prerequisite for the molecular apparatus handling the reabsorption of these proteins. Two multiligand endocytic receptors, megalin and cubilin, appear to be largely responsible for the proximal tubular reabsorption of protein (Christensen and Birn 2002). The two receptors are coexpressed along the endocytic and recycling pathway in these cells and apparently collaborate in the uptake of ligands (Christensen and Birn 2001; Verroust et al. 2002) (Table 1).

E. I. Christensen (✉) · R. Nielsen
University of Aarhus, Department of Cell Biology, Institute of Anatomy,
University Park, Building 234, 8000 Aarhus C, Denmark
e-mail: eic@ana.au.dk · Tel.: +45-8942-3057 · Fax: +45-8619-8664

Table 1 Ligands for megalin and cubilin

Megalin	Cubilin
Vitamin carrier proteins	
Folate binding protein (Birn et al. 2005)	
Retinol binding protein (Christensen et al. 1999)	
Transcobalamin-vitamin B_{12} (Moestrup et al. 1996)	Intrinsic factor-vitamin B_{12} (Seetharam et al. 1997)
Vitamin D binding protein (Nykjaer et al. 1999)	Vitamin D binding protein (Nykjaer et al. 2001)
Other carrier proteins	
Albumin (Cui et al. 1996)	Albumin (Birn et al. 2000a; Zhai et al. 2000)
Heavy metallothionein (Klassen et al. 2004)	
Hemoglobin (Gburek et al. 2002)	Hemoglobin (Gburek et al. 2002)
Myoglobin (Gburek et al. 2003)	Myoglobin (Gburek et al. 2003)
Lactoferrin (Willnow et al. 1992)	Transferrin (Kozyraki et al. 2001)
Liver-type fatty acid binding protein (Oyama et al. 2005)	
Neutrophil gelatinase-associated protein (Hvidberg et al. 2005)	
Odorant binding protein (Leheste et al. 1999)	
Sex hormone binding globulin (Hammes et al. 2005)	
Transthyretin (Sousa et al. 2000)	
Lipoproteins	
Apolipoprotein B (Stefansson et al. 1995a)	Apolipoprotein A1 (Kozyraki et al. 1999)
Apolipoprotein E (Willnow et al. 1992)	High-density lipoprotein (Hammad et al. 1999)
Apolipoprotein H (Moestrup et al. 1998b)	
Apolipoprotein J/clusterin (Kounnas et al. 1995)	
Apolipoprotein M (Faber et al. 2006)	
Hormones, hormone precursors, and signaling proteins	
Angiotensin II (Gonzalez-Villalobos et al. 2005)	
Bone morphogenic protein 4 (Spoelgen et al. 2005)	
Epidermal growth factor (Orlando et al. 1998)	
Insulin (Orlando et al. 1998)	
Leptin (Hama et al. 2004)	
Parathyroid hormone (Hilpert et al. 1999)	
Prolactin (Orlando et al. 1998)	
Sonic hedgehog protein (McCarthy et al. 2002)	
Thyroglobulin (Zheng et al. 1998)	
Enzymes and enzyme inhibitors	
α_1-Microglobulin (Leheste et al. 1999)	
α-Amylase (Birn et al. 2000b)	
Lipoprotein lipase (Kounnas et al. 1993)	
Lysozyme (Orlando et al. 1998)	
Plasminogen (Kanalas and Makker 1991)	
Plasminogen activator inhibitor type 1 (Stefansson et al. 1995b)	
Plasminogen activator inhibitor type 1-urokinase (Moestrup et al. 1993)	
Plasminogen activator inhibitor type 1-tissue plasminogen activator (Moestrup et al. 1993; Willnow et al. 1992)	
Pro-urokinase (Stefansson et al. 1995b)	

Table 1 (continued)

Megalin	Cubilin
Immune and stress response-related proteins	
Advanced glycation end products (Saito et al. 2003)	
β_2-Microglobulin (Orlando et al. 1998)	
Immunoglobulin light chains (Birn et al. 2003)	Immunoglobulin light chain (Batuman et al. 1998)
Pancreatitis-associated protein (Leheste et al. 1999)	Clara cell secretory protein (Burmeister et al. 2001)
Receptors and transmembrane proteins	
Cubilin (Moestrup et al. 1998a)	Megalin (Moestrup et al. 1998a)
Cation-independent mannose-6-phosphate receptor (Yammani et al. 2002)	Cation-independent mannose-6-phosphate receptor (Yammani et al. 2002)
Transcobalamin II-B_{12} receptor (Yammani et al. 2003)	AMN (Fyfe et al. 2004)
Drugs and toxins	
Aminoglycosides (Moestrup et al. 1995)	
Polymyxin B (Moestrup et al. 1995)	
Aprotinin (Moestrup et al. 1995)	
Trichosanthin (Chan et al. 2000)	
Others	
Ca^{2+} (Christensen et al. 1992)	
Cytochrome *c* (Orlando et al. 1998)	
Receptor-associated protein (RAP) (Christensen et al. 1992; Orlando et al. 1992; Willnow et al. 1992)	Receptor-associated protein (RAP) (Birn et al. 1997)
Seminal vesicle secretory protein II (Ranganathan et al. 1999)	

After binding of extracellular ligands at the apical membrane the receptors cluster into clathrin-coated pits; ligands are internalized into coated vesicles and subsequently into early and late endosomes. The interior of these vesicles is acidified because of proton pumps localized in the membrane causing dissociation of ligands from the receptors. The receptors are recycled to the apical membrane via so-called dense apical tubules (Christensen 1982), whereas the ligands are transferred to lysosomes for degradation of the protein. However, transport of megalin ligands from the apical to the basal pole (transcytosis) has been reported in other cell types like, for example, thyroglobulin in thyroid cells (Marino et al. 2000).

Evidence accumulated over the past decade suggests that the reabsorption of protein in the renal proximal tubule not only reduces the overall protein content of the final urine, but also provides uptake and thereby conservation of specific substances serving essential functions within the kidney and other tissues. This includes the recovery of several vitamins and ions filtered in complex with carrier proteins such as vitamin D binding protein (Nykjaer et al. 1999, 2001), retinol binding protein (Christensen et al. 1999), transcobalamin (Moestrup et al. 1996), and transferrin (Kozyraki et al. 2001), which not only minimizes urinary losses of the corresponding vitamins and iron but also provides substrate for renal metabolism of these substances, notably the renal activation of vitamin D.

Increased tubular reabsorption of specific proteins as seen in different diseases has been implicated in the progression of chronic renal disease characterized by proteinuria. Traditionally, proteinuria has been divided into glomerular and tubular proteinuria. Defects giving rise to tubular proteinuria originate in the proximal tubule and result in the excretion of proteins that normally are reabsorbed very efficiently by the proximal tubule by receptor-mediated endocytosis. Glomerular proteinuria diseases have their origin in defects of the glomerular filtration barrier and thereby introduce proteins, which are either normally not filtered or only filtered to a limited amount, into the glomerular ultrafiltrate. In other types of overload proteinuria, the pathogenesis is found outside the kidney. Examples include myoglobinuria, hemoglobinuria, and multiple myeloma. Although the capacity for reabsorption in the proximal tubule is very high, a high load of protein may result in the excretion not only of these proteins but also, because of competitive events for receptor binding, of carrier proteins and their substrates that are normally reabsorbed and processed in the proximal tubule. The extensive reabsorption of such an overload of filtered protein not only leads to tubular damage but may also subsequently induce tubular interstitial damage. Thus the focus is attracted toward the mechanisms that mediate and regulate the proximal tubule uptake of protein. This paper reviews the structure, expression, and functional properties of megalin and cubilin, focusing on their involvement in normal proximal tubule protein reabsorption as well as their implication in renal pathophysiology.

Molecular structure and ligands

Megalin

Megalin, also known as gp330, is a 600-kDa (nonglycosylated 517 kDa) endocytic receptor and was originally identified as the target antigen of Heymann nephritis (Kerjaschki and Farquhar 1982, 1983), which is an experimental model of membranous glomerulonephritis. The multifunctional endocytic receptor belongs to the low-density lipoprotein (LDL) receptor family (Raychowdhury et al. 1989), and determination of the primary structure revealed a 4600-aa protein that has an amino acid similarity of 77% between rat and human. The human megalin gene has been localized to chromosome 2q24–q31. Megalin (Fig. 1) is characterized by a large extracellular domain including four clusters of cysteine-rich, complement-type ligand binding motifs, a single transmembrane domain, and a short cytoplasmic tail (Hjälm et al. 1996; Saito et al. 1994). The extracellular ligand binding regions are flanked by growth factor repeats and cysteine-poor spacer regions containing YWTD motifs, involved in pH-dependent dissociation of receptor and ligands in acidic endosomal compartments (Davis et al. 1987). The carboxy terminal cytoplasmic tail contains two NPXY motifs and a NQNY motif (NPXY-like motif). The two NPXY motifs mediate clustering into coated pits and thereby initiate the endocytic process, and the NPXY-like motif is involved in apical sorting of the receptor (Takeda et al. 2003).

The existence of these three motifs as well as other cytoplasmic motifs, such as several phosphorylation sites and Src homology 3 (SH3) and Src homology 2 (SH2) binding motifs, make the cytoplasmic tail of megalin differ from the other members of the LDL

Fig. 1 Schematic drawing of cubilin (*left*) and megalin (*right*) with interacting proteins. Megalin has four ▶ extracellular binding motifs (complement type motifs) providing the binding of a huge number of ligands. Its cytoplasmic domain contains a number of protein interaction motifs as well as endocytosis motifs. Interaction via these motifs with ARH and Dab2 has been revealed to be important for endocytosis of glomerular filtered proteins. ARH and Dab2 further interact with AP-2 and clathrin, and assembly of these proteins is necessary for initial steps in endocytosis. Cubilin interacts with megalin via its 113-residue α-helical amino terminus, and this association supports internalization of cubilin, which lacks a transmembrane domain and an endocytosis motif. Amnionless is another partner of cubilin capable of mediating endocytosis of cubilin via its interaction with the EGF repeats in cubilin. The majority of cubilin is made up by CUB domains (acronym for complement Clr/Cls, Uegf, and bone morphogenic protein-1), and these are involved in ligand binding

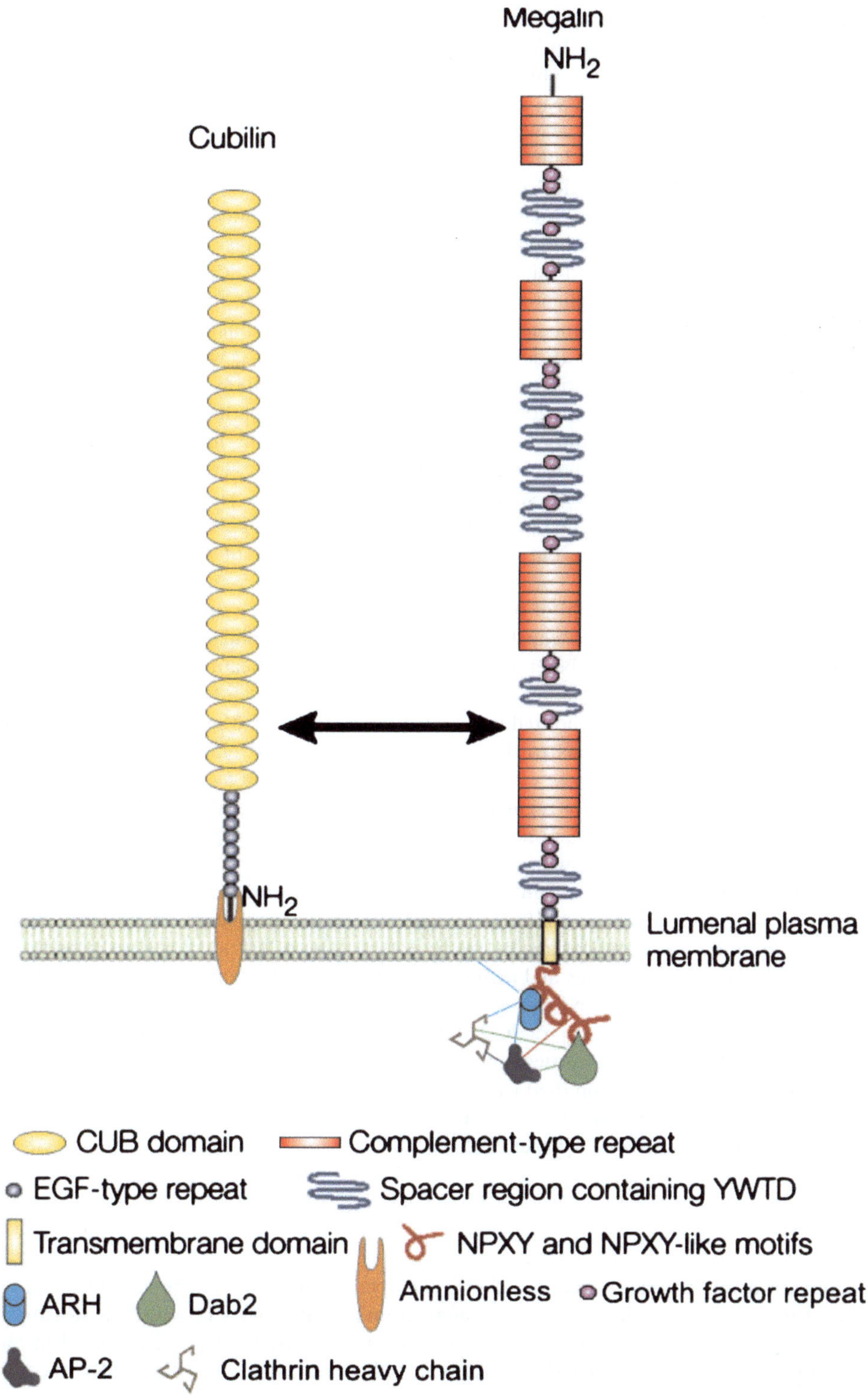
Megalin
NH2
Cubilin
NH2
Lumenal plasma membrane
CUB domain
Complement-type repeat
EGF-type repeat
Spacer region containing YWTD
Transmembrane domain
NPXY and NPXY-like motifs
ARH
Dab2
Amnionless
Growth factor repeat
AP-2
Clathrin heavy chain

receptor family, and the conception of megalin merely being a constitutively expressed recycling receptor is challenged. In line with the primary structure, several intracellular proteins involved in sorting and signaling have been demonstrated to bind to the megalin tail as, for example, JIP1 and -2, PDS-95, and MegBP, which further interact with transcriptional factors (Gotthardt et al. 2000; Larsson et al. 2003; Petersen et al. 2003).

Furthermore, apart from clathrin and adaptin-2 (AP-2) two adaptor proteins, Dab2 (Oleinikov et al. 2000) and ARH (Nagai et al. 2003), have been implicated directly in the endocytic process involving megalin. Through their PTB domains Dab2 binds to the second NPXY motif and ARH binds to the first NPXY motif (Nagai et al. 2003; Oleinikov et al. 2000). Dab2 gene-deleted mice as well as ARH-depleted mice (not published) have a mild tubular proteinuria implicating, together with the evidence for binding to megalin, their involvement in receptor-mediated endocytosis in the proximal tubule (Morris et al. 2002; Nagai et al. 2003, 2005). Binding of adaptor proteins such as Dab2 and ARH may be involved in controlling endocytosis, as well as signaling, by assembly of multimeric protein complexes through their interaction with other components of the cellular machinery, for example, through their ability to bind AP-2 and clathrin. The importance for adaptor proteins in the normal function of megalin is suggested by Muller et al. describing a patient with unknown etiology but with similar symptoms as megalin knockout mice, namely, holoprosencephalic syndrome (Willnow et al. 1996a) and renal tubular reabsorption defects seen as vitamin D deficiency, but without direct implications of megalin gene defects (Muller et al. 2001).

Besides having the potential of mediating signaling events through its intracellular domain, megalin is possibly also capable of signaling by extracellular signaling ligands as sonic hedgehog has shown to be endocytosed by megalin, implicating megalin in the regulation of sonic hedgehog signaling (McCarthy et al. 2002). Loss of sonic hedgehog leads to forebrain developmental abnormalities of varying degree (see McCarthy et al. 2002 for references), which is analogous to the seriously compromised forebrain development in megalin-knockout mice (Willnow et al. 1996a). Whether megalin-mediated endocytosis of sonic hedgehog is important for regulation of the levels of the signaling molecule, directly involved in hedgehog signaling, or necessary for long-range signaling of sonic hedgehog by transcytosis is at present unknown.

Generally megalin is present in apical membranes and in the endocytic apparatus of a number of tissues, therefore not being in contact with the circulation. Its role as a scavenger receptor has been emphasized by the broad spectrum of ligands it has been shown to bind. The ligands are numerous and include vitamin-binding proteins, enzyme and enzyme inhibitors, hormones, drugs and toxins, lipoproteins, calcium, albumin, hemoglobin, myoglobin, and receptor-associated protein (RAP) (Christensen and Birn 2002). The importance of the reabsorption mediated by megalin in the renal proximal tubule is observed in megalin-deficient mice and best exemplified by the ligand vitamin D binding protein. The inability of these mice to reabsorb and thereby metabolize filtered vitamin D results in vitamin D deficiency and bone disease (Nykjaer et al. 1999).

Cubilin

Cubilin, is a 460-kDa (nonglycosylated 400 kDa) peripheral membrane protein identical to the intrinsic factor-vitamin B_{12} receptor known from the small intestine with little structural homology to known endocytic receptors. Its complete cDNA has been determined in rat (Moestrup et al. 1998a), human (Kozyraki et al. 1998), and canine (Xu et al. 1999). The human gene is located to chromosome 10p12.33–p13. Cubilin has obtained its name because of the majority of the protein being occupied by CUB domains, which are 120-amino acid ligand binding modules originally identified in the complement C1r/C1s, Uegf (epidermal growth factor-related sea urchin protein), and bone morphogenic protein-1. These modules are present in a diverse array of proteins, many of which are involved in developmental processes (e.g., BMP-1, tolloid, and TSG6)(Bork and Beckmann 1993). This accumulation of CUB domains suggests that cubilin may interact with a wide variety of ligands. The protein (Fig. 1) consists of a 110-amino acid amino-terminal stretch (Moestrup et al. 1998a) and 27 CUB domains. CUB domains 5–8 and 13–14 have been shown to contain binding sites for intrinsic factor-vitamin B_{12} and the receptor-associated protein (RAP), respectively (Kristiansen et al. 1999). The membrane association of cubilin depends on the 110 amino acids at the amino-terminus stretch (Kristiansen et al. 1999) and may involve a putative amphipathic helix as well as palmitoylation. Biochemical and immunomorphological data suggest that the internalization of cubilin is, at least in part, carried out by megalin (Kozyraki et al. 2001; Moestrup et al. 1998a), possibly by binding of megalin to cubilin within the amino-terminal region including CUB domains 1 and 2 (Yammani et al. 2001).

Expression

The two receptors are heavily expressed and colocalize in the kidney proximal tubule brush border and the luminal endocytic apparatus and to some extent in lysosomal structures (Figs. 2, 3, and 4) (Bachinsky et al. 1993; Chatelet et al. 1986; Christensen et al. 1992, 1995; Kerjaschki and Farquhar 1982; Sahali et al. 1988, 1992; Seetharam et al. 1997). Megalin has also been identified in glomerular podocytes of Lewis rats (Kerjaschki and Farquhar 1983). The two receptors also colocalize in several extrarenal tissues, in particular absorptive epithelia such as the visceral yolk sac (Sahali et al. 1988), the epithelium of the small intestine (Birn et al. 1997; Levine et al. 1984), and the placenta (Hammad et al. 2000; Lundgren et al. 1997; Zheng et al. 1994). In addition, megalin is expressed in the choroid plexus, ependymal cells, epididymis, oviduct, thyroid cells, labyrinthic cells of the inner ear, type II pneumocytes, the PTH-secreting cells of the parathyroid gland, the endometrium, the ciliary epithelium of the eye, and embryonic tissues such as the trophoectodermic cells and the neuroectoderm (Zheng et al. 1994) (reviewed in Christensen et al. 1998). The expression of cubilin at present appears more restricted, although the receptor has been identified in other tissues including thymus (Hammad et al. 2000).

The normal expression of megalin is dependent on receptor-associated protein (RAP) (Birn et al. 2000b) serving as a chaperone protecting newly synthesized receptor from the early binding of ligands and possibly also involved in folding of the receptor (Bu et al. 1995; Bu and Rennke 1996; Willnow et al. 1995, 1996b; Willnow 1998). RAP binds megalin with high affinity within the RER and functions as an intracellular ligand inhibiting the binding of most other ligands to megalin. RAP also binds other members of the LDL receptor family with a similar function. Because RAP specifically inhibits binding to megalin, it serves as an important tool for the study of megalin function. RAP also binds to cubilin (Birn et al. 1997; Kristiansen et al. 1999), although its role for the posttranslational processing of this receptor is unknown. Normal expression of cubilin is dependent on amnionless (AMN), a 45-kDa transmembrane protein identified as an important factor for the normal development of the middle portion of the primitive streak in mice (Kalantry et al. 2001). AMN colocalizes with cubilin in kidney proximal tubule (Fyfe et al. 2004), interacts with the EGF-type repeats of

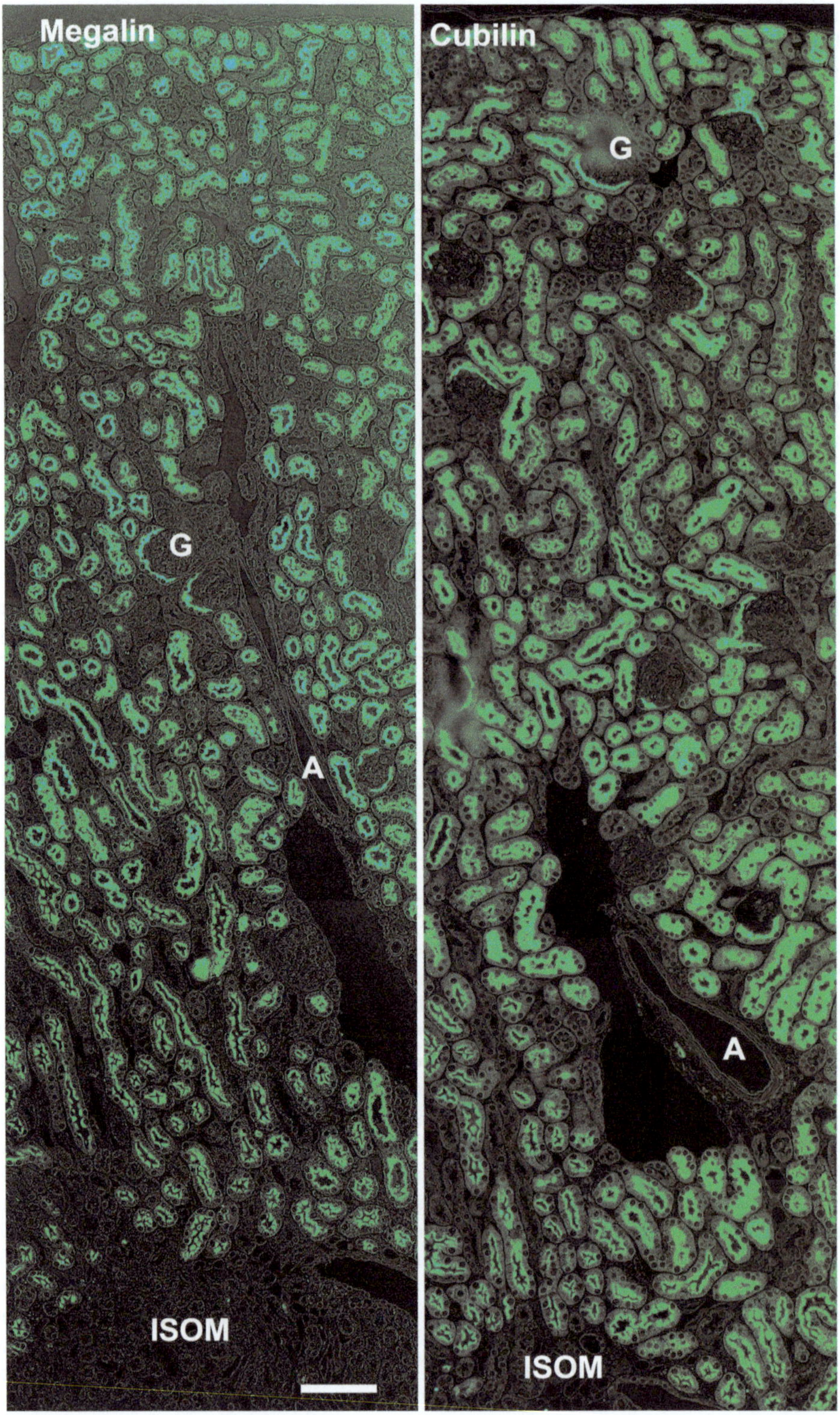
Megalin
G
A
ISOM
Cubilin
G
A
ISOM

◂ **Fig. 2** Immunofluorescence for megalin and cubilin of paraffin section from mouse kidney from the renal capsule (*top*) through the cortex and the outer stripe of the outer medulla. Only proximal tubules, including proximal tubule cells constituting part of Bowman's capsule of the glomeruli (*G*) are labeled. *A*, arteries; *ISOM*, inner stripe of outer medulla; *bar*, 100 μm

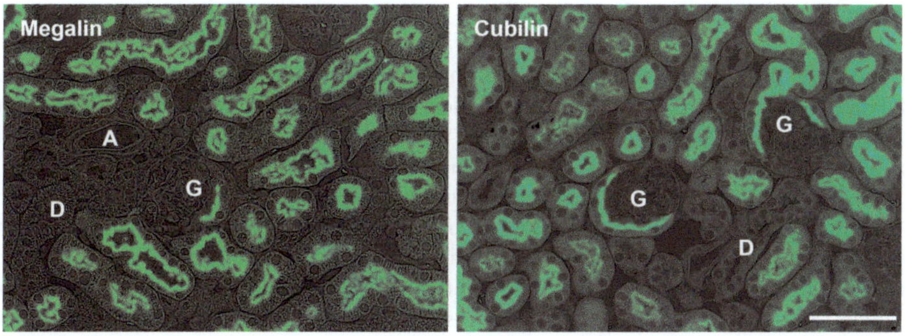

Fig. 3 Immunofluorescence for megalin and cubilin of paraffin section from mouse kidney cortex. Only proximal tubules are labeled. Glomerular cells (*G*) and distal tubules (*D*) are unlabeled. *A*, Arterial branch; *bar*, 100 μm

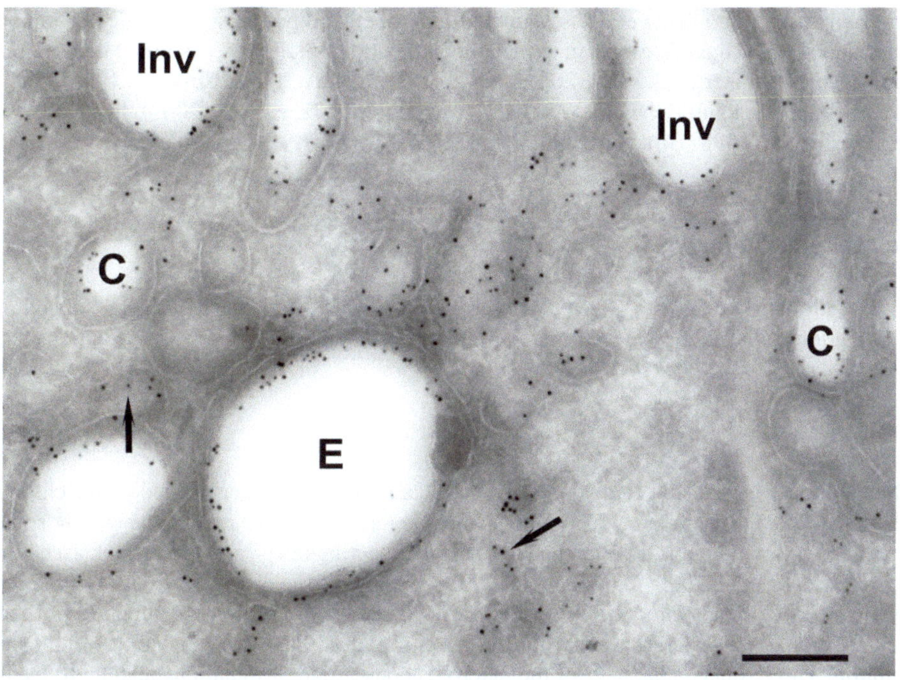

Fig. 4 Immunogold double labeling for megalin and cubilin of mouse renal proximal tubule cell, megalin (10-nm gold particles) and cubilin (5-nm gold particles). Labeling for both receptors is seen along the membrane of apical invaginations (*Inv*), coated pits/vesicles (*C*) and endosomes (E). *Bar*, 0.5 μm

cubilin, and is essential for normal translocation of the cubilin-AMN complex from the ER to the plasma membrane and for subsequent endocytosis (Coudroy et al. 2005; Fyfe et al. 2004).

Interaction between megalin and cubilin

At the time when cubilin was described as an endocytic receptor lacking a cytoplasmic tail and thereby internalization motifs, it was evident that it had to interact with other proteins to fulfill its role. In 1998 Moestrup et al. showed that cubilin binds megalin with high affinity in vitro (Moestrup et al. 1998a), and based on this observation and their colocalization (Figs. 2, 3, and 4) it was suggested that megalin mediates the endocytosis and intracellular trafficking of cubilin (Moestrup et al. 1998a). Several other findings supported this concept. Among others, in vitro uptake studies of the cubilin ligands transferrin and apolipoprotein A-I (apo A-I)/high-density lipoprotein (HDL) showed that uptake was decreased by anti-megalin antibodies (Kozyraki et al. 1999, 2001) as well as by megalin antisense oligonucleotides (Hammad et al. 2000). Furthermore, in megalin-deficient mice endogenous transferrin accumulates on the apical surface of the proximal tubule cells without being internalized (Kozyraki et al. 2001). Later, the cubilin-AMN interaction, mentioned above, was discovered, and as the cubilin-AMN complex is internalized in cells lacking megalin (Fyfe et al. 2004) cubilin may have other interaction partners to mediate its internalization. Whether megalin is preferentially used in epithelia expressing megalin or whether a potential preference pertains to internalization of ligands common to megalin and cubilin, including vitamin D binding protein (DBP) (Nykjaer et al. 1999, 2001), albumin (Birn et al. 2000a; Cui et al. 1996; Zhai et al. 2000), hemoglobin (Gburek et al. 2002), myoglobin (Gburek et al. 2003), Ig light chains (Batuman et al. 1998; Birn et al. 2003), and RAP (Birn et al. 1997) is unknown.

Tubular proteinuria

In the following we discuss selected diseases in which changes in the expression/subcellular distribution of the endocytic receptors has either been demonstrated or may be anticipated.

Dent's disease

Dent's disease is an X-linked disorder caused by mutations in the gene encoding the chloride channel CLC-5 (Scheinman 1998). The channel is localized to the membrane of endosomes in renal proximal tubules, in intercalated cells of collecting ducts, and in the thick ascending limb (Devuyst et al. 1999; Günther et al. 1998). In patients the disease presents as a Fanconi syndrome including aminoaciduria, glucosuria, phosphaturia, kaliuresis, impaired urinary acidification, hypercalciuria, nephrolithiasis, renal failure, and low-molecular-weight proteinuria. The X-linked character of the heterogeneous group of mutations leading to Dent's disease causes a milder phenotype in heterozygous female patients (Wrong et al. 1994). Their phenotype is based on random X inactivation whereby heterozygous females achieve cells expressing either wild-type or mutant gene product and the differing expression is reflected in proximal tubular protein uptake (Piwon et al. 2000). Several mouse models for Dent's disease have been developed that show some of the same symptoms as Dent's disease patients (Luyckx et al. 1999; Piwon et al. 2000; Wang et al. 2000). The low-molecular-weight proteinuria detected in these models was found to be caused by defects in early endosomal function, whereby endocytosis of filtered proteins are impaired in proximal tubule cells (Piwon et al. 2000; Wang et al. 2000).

The low-molecular-weight proteinuria as detected in Dent's disease patients is most likely a result of decreased expression of megalin as described in ClC-5-knockout mice (Piwon et al. 2000) and the additional, even more pronounced decrease in expression of cubilin (25% of wild type) demonstrated recently (Christensen et al. 2003). This reduction was visualized as a virtually total disappearance of the two receptors from the microvilli (Fig. 5), which suggests an intracellular trafficking defect of megalin and cubilin.

The underlying cause for altered receptor trafficking in CLC-5-knockout mice has been suggested to be reduced acidification of apical endosomes in these cells because ClC-5 colocalizes with H^+-ATPase and has been thought to act as an electrical shunt for H^+ secretion into early endosomes (Marshansky et al. 2002; Piwon et al. 2000; Wang et al. 2000). Recently, it has, however, been evidenced that CLC-5 is a chloride/proton antiporter and therefore *per se* is a direct player in endosomal acidification (Picollo and Pusch 2005; Scheel et al. 2005). Perturbed acidification results in lack of dissociation of receptor-ligand complexes, but recently it has furthermore been shown that recruitment of Arf6 and the GTP/GDP exchange factor (ARNO) is dependent on endosome pH and that these proteins partially colocalize with V-ATPase in proximal tubule endosomes (Maranda et al. 2001; Marshansky et al. 2002).

Arf small GTPases are regulators of membrane trafficking by coat recruitment (COPI, AP1, Ap3, AP4), and all Arfs can activate phospholipase D (PLD) and phosphatidylinositol 4-phosphate 5-kinase (PIP5K) to generate phosphatidic acid and phosphatidylinositol 4,5-bisphosphate (PIP_2), respectively (Donaldson and Honda 2005). In fact, the influence of Arf6 on PIP_2 levels has been shown to be important for AP2 and clathrin assembly in synaptic membranes (Krauss et al. 2003). Therefore, it was suggested that pH-dependent recruitment of Arf6 and ARNO is required for normal endocytosis through recruitment of coat proteins in the proximal tubule. Thus lack of acidification might perturb both receptor-ligand dissociation as well as recruitment of Arfs. Furthermore, ClC-5 associates with cofilin and thereby alters the actin cytoskeleton, and albumin uptake is dependent on ubiquination of ClC-5 by Nedd-4-2 in proximal tubule cells (Hryciw et al. 2003, 2004).

Thus whether the cause of altered megalin and cubilin trafficking, when ClC-5 is disturbed, is mediated by a lack of gathering of coat proteins such as Arf6 and ARNO, the possible involvement of ClC-5 in endosome acidification *per se*, or the involvement of ClC-5 in protein-protein interactions required for receptor-mediated endocytosis is unknown. Thus megalin/cubilin trafficking and endocytosis of albumin may require assembly of a multimeric complex containing several proteins besides the known adaptor proteins.

The reduction of the receptors in the brush border further results in significantly decreased urinary excretion of megalin and unchanged excretion of cubilin (Christensen et al. 2003). A similar pattern of decreased urinary excretion of megalin and unchanged excretion of cubilin was found in Dent's disease patients (Norden et al. 2002). Interestingly, patients with Lowe syndrome exhibited a similar pattern of urinary excretion of the two receptors (Norden et al. 2002). The oculocerebrorenal syndrome of Lowe is also an X-linked disease characterized by mental retardation, cataracts, and renal Fanconi syndrome (Lowe et al. 1952). The affected gene, *OCRL1*, encodes a phosphatidylinositol 4,5 bisphosphate 5-phosphatase localized to the trans-Golgi network (Dressman et al. 2000; Suchy et al. 1995) and suggested to be involved in TGN sorting. Whether or not Lowe syndrome patients have a defect in sorting of megalin and/or cubilin remains to be determined. However, recently

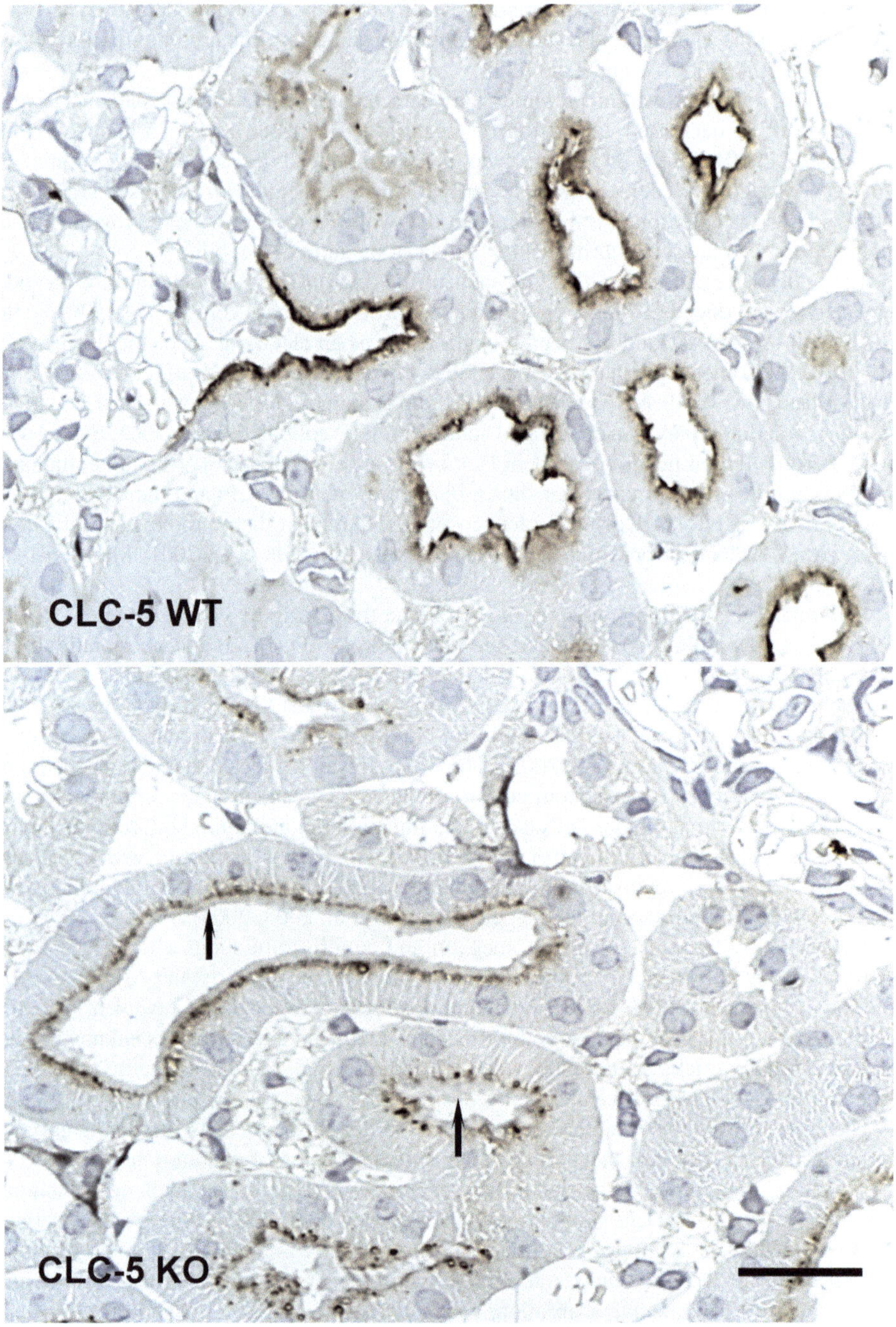

Fig. 5 Immunoperoxidase labeling for megalin of cryosections from renal cortex of CLC-5 wild-type and knockout mice. In the CLC-5-knockout mouse megalin has virtually totally disappeared from the microvilli (*arrows*). *Bar*, 40 μm

a direct link to Dent's disease has become evident from a study demonstrating patients with mutations in *OCRL1* who have the isolated renal phenotype of Dent's disease (Hoopes et al. 2005).

Recently it has also become evident that besides megalin, cubilin, and ClC-5, other ion transport proteins are implicated in albumin endocytosis occurring in the renal proximal tubule (Marshansky et al. 2002). As mentioned above the v-H^+-ATPase is involved in endosome acidification (Marshansky et al. 2002) and NHE3, which is implicated in sodium retention and endosome acidification in the proximal tubule, seems to play a role as well (Gekle et al. 1999, 2002). Inhibition of NHE3 has been shown directly to decrease albumin uptake in proximal tubule cells (Gekle et al. 2002), and this has been confirmed in vivo by pharmacological inhibition of NHE3 resulting in decreased megalin-mediated endocytosis of cytochrome *c* as well as the detection of tubular proteinuria in NHE3-knockout mice (Gekle et al. 2004). Furthermore, besides a role in Na^+ homeostasis and acidification, NHE3 may be able to interact directly with megalin (Biemesderfer et al. 1999).

Imerslund-Gräsbeck syndrome

Selective intestinal malabsorption of cobalamin (vitamin B_{12}) accompanied by tubular proteinuria, known as Imerslund-Gräsbeck syndrome (I-GS) or megaloblastic anemia 1 (MGA1), is a rare autosomal recessive disorder with 200 patients identified so far (Gräsbeck et al. 1960; Imerslund 1960). MGA1 occurs worldwide, but it has a higher prevalence in several Middle Eastern countries and in Norway, and highest in Finland (Rosenblatt and Fenton 1999).

One of cubilin's ligands not shared with megalin is the intrinsic factor-vitamin-B_{12} complex (IF-B_{12}). Cubilin is thus responsible for uptake of IF-B_{12} (Seetharam et al. 1997) in the small intestine. Defects in cubilin-mediated IF-B_{12} absorption were therefore likely to be involved in I-GS, and accordingly linkage studies have shown I-GS to be caused by mutation in a region designated MGA1 (megaloblastic anemia-1), located on 10p, where cubilin has been mapped as well (Kozyraki et al. 1998).

The etiology in the families identified in different countries seems to be different. In the Finnish families two distinct mutations of the cubilin gene have been identified (Aminoff et al. 1999). The first mutation (FM1) consists of a point mutation in CUB domain 8 that binds IF-B_{12}. The FM2 mutation, so far detected only in a single patient, is an intronic mutation within CUB domain 6 that probably results in the synthesis of a truncated and/or rapidly degraded protein. Diversity in mutations may explain why only some patients with selective IF-B_{12} malabsorption have proteinuria. Mutations in the cubilin gene affecting more binding sites or resulting in the absence of functional receptors are more likely to give proteinuria than mutations affecting only the IF-B_{12} binding site. The FM2 patient has pronounced proteinuria of cubilin ligands (Nykjaer et al. 2001; Wahlstedt-Froberg et al. 2003); however, patients with FM1 may also have proteinuria (Wahlstedt-Froberg et al. 2003).

The Norwegian I-GS patients (Tanner et al. 2004) do not have mutations in the cubilin gene, but, as some patients from Turkey, they have mutations in the amnionless (AMN) gene (Tanner et al. 2004).

Furthermore, inbred dogs with failure to insert cubilin in the apical membrane (Fyfe et al. 1991) showed highly significant linkage to the same locus as the Norwegian patients (He et al. 2003), and two distinct *AMN* mutations in two unrelated canine kindreds with I-GS have now been identified (He et al. 2005). The dogs also have evidence of B_{12} deficiency and proteinuria of cubilin ligands (Birn et al. 2000a; Kozyraki et al. 2001; Nykjaer et al.

2001). This proteinuria, however, is less pronounced than that seen in megalin gene-deleted mice (Leheste et al. 1999) because the dogs express normal levels of megalin and therefore reabsorb ligands to megalin normally and to some extent also ligands shared by the two receptors (Nykjaer et al. 2001).

Fabry disease and cystinosis

We will just briefly mention two other diseases, because they involve the endocytic/lysosomal compartments in the kidney and especially the proximal tubule. Anderson-Fabry disease is an X-linked lysosomal storage disease caused by deficiency of α-galactosidase A. So far, the only effective treatment is enzyme replacement therapy with recombinant α-galactosidase A (for a comprehensive review see (Grabowski and Hopkin 2003). Although enzyme replacement therapy appears to effectively reduce glycolipid deposits in vascular endothelial cells, the effect on, for example, proximal tubular cells (Branton et al. 2002) and distal tubule cells (Thurberg et al. 2002) appears less pronounced, which makes sense, considering the accessibility of the enzyme to the cells. Besides the vascular endothelial cells the glomerular podocyte appears to be the most severely affected cell in the kidney, and many patients have a significant proteinuria. How this disease affects the apical endocytic receptors in the proximal tubule is not known.

Cystinosis is the most common inherited form of the Fanconi syndrome. The causative gene, *CTNS*, found by positional cloning (Town et al. 1998) encodes cystinosin, a lysosomal membrane protein responsible for lysosomal transport of cystine into the cytoplasm (for a review see Kalatzis and Antignac 2003). The patients show lysosomal accumulation of cystine due to lack of transport out of the lysosomes (Kalatzis and Antignac 2003). Because these patients develop a severe proteinuria, it would be interesting to study possible changes in expression and subcellular localization of megalin and cubilin in the proximal tubule.

Nephrotoxicity induced by megalin-mediated aminoglycoside uptake

Aminoglycosides are of low molecular weight and assumed to be freely filtered in the renal glomeruli. The majority is lost by urinary excretion; however; 10% of the dose enters the renal cortex (Mingeot-Leclercq and Tulkens 1999). More specifically, they have been shown to accumulate in the endocytic apparatus of the proximal tubule (Inui et al. 1988; Silverblatt and Kuehn 1979; Vandewalle et al. 1981). Here aminoglycosides are accumulated with a long half-life and influence the plasma membrane, mitochondria, and lysosomes eventually leading to renal damage (Mingeot-Leclercq and Tulkens 1999). In 1995 it was shown that megalin interacts with aminoglycosides (Moestrup et al. 1995). The three different aminoglycosides tested were able to inhibit megalin binding to PAI-1 complexes. Similarly, gentamicin is able to inhibit albumin binding to megalin in the proximal tubule as observed by microperfusion experiments (Cui et al. 1996). With gentamicin being intravenously administered other megalin ligands such as vitamin D binding protein and calcium are increasingly secreted in the urine, and competition on Western blots between gentamicin and calcium has been observed (Nagai et al. 2001). Support of the role of megalin in aminoglycoside-induced nephrotoxicity has been obtained by intraperitoneal administration of gentamicin to megalin-knockout mice. These studies showed a marked decrease of gentamicin accumulation in the kidney of the knockout mice (10% to 0.6% of the dose) (Schmitz et al. 2002). As aminoglycosides apparently also bind to LRP1, another member of the LDL receptor family present in hepatocytes, it is quite striking that no accumulation is observed in the liver, even under conditions where gentamicin plasma levels are kept high for extended periods of time. It could be speculated that other factors such as megalin-mediated uptake of cubilin are involved in the renal accumulation.

Glomerular proteinuria

We will not go into a detailed discussion of glomerular proteinuria and overload proteinuria. It should, however, be emphasized that any increased exposure of the endocytic receptors in the proximal tubule to protein will undoubtedly influence the uptake of proteins, vitamins, and other trace elements normally reabsorbed here. Because the tubular fluid after the proximal tubule, at least in humans, is virtually devoid of filtered proteins, it should also be emphasized that the remaining part of the nephron and the collecting duct under proteinuric conditions may be exposed to high concentrations of plasma proteins. The uptake of these proteins, probably by nonspecific/fluid-phase endocytosis, which can take place in both distal tubules and collecting ducts (Christensen et al. 1981; Madsen et al. 1982; Straus 1964), but also simply their presence in the tubular fluid may interfere with important processes taking place here and thereby add to the toxicity and subsequent events leading to interstitial inflammation, fibrosis, and other functional disturbances, notably reduced activation of vitamin D due to lack of reabsorption of DBP.

Summary and future directions

Megalin and cubilin are well-established scavenger receptors in the renal proximal tubule, and it is now obvious that tubular proteinuria develops if either of the receptors is not trafficked properly or absent. Thus other syndromes involving tubular proteinuria as part of a Fanconi syndrome might have defective megalin and/or cubilin receptor function, mediated by yet unidentified cellular pathways. Future studies aimed at elucidating cellular mechanisms involved in megalin and cubilin trafficking are therefore needed to approach the origin of Fanconi syndromes as observed in a variety of renal diseases.

Besides this important role in clearance of the primary filtrate it seems evident that megalin and cubilin also serve as an important link in renal vitamin metabolism and preservation of certain vitamins. In addition, data implicating megalin in cellular signaling via intracellular protein interactions are accumulating. These findings indicate that megalin interacts in more refined processes than merely bulk uptake of molecules destined for degradation. The goal is now to resolve the potential biochemical pathways induced by ligand binding and the cellular response they regulate. Such future studies will reveal whether interference with their receptor function leads to detrimental effects other than tubular proteinuria during renal disease.

Acknowledgements. This work was supported by grants from the Danish Medical Research Council, the Novo Nordic Foundation, the Danish Biotechnology Program, the Velux Foundation, the University of Aarhus Research Foundation, and the European Commission (EU Framework Program 6, EureGene, contract number 05085). We thank H. Sidelmann and I. Blenker Kristoffersen for excellent technical assistance.

References

Aminoff M, Carter JE, Chadwick RB, Johnson C, Grasbeck R, Abdelaal MA, Broch H, Jenner LB, Verroust PJ, Moestrup SK, de la Chapelle A, Krahe R (1999) Mutations in CUBN, encoding the intrinsic factor-vitamin B12 receptor, cubilin, cause hereditary megaloblastic anaemia 1. Nat Genet 21:309–313

Bachinsky DR, Zheng G, Niles JL, McLaughlin M, Abbate M, Andres G, Brown D, McCluskey RT (1993) Detection of two forms of GP330. Their role in Heymann nephritis. Am J Pathol 143:598–611

Batuman V, Verroust PJ, Navar GL, Kaysen JH, Goda FO, Campbell WC, Simon E, Pontillon F, Lyles M, Bruno J, Hammond TG (1998) Myeloma light chains are ligands for cubilin (gp280). Am J Physiol Renal Physiol 275:F246–F254

Biemesderfer D, Nagy T, DeGray B, Aronson PS (1999) Specific association of megalin and the Na^+/H^+ exchanger isoform NHE3 in the proximal tubule. J Biol Chem 274:17518–17524

Birn H, Fyfe JC, Jacobsen C, Mounier F, Verroust PJ, Orskov H, Willnow TE, Moestrup SK, Christensen EI (2000a) Cubilin is an albumin binding protein important for renal tubular albumin reabsorption. J Clin Invest 105:1353–1361

Birn H, Leboulleux M, Moestrup SK, Ronco PM, Aucouturier P, Christensen EI (2003) Megalin, cubilin and immunoglobulin light chains: receptor-mediated uptake of light chains in kidney proximal tubule. In: Touchard G, Aucouturier P, Hermine O, Ronco PM (eds) Monoclonal Gammopathies and the Kidney. Kluwer, The Netherlands, pp 37–48

Birn H, Verroust PJ, Nexø E, Hager H, Jacobsen C, Christensen EI, Moestrup SK (1997) Characterization of an epithelial approximately 460-kDa protein that facilitates endocytosis of intrinsic factor-vitamin B12 and binds receptor-associated protein. J Biol Chem 272:26497–26504

Birn H, Vorum H, Verroust PJ, Moestrup SK, Christensen EI (2000b) Receptor-associated protein is important for normal processing of megalin in kidney proximal tubules. J Am Soc Nephrol 11:191–202

Birn H, Zhai X, Holm J, Hansen SI, Jacobsen C, Christensen EI, Moestrup SK (2005) Megalin binds and mediates cellular internalization of folate binding protein. FEBS J 272:4423–4430

Bork P, Beckmann G (1993) The CUB domain. A widespread module in developmentally regulated proteins. J Mol Biol 231:539–545

Branton M, Schiffmann R, Kopp JB (2002) Natural history and treatment of renal involvement in Fabry disease. J Am Soc Nephrol 13(Suppl 2):S139–S143

Bu G, Geuze HJ, Strous GJ, Schwartz AL (1995) 39 kDa receptor-associated protein is an ER resident protein and molecular chaperone for LDL receptor-related protein. EMBO J 14:2269–2280

Bu G, Rennke S (1996) Receptor-associated protein is a folding chaperone for low density lipoprotein receptor-related protein. J Biol Chem 271:22218–22224

Burmeister R, Boe IM, Nykjaer A, Jacobsen C, Moestrup SK, Verroust P, Christensen EI, Lund J, Willnow TE (2001) A two-receptor pathway for catabolism of Clara cell secretory protein in the kidney. J Biol Chem 276:13295–13301

Chan WL, Shaw PC, Tam SC, Jacobsen C, Gliemann J, Nielsen MS (2000) Trichosanthin interacts with and enters cells via LDL receptor family members. Biochem Biophys Res Commun 270(2):453–457

Chatelet F, Brianti E, Ronco P, Roland J, Verroust P (1986) Ultrastructural localization by monoclonal antibodies of brush border antigens expressed by glomeruli. I. Renal distribution. Am J Pathol 122:500–511

Christensen EI (1982) Rapid membrane recycling in renal proximal tubule cells. Eur J Cell Biol 29:43–49

Christensen EI, Birn H (2002) Megalin and cubilin: multifunctional endocytic receptors. Nat Rev Mol Cell Biol 3:256–266

Christensen EI, Birn H (2001) Megalin and cubilin: synergistic endocytic receptors in renal proximal tubule. Am J Physiol Renal Physiol 280:F562–F573

Christensen EI, Birn H, Verroust P, Moestrup SK (1998) Membrane receptors for endocytosis in the renal proximal tubule. Int Rev Cytol 180:237–284

Christensen EI, Carone FA, Rennke HG (1981) Effect of molecular charge on endocytic uptake of ferritin in renal proximal tubule cells. Lab Invest 44:351–358

Christensen EI, Devuyst O, Dom G, Nielsen R, Van Der Smissen P, Verroust P, Leruth M, Guggino WB, Courtoy PJ (2003) Loss of chloride channel ClC-5 impairs endocytosis by defective trafficking of megalin and cubilin in kidney proximal tubules. Proc Natl Acad Sci USA 100:8472–8477

Christensen EI, Gliemann J, Moestrup SK (1992) Renal tubule gp330 is a calcium binding receptor for endocytic uptake of protein. J Histochem Cytochem 40:1481–1490

Christensen EI, Moskaug JO, Vorum H, Jacobsen C, Gundersen TE, Nykjaer A, Blomhoff R, Willnow TE, Moestrup SK (1999) Evidence for an essential role of megalin in transepithelial transport of retinol. J Am Soc Nephrol 10:685–695

Christensen EI, Nielsen S, Moestrup SK, Borre C, Maunsbach AB, de Heer E, Ronco P, Hammond TG, Verroust P (1995) Segmental distribution of the endocytosis receptor gp330 in renal proximal tubules. Eur J Cell Biol 66:349–364

Coudroy G, Gburek J, Kozyraki R, Madsen M, Trugnan G, Moestrup SK, Verroust PJ, Maurice M (2005) Contribution of cubilin and amnionless to processing and membrane targeting of cubilin-amnionless complex. J Am Soc Nephrol 16:2330–2337

Cui S, Verroust PJ, Moestrup SK, Christensen EI (1996) Megalin/gp330 mediates uptake of albumin in renal proximal tubule. Am J Physiol 271:F900–F907

Davis CG, Goldstein JL, Sudhof TC, Anderson RG, Russell DW, Brown MS (1987) Acid-dependent ligand dissociation and recycling of LDL receptor mediated by growth factor homology region. Nature 326:760–765

Devuyst O, Christie PT, Courtoy PJ, Beauwens R, Thakker RV (1999) Intra-renal and subcellular distribution of the human chloride channel, CLC-5, reveals a pathophysiological basis for Dent's disease. Hum Mol Genet 8:247–257

Donaldson JG, Honda A (2005) Localization and function of Arf family GTPases. Biochem Soc Trans 33:639–642

Dressman MA, Olivos-Glander IM, Nussbaum RL, Suchy SF (2000) Ocrl1, a PtdIns(4,5)P(2) 5-phosphatase, is localized to the trans-Golgi network of fibroblasts and epithelial cells. J Histochem Cytochem 48:179–190

Faber K, Hvidberg V, Moestrup SK, Dahlback B, Nielsen LB (2006) Megalin is a receptor for apolipoprotein M, and kidney-specific megalin-deficiency confers urinary excretion of apolipoprotein M. Mol Endocrinol 20:212–218

Fyfe JC, Madsen M, Hojrup P, Christensen EI, Tanner SM, de la Chapelle A, He Q, Moestrup SK (2004) The functional cobalamin (vitamin B12)-intrinsic factor receptor is a novel complex of cubilin and amnionless. Blood 103:1573–1579

Fyfe JC, Ramanujam KS, Ramaswamy K, Patterson DF, Seetharam B (1991) Defective brush-border expression of intrinsic factor-cobalamin receptor in canine inherited intestinal cobalamin malabsorption. J Biol Chem 266:4489–4494

Gburek J, Birn H, Verroust PJ, Goj B, Jacobsen C, Moestrup SK, Willnow TE, Christensen EI (2003) Renal uptake of myoglobin is mediated by the endocytic receptors megalin and cubilin. Am J Physiol Renal Physiol 285:F451–F458

Gburek J, Verroust PJ, Willnow TE, Fyfe JC, Nowacki W, Jacobsen C, Moestrup SK, Christensen EI (2002) Megalin and cubilin are endocytic receptors involved in renal clearance of hemoglobin. J Am Soc Nephrol 13:423–430

Gekle M, Drumm K, Mildenberger S, Freudinger R, Gassner B, Silbernagl S (1999) Inhibition of Na^+-H^+ exchange impairs receptor-mediated albumin endocytosis in renal proximal tubule-derived epithelial cells from opossum. J Physiol 520:709–721

Gekle M, Serrano OK, Drumm K, Mildenberger S, Freudinger R, Gassner B, Jansen HW, Christensen EI (2002) NHE3 serves as a molecular tool for cAMP-mediated regulation of receptor-mediated endocytosis. Am J Physiol Renal Physiol 283:F549–F558

Gekle M, Volker K, Mildenberger S, Freudinger R, Shull GE, Wiemann M (2004) NHE3 Na^+/H^+ exchanger supports proximal tubular protein reabsorption in vivo. Am J Physiol Renal Physiol 287:F469–F473

Gonzalez-Villalobos R, Klassen RB, Allen PL, Navar LG, Hammond TG (2005) Megalin binds and internalizes angiotensin II. Am J Physiol Renal Physiol 288:F420–F427

Gotthardt M, Trommsdorff M, Nevitt MF, Shelton J, Richardson JA, Stockinger W, Nimpf J, Herz J (2000) Interactions of the low density lipoprotein receptor gene family with cytosolic adaptor and scaffold proteins suggest diverse biological functions in cellular communication and signal transduction. J Biol Chem 275:25616–25624

Grabowski GA, Hopkin RJ (2003) Enzyme therapy for lysosomal storage disease: principles, practice, and prospects. Annu Rev Genomics Hum Genet 4:403–436

Gräsbeck R, Gordin R, Kantero I, Kuhlbäck B (1960) Selective vitamin B12 malabsorption and proteinuria in young people. Acta Med Scand 167:289–296

Günther W, Lüchow A, Cluzeaud F, Vandewalle A, Jentsch TJ (1998) ClC-5, the chloride channel mutated in Dent's disease, colocalizes with the proton pump in endocytotically active kidney cells. Proc Natl Acad Sci USA 95:8075–8080

Hama H, Saito A, Takeda T, Tanuma A, Xie Y, Sato K, Kazama JJ, Gejyo F (2004) Evidence indicating that renal tubular metabolism of leptin is mediated by megalin but not by the leptin receptors. Endocrinology 145:3935–3940

Hammad SM, Barth JL, Knaak C, Argraves WS (2000) Megalin acts in concert with cubilin to mediate endocytosis of high density lipoproteins. J Biol Chem 275:12003–12008

Hammad SM, Stefansson S, Twal WO, Drake CJ, Fleming P, Remaley A, Brewer HBJ, Argraves WS (1999) Cubilin, the endocytic receptor for intrinsic factor-vitamin B_{12} complex, mediates high-density lipoprotein holoparticle endocytosis. Proc Natl Acad Sci USA 96:10158–10163

Hammes A, Andreassen TK, Spoelgen R, Raila J, Hubner N, Schulz H, Metzger J, Schweigert FJ, Luppa PB, Nykjaer A, Willnow TE (2005) Role of endocytosis in cullular uprake of sex steroids. Cell 122:751–762

He Q, Fyfe JC, Schäffer AA, Kilkenney A, Werner P, Kirkness EF, Henthorn PS (2003) Canine Imerslund-Gräsbeck syndrome maps to a region orthologous to HSA14q. Mamm Genome 14:765–777

He Q, Madsen M, Kilkenney A, Gregory B, Christensen EI, Vorum H, Hojrup P, Schaffer AA, Kirkness EF, Tanner SM, de la Chapelle A, Giger U, Moestrup SK, Fyfe JC (2005) Amnionless function is required for cubilin brush-border expression and intrinsic factor-cobalamin (vitamin B12) absorption in vivo. Blood 106:1447–1453

Hilpert J, Nykjaer A, Jacobsen C, Wallukat G, Nielsen R, Moestrup SK, Haller H, Luft FC, Christensen EI, Willnow TE (1999) Megalin antagonizes activation of the parathyroid hormone receptor. J Biol Chem 274:5620–5625

Hjälm G, Murray E, Crumley G, Harazim W, Lundgren S, Onyango I, Ek B, Larsson M, Juhlin C, Hellman P, Davis H, Åkerström G, Rask L, Morse B (1996) Cloning and sequencing of human gp330, a Ca^{2+}-binding receptor with potential intracellular signaling properties. Eur J Biochem 239:132–137

Hoopes RR, Jr., Shrimpton AE, Knohl SJ, Hueber P, Hoppe B, Matyus J, Simckes A, Tasic V, Toenshoff B, Suchy SF, Nussbaum RL, Scheinman SJ (2005) Dent disease with mutations in OCRL1. Am J Hum Genet 76:260–267

Hryciw DH, Ekberg J, Lee A, Lensink IL, Kumar S, Guggino WB, Cook DI, Pollock CA, Poronnik P (2004) Nedd4-2 functionally interacts with ClC-5: involvement in constitutive albumin endocytosis in proximal tubule cells. J Biol Chem 279:54996–55007

Hryciw DH, Wang Y, Devuyst O, Pollock CA, Poronnik P, Guggino WB (2003) Cofilin interacts with ClC-5 and regulates albumin uptake in proximal tubule cell lines. J Biol Chem 278:40169–40176

Hvidberg V, Jacobsen C, Strong RK, Cowland JB, Moestrup SK, Borregaard N (2005) The endocytic receptor megalin binds the iron transporting neutrophil-gelatinase-associated lipocalin with high affinity and mediates its cellular uptake. FEBS Lett 579:773–777

Imerslund O (1960) Idiopathic chronic megaloblastic anemia in children. Acta Paediatr Scand 49(suppl.119):1–115

Inui K, Saito H, Iwata T, Hori R (1988) Aminoglycoside-induced alterations in apical membranes of kidney epithelial cell line (LLC-PK1). Am J Physiol Cell Physiol 254:C251–C257

Kalantry S, Manning S, Haub O, Tomihara-Newberger C, Lee HG, Fangman J, Disteche CM, Manova K, Lacy E (2001) The amnionless gene, essential for mouse gastrulation, encodes a visceral-endoderm-specific protein with an extracellular cysteine-rich domain. Nat Genet 27:412–416

Kalatzis V, Antignac C (2003) New aspects of the pathogenesis of cystinosis. Pediatr Nephrol 18:207–215

Kanalas JJ, Makker SP (1991) Identification of the rat Heymann nephritis autoantigen (GP330) as a receptor site for plasminogen. J Biol Chem 266:10825–10829

Kerjaschki D, Farquhar MG (1982) The pathogenic antigen of Heymann nephritis is a membrane glycoprotein of the renal proximal tubule brush border. Proc Natl Acad Sci USA 79:5557–5561

Kerjaschki D, Farquhar MG (1983) Immunocytochemical localization of the Heymann nephritis antigen (GP330) in glomerular epithelial cells of normal Lewis rats. J Exp Med 157:667–686

Klassen RB, Crenshaw K, Kozyraki R, Verroust PJ, Tio L, Atrian S, Allen PL, Hammond TG (2004) Megalin mediates renal uptake of heavy metal metallothionein complexes. Am J Physiol Renal Physiol 287:F393–F403

Kounnas MZ, Chappell DA, Strickland DK, Argraves WS (1993) Glycoprotein 330, a member of the low density lipoprotein receptor family, binds lipoprotein lipase in vitro. J Biol Chem 268:14176–14181

Kounnas MZ, Loukinova EB, Stefansson S, Harmony JA, Brewer BH, Strickland DK, Argraves WS (1995) Identification of glycoprotein 330 as an endocytic receptor for apolipoprotein J/clusterin. J Biol Chem 270:13070–13075

Kozyraki R, Fyfe J, Kristiansen M, Gerdes C, Jacobsen C, Cui S, Christensen EI, Aminoff M, de la Chapelle A, Krahe R, Verroust PJ, Moestrup SK (1999) The intrinsic factor-vitamin B12 receptor, cubilin, is a high-affinity apolipoprotein A-I receptor facilitating endocytosis of high-density lipoprotein. Nat Med 5:656–661

Kozyraki R, Fyfe J, Verroust PJ, Jacobsen C, Dautry-Varsat A, Gburek J, Willnow TE, Christensen EI, Moestrup SK (2001) Megalin-dependent cubilin-mediated endocytosis is a major pathway for the apical uptake of transferrin in polarized epithelia. Proc Natl Acad Sci USA 98:12491–12496

Kozyraki R, Kristiansen M, Silahtaroglu A, Hansen C, Jacobsen C, Tommerup N, Verroust PJ, Moestrup SK (1998) The human intrinsic factor-vitamin B12 receptor, cubilin: molecular characterization and chromosomal mapping of the gene to 10p within the autosomal recessive megaloblastic anemia (MGA1) region. Blood 91:3593–3600

Krauss M, Kinuta M, Wenk MR, De Camilli P, Takei K, Haucke V (2003) ARF6 stimulates clathrin/AP-2 recruitment to synaptic membranes by activating phosphatidylinositol phosphate kinase type Iγ. J Cell Biol 162:113–124

Kristiansen M, Kozyraki R, Jacobsen C, Nexø E, Verroust PJ, Moestrup SK (1999) Molecular dissection of the intrinsic factor-vitamin B12 receptor, cubilin, discloses regions important for membrane association and ligand binding. J Biol Chem 274:20540–20544

Larsson M, Hjalm G, Sakwe AM, Engstrom A, Hoglund AS, Larsson E, Robinson RC, Sundberg C, Rask L (2003) Selective interaction of megalin with postsynaptic density-95 (PSD-95)-like membrane-associated guanylate kinase (MAGUK) proteins. Biochem J 373:381–391

Leheste JR, Rolinski B, Vorum H, Hilpert J, Nykjaer A, Jacobsen C, Aucouturier P, Moskaug J, Otto A, Christensen EI, Willnow TE (1999) Megalin knockout mice as an animal model of low molecular weight proteinuria. Am J Pathol 155:1361–1370

Levine JS, Allen RH, Alpers DH, Seetharam B (1984) Immunocytochemical localization of the intrinsic factor-cobalamin receptor in dog ileum: distribution of intracellular receptor during cell maturation. J Cell Biol 98:1111–1118

Lowe CU, Terrey M, MacLachlan EA (1952) Organic aciduria, decreased renal ammonia production, hydrophthalmos, and mental retardation; a clinical entity. AMA Am J Dis Child 83:164–184

Lundgren S, Carling T, Hjälm G, Juhlin C, Rastad J, Pihlgren U, Rask L, Åkerström G, Hellman P (1997) Tissue distribution of human gp330/megalin, a putative Ca^{2+}-sensing protein. J Histochem Cytochem 45:383–392

Luyckx VA, Leclercq B, Dowland LK, Yu AS (1999) Diet-dependent hypercalciuria in transgenic mice with reduced CLC5 chloride channel expression. Proc Natl Acad Sci USA 96:12174–12179

Madsen KM, Harris RH, Tisher CC (1982) Uptake and intracellular distribution of ferritin in the rat distal convoluted tubule. Kidney Int 21:354–361

Maranda B, Brown D, Bourgoin S, Casanova JE, Vinay P, Ausiello DA, Marshansky V (2001) Intra-endosomal pH-sensitive recruitment of the Arf-nucleotide exchange factor ARNO and Arf6 from cytoplasm to proximal tubule endosomes. J Biol Chem 276:18540–18550

Marino M, Zheng G, Chiovato L, Pinchera A, Brown D, Andrews D, McCluskey RT (2000) Role of megalin (gp330) in transcytosis of thyroglobulin by thyroid cells. A novel function in the control of thyroid hormone release. J Biol Chem 275:7125–7137

Marshansky V, Ausiello DA, Brown D (2002) Physiological importance of endosomal acidification: potential role in proximal tubulopathies. Curr Opin Nephrol Hypertens 11:527–537

McCarthy RA, Barth JL, Chintalapudi MR, Knaak C, Argraves WS (2002) Megalin functions as an endocytic sonic hedgehog receptor. J Biol Chem 277:25660–25667

Mingeot-Leclercq MP, Tulkens PM (1999) Aminoglycosides: nephrotoxicity. Antimicrob Agents Chemother 43:1003–1012

Moestrup SK, Birn H, Fischer PB, Petersen CM, Verroust PJ, Sim RB, Christensen EI, Nexø E (1996) Megalin-mediated endocytosis of transcobalamin-vitamin-B12 complexes suggests a role of the receptor in vitamin-B12 homeostasis. Proc Natl Acad Sci USA 93:8612–8617

Moestrup SK, Cui S, Vorum H, Bregengård C, Bjørn SE, Norris K, Gliemann J, Christensen EI (1995) Evidence that epithelial glycoprotein 330/megalin mediates uptake of polybasic drugs. J Clin Invest 96:1404–1413

Moestrup SK, Kozyraki R, Kristiansen M, Kaysen JH, Rasmussen HH, Brault D, Pontillon F, Goda FO, Christensen EI, Hammond TG, Verroust PJ (1998a) The intrinsic factor-vitamin B12 receptor and target of teratogenic antibodies is a megalin-binding peripheral membrane protein with homology to developmental proteins. J Biol Chem 273:5235–5242

Moestrup SK, Nielsen S, Andreasen P, Jørgensen KE, Nykjaer A, Røigaard H, Gliemann J, Christensen EI (1993) Epithelial glycoprotein-330 mediates endocytosis of plasminogen activator-plasminogen activator inhibitor type-1 complexes. J Biol Chem 268:16564–16570

Moestrup SK, Schousboe I, Jacobsen C, Leheste JR, Christensen EI, Willnow TE (1998b) β2-Glycoprotein-I (apolipoprotein H) and β2-glycoprotein-I-phospholipid complex harbor a recognition site for the endocytic receptor megalin. J Clin Invest 102:902-909

Morris SM, Tallquist MD, Rock CO, Cooper JA (2002) Dual roles for the Dab2 adaptor protein in embryonic development and kidney transport. EMBO J 21:1555–1564

Muller D, Ankermann T, Stephani U, Kirschstein M, Szelestei T, Luft FC, Willnow TE (2001) Holoprosencephaly and low molecular weight proteinuria: the human homologue of murine megalin deficiency. Am J Kidney Dis 37:624–628

Nagai J, Christensen EI, Morris SM, Willnow TE, Cooper JA, Nielsen R (2005) Mutually dependent localization of megalin and Dab2 in the renal proximal tubule. Am J Physiol Renal Physiol 289:F569–F576

Nagai J, Tanaka H, Nakanishi N, Murakami T, Takano M (2001) Role of megalin in renal handling of aminoglycosides. Am J Physiol Renal Physiol 281:F337–F344

Nagai M, Meerloo T, Takeda T, Farquhar MG (2003) The adaptor protein ARH escorts megalin to and through endosomes. Mol Biol Cell 14:4984–4996

Norden AG, Lapsley M, Igarashi T, Kelleher CL, Lee PJ, Matsuyama T, Scheinman SJ, Shiraga H, Sundin DP, Thakker RV, Unwin RJ, Verroust P, Moestrup SK (2002) Urinary megalin deficiency implicates abnormal tubular endocytic function in Fanconi syndrome. J Am Soc Nephrol 13:125–133

Nykjaer A, Dragun D, Walther D, Vorum H, Jacobsen C, Herz J, Melsen F, Christensen EI, Willnow TE (1999) An endocytic pathway essential for renal uptake and activation of the steroid 25-(OH) vitamin D3. Cell 96:507–515

Nykjaer A, Fyfe JC, Kozyraki R, Leheste JR, Jacobsen C, Nielsen MS, Verroust PJ, Aminoff M, de la Chapelle A, Moestrup SK, Ray R, Gliemann J, Willnow TE, Christensen EI (2001) Cubilin dysfunction causes abnormal metabolism of the steroid hormone 25(OH) vitamin D_3. Proc Natl Acad Sci USA 98:13895–13900

Oleinikov AV, Zhao J, Makker SP (2000) Cytosolic adaptor protein Dab2 is an intracellular ligand of endocytic receptor gp600/megalin. Biochem J 347:613–621

Orlando RA, Kerjaschki D, Kurihara H, Biemesderfer D, Farquhar MG (1992) gp330 associates with a 44-kDa protein in the rat kidney to form the Heymann nephritis antigenic complex. Proc Natl Acad Sci USA 89:6698–6702

Orlando RA, Rader K, Authier F, Yamazaki H, Posner BI, Bergeron JJ, Farquhar MG (1998) Megalin is an endocytic receptor for insulin. J Am Soc Nephrol 9:1759–1766

Oyama Y, Takeda T, Hama H, Tanuma A, Iino N, Sato K, Kaseda R, Ma M, Yamamoto T, Fujii H, Kazama JJ, Odani S, Terada Y, Mizuta K, Gejyo F, Saito A (2005) Evidence for megalin-mediated proximal tubular uptake of L-FABP, a carrier of potentially nephrotoxic molecules. Lab Invest 85:522–531

Petersen HH, Hilpert J, Militz D, Zandler V, Jacobsen C, Roebroek AJ, Willnow TE (2003) Functional interaction of megalin with the megalin binding protein (MegBP), a novel tetratrico peptide repeat-containing adaptor molecule. J Cell Sci 116:453–461

Picollo A, Pusch M (2005) Chloride/proton antiporter activity of mammalian CLC proteins ClC-4 and ClC-5. Nature 436:420–423

Piwon N, Gunther W, Schwake M, Bosl MR, Jentsch TJ (2000) ClC-5 Cl^--channel disruption impairs endocytosis in a mouse model for Dent's disease. Nature 408:369–373

Ranganathan S, Knaak C, Morales CR, Argraves WS (1999) Identification of low density lipoprotein receptor-related protein-2/megalin as an endocytic receptor for seminal vesicle secretory protein II. J Biol Chem 274:5557–5563

Raychowdhury R, Niles JL, McCluskey RT, Smith JA (1989) Autoimmune target in Heymann nephritis is a glycoprotein with homology to the LDL receptor. Science 244:1163–1165

Rosenblatt DS, Fenton WA (1999) Inborn errors of cobalamin metabolism. In: Banerjee R (ed) Chemistry and Biochemistry of Vitamin B12. Wiley and Sons, New York, pp 367–384

Sahali D, Mulliez N, Chatelet F, Dupuis R, Ronco P, Verroust P (1988) Characterization of a 280-kD protein restricted to the coated pits of the renal brush border and the epithelial cells of the yolk sac. Teratogenic effect of the specific monoclonal antibodies. J Exp Med 167:213–218

Sahali D, Mulliez N, Chatelet F, Laurent Winter C, Citadelle D, Roux C, Ronco P, Verroust P (1992) Coexpression in humans by kidney and fetal envelopes of a 280 kDa-coated pit-restricted protein. Similarity with the murine target of teratogenic antibodies. Am J Pathol 140:33–44

Saito A, Nagai R, Tanuma A, Hama H, Cho K, Takeda T, Yoshida Y, Toda T, Shimizu F, Horiuchi S, Gejyo F (2003) Role of megalin in endocytosis of advanced glycation end products: implications for a novel protein binding to both megalin and advanced glycation end products. J Am Soc Nephrol 14:1123–1131

Saito A, Pietromonaco S, Loo AK, Farquhar MG (1994) Complete cloning and sequencing of rat gp330/"megalin," a distinctive member of the low density lipoprotein receptor gene family. Proc Natl Acad Sci USA 91:9725–9729

Scheel O, Zdebik AA, Lourdel S, Jentsch TJ (2005) Voltage-dependent electrogenic chloride/proton exchange by endosomal CLC proteins. Nature 436:424–427

Scheinman SJ (1998) X-linked hypercalciuric nephrolithiasis: clinical syndromes and chloride channel mutations. Kidney Int 53:3-17

Schmitz C, Hilpert J, Jacobsen C, Boensch C, Christensen EI, Luft FC, Willnow TE (2002) Megalin deficiency offers protection from renal aminoglycoside accumulation. J Biol Chem 277:618–622

Seetharam B, Christensen EI, Moestrup SK, Hammond TG, Verroust PJ (1997) Identification of rat yolk sac target protein of teratogenic antibodies, gp280, as intrinsic factor-cobalamin receptor. J Clin Invest 99:2317–2322

Silverblatt FJ, Kuehn C (1979) Autoradiography of gentamicin uptake by the rat proximal tubule cell. Kidney Int 15:335–345

Sousa MM, Norden AG, Jacobsen C, Willnow TE, Christensen EI, Thakker RV, Verroust PJ, Moestrup SK, Saraiva MJ (2000) Evidence for the role of megalin in renal uptake of transthyretin. J Biol Chem 275:38176–38181

Spoelgen R, Hammes A, Anzenberger U, Zechner D, Andersen OM, Jerchow B, Willnow TE (2005) LRP2/megalin is required for patterning of the ventral telencephalon. Development 132:405–414

Stefansson S, Chappell DA, Argraves KM, Strickland DK, Argraves WS (1995a) Glycoprotein 330/low density lipoprotein receptor-related protein-2 mediates endocytosis of low density lipoproteins via interaction with apolipoprotein B100. J Biol Chem 270:19417–19421

Stefansson S, Kounnas MZ, Henkin J, Mallampalli RK, Chappell DA, Strickland DK, Argraves WS (1995b) gp330 on type II pneumocytes mediates endocytosis leading to degradation of pro-urokinase, plasminogen activator inhibitor-1 and urokinase-plasminogen activator inhibitor-1 complex. J Cell Sci 108:2361–2368

Straus W (1964) Occurrence of phagosomes and phago-lysosomes in different segments of the nephron in relation to the reabsorption, transport, digestion and extrusion of intravenously injected horseradish peroxidase. J Cell Biol 21:295–308

Suchy SF, Olivos-Glander IM, Nussabaum RL (1995) Lowe syndrome, a deficiency of phosphatidylinositol 4,5-bisphosphate 5-phosphatase in the Golgi apparatus. Hum Mol Genet 4:2245–2250

Takeda T, Yamazaki H, Farquhar MG (2003) Identification of an apical sorting determinant in the cytoplasmic tail of megalin. Am J Physiol Cell Physiol 284:C1105–C1113

Tanner SM, Li Z, Bisson R, Acar C, Oner C, Oner R, Cetin M, Abdelaal MA, Ismail EA, Lissens W, Krahe R, Broch H, Grasbeck R, de la Chapelle A (2004) Genetically heterogeneous selective intestinal malabsorption of vitamin B12: founder effects, consanguinity, and high clinical awareness explain aggregations in Scandinavia and the Middle East. Hum Mutat 23:327–333

Thurberg BL, Rennke H, Colvin RB, Dikman S, Gordon RE, Collins AB, Desnick RJ, O'Callaghan M (2002) Globotriaosylceramide accumulation in the Fabry kidney is cleared from multiple cell types after enzyme replacement therapy. Kidney Int 62:1933–1946

Town M, Jean G, Cherqui S, Attard M, Forestier L, Whitmore SA, Callen DF, Gribouval O, Broyer M, Bates GP, van't Hoff W, Antignac C (1998) A novel gene encoding an integral membrane protein is mutated in nephropathic cystinosis. Nat Genet 18:319–324

Vandewalle A, Farman N, Morin JP, Fillastre JP, Hatt PY, Bonvalet JP (1981) Gentamicin incorporation along the nephron: autoradiographic study on isolated tubules. Kidney Int 19:529–539

Verroust PJ, Birn H, Nielsen R, Kozyraki R, Christensen EI (2002) The tandem endocytic receptors megalin and cubilin are important proteins in renal pathology. Kidney Int 62:745–756

Wahlstedt-Froberg V, Pettersson T, Aminoff M, Dugue B, Grasbeck R (2003) Proteinuria in cubilin-deficient patients with selective vitamin B12 malabsorption. Pediatr Nephrol 18:417–421

Wang SS, Devuyst O, Courtoy PJ, Wang XT, Wang H, Wang Y, Thakker RV, Guggino S, Guggino WB (2000) Mice lacking renal chloride channel, CLC-5, are a model for Dent's disease, a nephrolithiasis disorder associated with defective receptor-mediated endocytosis. Hum Mol Genet 9:2937–2945

Willnow TE (1998) Receptor-associated protein (RAP): a specialized chaperone for endocytic receptors. Biol Chem 379:1025–1031

Willnow TE, Armstrong SA, Hammer RE, Herz J (1995) Functional expression of low density lipoprotein receptor- related protein is controlled by receptor-associated protein in vivo. Proc Natl Acad Sci USA 92:4537–4541

Willnow TE, Goldstein JL, Orth K, Brown MS, Herz J (1992) Low density lipoprotein receptor-related protein and gp330 bind similar ligands, including plasminogen activator-inhibitor complexes and lactoferrin, an inhibitor of chylomicron remnant clearance. J Biol Chem 267:26172–26180

Willnow TE, Hilpert J, Armstrong SA, Rohlmann A, Hammer RE, Burns DK, Herz J (1996a) Defective forebrain development in mice lacking gp330/megalin. Proc Natl Acad Sci USA 93:8460–8464

Willnow TE, Rohlmann A, Horton J, Otani H, Braun JR, Hammer RE, Herz J (1996b) RAP, a specialized chaperone, prevents ligand-induced ER retention and degradation of LDL receptor-related endocytic receptors. EMBO J 15:2632–2639

Wrong OM, Norden AG, Feest TG (1994) Dent's disease; a familial proximal renal tubular syndrome with low-molecular-weight proteinuria, hypercalciuria, nephrocalcinosis, metabolic bone disease, progressive renal failure and a marked male predominance. QJM 87:473–493

Xu D, Kozyraki R, Newman TC, Fyfe JC (1999) Genetic evidence of an accessory activity required specifically for cubilin brush-border expression and intrinsic factor-cobalamin absorption. Blood 94:3604–3606

Yammani RR, Seetharam S, Dahms NM, Seetharam B (2003) Transcobalamin II receptor interacts with megalin in the renal apical brush border membrane. J Membr Biol 193:57–66

Yammani RR, Seetharam S, Seetharam B (2001) Cubilin and megalin expression and their interaction in the rat intestine: effect of thyroidectomy. Am J Physiol Endocrinol Metab 281:E900–E907

Yammani RR, Sharma M, Seetharam S, Moulder JE, Dahms NM, Seetharam B (2002) Loss of albumin and megalin binding to renal cubilin in rats results in albuminuria after total body irradiation. Am J Physiol Regul Integr Comp Physiol 283:R339–R346

Zhai XY, Nielsen R, Birn H, Drumm K, Mildenberger S, Freudinger R, Moestrup SK, Verroust PJ, Christensen EI, Gekle M (2000) Cubilin- and megalin-mediated uptake of albumin in cultured proximal tubule cells of opossum kidney. Kidney Int 58:1523–1533

Zheng G, Bachinsky DR, Stamenkovic I, Strickland DK, Brown D, Andres G, McCluskey RT (1994) Organ distribution in rats of two members of the low-density lipoprotein receptor gene family, gp330 and LRP/α2MR, and the receptor-associated protein (RAP). J Histochem Cytochem 42:531–542

Zheng G, Marino, Zhao J, McCluskey RT (1998) Megalin (gp330): a putative endocytic receptor for thyroglobulin (Tg). Endocrinology 139:1462–1465

Rev Physiol Biochem Pharmacol (2007)
DOI 10.1007/112_2006_0605

G. Zifarelli · M. Pusch

CLC chloride channels and transporters: a biophysical and physiological perspective

Published online: 5 December 2006

Abstract Chloride-transporting proteins play fundamental roles in many tissues in the plasma membrane as well as in intracellular membranes. They have received increasing attention in the last years because crucial, and often unexpected and novel, physiological functions have been disclosed with gene-targeting approaches, X-ray crystallography, and biophysical analysis. CLC proteins form a gene family that comprises nine members in mammals, at least four of which are involved in human genetic diseases. The X-ray structure of the bacterial CLC homolog, ClC-ec1, revealed a complex fold and confirmed the anticipated homodimeric double-barreled architecture of CLC-proteins with two separate Cl^- ion transport pathways, one in each subunit. Four of the mammalian CLC proteins, ClC-1, ClC-2, ClC-Ka, and ClC-Kb, are chloride ion channels that fulfill their functional roles—stabilization of the membrane potential, transepithelial salt transport, and ion homeostasis—in the plasma membrane. The other five CLC proteins are predominantly expressed in intracellular organelles like endosomes and lysosomes, where they are probably important for a proper luminal acidification, in concert with the V-type H^+-ATPase. Surprisingly, ClC-4, ClC-5, and probably also ClC-3, are not Cl^- ion channels but exhibit significant Cl^-/H^+ antiporter activity, as does the bacterial homolog ClC-ec1 and the plant homolog AtCLCa. The physiological significance of the Cl^-/H^+ antiport activity remains to be established.

Overview and scope

The lipid bilayer that surrounds all living cells and the organelles inside eukaryotic cells presents, by virtue of its fatty nature, an insurmountable electrostatic barrier for the diffusive passage of small inorganic ions like Na^+, K^+, Ca^{2+}, Cl^- and also small organic ions like amino acids or HCO_3^-. To overcome this barrier and to allow the exchange of these substrates across the lipid bilayer in a controlled manner, nature has invented an incredible

G. Zifarelli · M. Pusch (✉)
CNR, Istituto di Biofisica,
Via De Marini 6, 16149 Genova, Italy
e-mail: pusch@ge.ibf.cnr.it · Tel.: +39-0106475-561/522 · Fax: +39-0106475-500

variety of different ion-transporting proteins, most of which allow the specific passage of only a very limited subset of ions. Transport proteins can be grossly subdivided into passive transporters and active transporters. Conceptually, passive transporters can be regarded as enzymes that lower the activation energy for passive diffusion across the lipid bilayer. The most important example of passive transporters are ion channels, which provide a selective pore that allows a high-throughput transport, close to the diffusion limit in some cases, while maintaining exquisite selectivity. Active transporters couple the energy of the translocation of one substrate, or other energy sources such as ATP hydrolysis, to the transport of another substrate, often in a strictly stoichiometric manner. One prominent example of this class of proteins are the familiar P-type ion pumps and ion cotransporters. Active transport is generally associated with the picture of an alternating access model of transport in which the transporter exposes its ion binding sites alternatively to one or the other side of the membrane (see Tanford 1983). According to this mechanism, one or a few substrate molecules are translocated for each transport cycle, leading to the slow transport rates seen for active transporters, compared to those of ion channels (Hille 2001). As a consequence, in general, the architecture of active transporter proteins (see, e.g., Abramson et al. 2003; Toyoshima et al. 2000) is quite different from that of ion channels (see, e.g., Doyle et al. 1998; Miyazawa et al. 2003).

The present review focuses on anion-selective channels and, in particular, on Cl^- channels from the CLC family (Jentsch et al. 2002). However, as described below, the same basic architecture in the CLC family of proteins (Jentsch et al. 2005c) can be used to produce either active transporters (Accardi and Miller 2004; De Angeli et al. 2006; Picollo and Pusch 2005; Scheel et al. 2005) or passive chloride channels (Bauer et al. 1991). Since a full appreciation of the physiological role of CLC proteins requires a molecular comprehension of their mechanism of transport, we will have to consider passive channel-mediated diffusion as well as the active antiport of protons and Cl^- ions.

It is important to note that the CLC family represents only one of several classes of proteins carrying out Cl^- transport. A detailed treatment of such a vast and variegated array is beyond the scope of this review, but we nevertheless provide a brief overview of the physiological roles of Cl^- channels not belonging to the CLC branch.

The transport of Cl^- (or any other ion) across the plasma membrane has two distinct consequences: transport of the substrate and transport of electrical charge. The transport of charge is fundamental for the regulation of excitability in nerve and muscle, whereas the transport of substrate is of paramount importance for epithelial physiology. In neurons and muscle cells the membrane potential, V_m, is one of the most critical physiological variables. The activation of closed Cl^- channels, or the inactivation of active Cl^- channels, changes V_m according to the equilibrium potential for Cl^-, E_{Cl}. In most cases, the intracellular Cl^- concentration ($[Cl^-]_{int}$) is low, such that E_{Cl} is very negative and close to or even more negative than E_K. Low $[Cl^-]_{int}$ is achieved by secondary active KCl cotransport proteins (Hübner et al. 2001). Thus Cl^- channel activity in nerve and muscle generally dampens excitability, stabilizing a negative membrane potential. For the dampening and stabilization of the membrane potential not only is the value of E_{Cl} important, but also the chloride conductance, g_{Cl}, relative to other conductances, that is, a large g_{Cl} associated with a slightly depolarized E_{Cl} will nevertheless impede strong depolarization caused by a (relatively) small depolarizing conductance. A typical example, the skeletal muscle Cl^- conductance that is provided by the ClC-1 Cl^- channel, is described in more detail below. In neurons, postsynaptic GABA and glycine receptors are the most important anion channels in the plasma membrane (Jentsch et al. 2002). The traditional view is that their activation suppresses excitation (i.e., action potential firing) of the postsynaptic cell. It is clearly beyond the scope of this review to describe these neuronal channels in detail. However, we would like to mention that activation of GABA and glycine receptors is not always inhibitory: In the developing nervous system and in some specialized neuronal structures, $[Cl^-]_{int}$ is relatively high, leading to a paradoxical excitatory effect of receptor activation (Marty and Llano 2005; Misgeld et al. 1986). GABA and glycine receptors are poorly selective for Cl^-, showing a significant permeability even to cations (Wotring et al. 2003). Physiologically, the permeability to bicarbonate (HCO_3^-) seems to be of particular relevance as it significantly contributes to a rise of $[Cl^-]_{int}$ after GABA stimulation (see Marty and Llano 2005).

Apart from CLC proteins and GABA/glycine receptors, the only molecularly identified Cl^- channel is the "cystic fibrosis transmembrane conductance regulator," CFTR (Riordan et al. 1989). CFTR is a widely expressed, but mostly epithelial, Cl^- channel. Mutations in the gene coding for CFTR cause cystic fibrosis (Tsui 1991), one of the most common lethal genetic diseases. Structurally, CFTR belongs to the very large class of ABC transporters, but it seems to be the only channel member of this family of active transport proteins. Despite extensive research in the 15 years since its cloning, the molecular mechanisms of channel gating by protein kinase A and intracellular ATP and also its physiological role are still relatively unclear. Excellent reviews about many aspects of CFTR have been published recently (Guggino 2004; Hanrahan and Wioland 2004; Riordan 2005).

Several important anion conductances have been described in various mammalian cell types whose molecular identity is still unknown or in dispute. The most typical examples are the swelling-activated Cl^- channel, also known as VRAC (volume-regulated anion channel) (Eggermont et al. 2001), and various types of calcium-activated Cl^- channels. VRAC is probably present in all animal cells and is activated by cell swelling, but the molecular mechanism leading to its activation is unknown (Eggermont et al. 2001). This channel is also permeable to small organic solutes and has been proposed to be important for a process called regulatory volume decrease (RVD). Cellular volume regulation is essential for all cell types to respond to osmotic challenges caused by changes of the extracellular medium as well as to metabolically induced changes in intracellular osmolarity. The functional properties of VRAC have been extensively studied, and several proteins have been proposed as molecular correlates of VRAC, but none of these is generally accepted (see Jentsch et al. 2002).

Ca^{2+} activated Cl^- channels, CaCCs, are also found in many different cell systems including smooth muscle, epithelia, and olfactory receptors. Their activation, via an increase of intracellular $[Ca^{2+}]$, generally leads to cell depolarization and thus, for example, smooth muscle contraction or amplification of olfactory sensation (Hartzell et al. 2005). In epithelia, CaCC activation is responsible for transient Cl^- (and water) secretion, for example, in salivary glands. Similar to VRAC, several proteins have been proposed as molecular correlates of CaCCs, none of them being as yet fully accepted. Currently, the family of bestrophin proteins is under intense study as CaCC candidates (Hartzell et al. 2005), even though a definite proof of their identity is still missing (see, e.g., Rosenthal et al. 2006).

Another example of a Cl^- conductance for which the molecular association with a membrane protein is still lacking is the hyperpolarization- and cAMP-activated Cl^- current measured in choroid plexus cells (Kibble et al. 1996). This current superficially resembles ClC-2 currents, but is found unaltered in ClC-2 knockout mice (Speake et al. 2002). Other examples include an ATP-activated Cl^- current described in mouse parotid acinar cells (Arreola and Melvin 2003), and a proton-activated Cl^- channel (Nobles et al. 2004), both sharing some characteristics with VRAC.

Epithelial ion transporters are designed to allow massive but specific translocation of salts across the epithelial cell sheet. To allow for vectorial ion movement, transporters must be expressed in a polarized manner. For example, the Na^+-K^+-ATPase is usually expressed on the basolateral membrane in epithelial cells. Thus it is important to understand the mechanisms underlying the correct targeting of chloride channels and transporters to the apical versus basolateral membrane. Very little is known about the targeting of the molecularly identified Cl^- channels (CLC channels, GABA/glycine receptors, CFTR), even though several putative partner proteins of CLC channels, possibly important for targeting, have been identified in recent years (Dhani and Bear 2006) and are described in some detail below.

Cl^- channels are not restricted to the plasma membrane but are also found in intracellular organelles. Relatively little is known about the intracellular Cl^- channels from in situ studies. This is largely explained by the inaccessibility of the small intracellular organelles to standard patch clamp techniques. As discussed in detail in later sections of this review, five of the nine mammalian CLC homologs reside in intracellular membranes, and their study thus opens new and promising perspectives for the understanding of the role of intracellular Cl^- channels and transporters.

The present review first describes the general mechanism underlying the function of CLC proteins and then focuses on the biophysical properties and physiological and pathophysiological roles of mammalian, and in particular human, CLC members. For the mechanistic aspects, two "model" CLCs have been most extensively studied. One is the *Torpedo* channel ClC-0, which, compared to many, physiologically more relevant, channels, has favorable biophysical properties, for example, a relatively large single-channel conductance, and whose mechanisms of gating are best understood. The other model CLC is, of course, the bacterial ClC-ec1, for which we have detailed structural information and which can also be studied functionally. For reasons of space we do not attempt to cover the research on CLC proteins in other organisms like plants (Barbier-Brygoo et al. 2000; De Angeli et al. 2006), *Caenorhabditis elegans* (Strange 2003), or other model organisms or pathogens (see, e.g., Salas-Casas et al. 2006).

Introduction: The CLC family of chloride-transporting proteins

The research in the CLC chloride channel field has always been accompanied, right from its very beginning, by a great number of unexpected findings and surprises. Already the first step in the field, the identification of the *Torpedo* channel by Miller and coworkers (White and Miller 1979), was a sort of accident (or artifact) in the quest of the authors to investigate acetylcholine-gated cation channels.

The basic properties of the *Torpedo* chloride channel were established by Miller and coworkers in a series of experiments on reconstituted channels from the electroplax of *Torpedo californica* (Miller and Richard 1990). This organ constitutes an internal battery that the fish use as a source of electric current to stun their prey. The plasma membrane of the electrocytes is extremely rich in a specific type of Cl^- channel (later named ClC-0). Miller and colleagues reconstituted the channel in lipid bilayers and analyzed both macroscopic and single-channel currents (Miller and White 1980; White and Miller 1979). At the single-channel level, an unusual gating behavior was observed, with bursts of channel activity separated by periods in which the channel was closed (Fig. 1). Interestingly, the bursting events had a characteristic pattern with three different, equally spaced levels of conductance (0, 11, and 22 pS, respectively). This behavior was found in many different measuring conditions (Hanke and Miller 1983; Miller 1982) and was immediately interpreted as suggesting that the chloride channel was a functional dimer. In this view, the three substates during

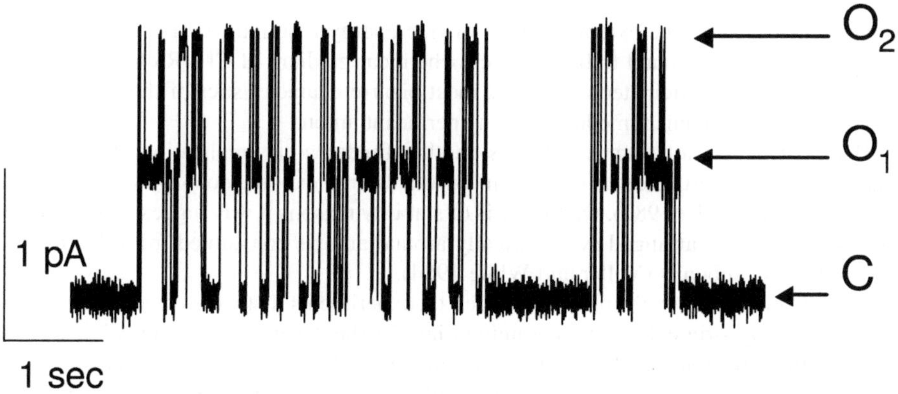

Fig. 1 Schematic (simulated) single-channel trace of the *Torpedo* channel ClC-0. Channel activity occurs in bursts that are separated by long closed periods. Within each bursts two open conductance levels (*O1* and *O2*) are seen, where *O2* has exactly twice the conductance of *O1*

the bursts would represent the independent opening and closing of two identical Cl^- diffusion pathways, called protochannels; the dimeric channel complex may exist with both protochannels simultaneously open, with one open and one closed, or with both closed, generating the three conductance substates.

At all voltages tested, the frequency of substates during a burst followed a binomial distribution as predicted for two independently opening and closing protochannels. Moreover, the probability of a single protopore to be in its conducting state depended on voltage according to a Boltzmann distribution, as expected for a two-state mechanism. This is in agreement with the presence of two independently opening and closing Cl^- pathways and in contrast with the presence of a single channel with different subconductance levels (Miller 1982).

This model was strengthened by a study of DIDS (4,4′-diisothiocyanatostilbene-2,2′-disulfonate) inhibition of single-channel currents (Miller and White 1984). Addition of 10 μM DIDS to the *cis* side of the chamber eliminated first the 22-pS conductance level and, subsequently, the 11-pS conductance level, that is, the bursting activity disappeared. The authors interpreted the finding as being due to the binding (and inhibition) of DIDS first to one and then the other protopore. This strongly supported a model with two separated diffusion pathways (pores) each with a single open state rather than a single Cl^- diffusion pathway with multiple conductance states.

Incidentally, the fact that DIDS inhibited the oriented channels only if added to the *cis* side of the preparation implied that the two protopores had the same orientation in the channel complex.

The fact that the channel activity presented periods of activity (bursts) and periods of no activity (Fig. 1) indicated that the two protochannels were not completely independent from each other. Therefore, it was suggested that there is an inactivating process that closes both protochannels simultaneously and on a slower time scale (which was later defined as a common gate or slow gate) compared to the closing events within a burst (which were attributed to what was later named fast gate) (Miller and Richard 1990).

Another peculiar feature of ClC-0 emerged from the inspection of the beginnings and the endings of the bursts. Burst activity tended to begin with both protopores open and ended more often with only one protopore open (Richard and Miller 1990). This time asymmetry implies that the transitions between the possible states of the protopores are not in thermodynamic equilibrium. The external source of free energy required to drive the irreversible gating transitions was found to be the electrochemical gradient of Cl^- (Richard and Miller 1990). This finding anticipated one of the most bizarre characteristics of the CLC channel family, a gating mechanism mediated by the permeant anion.

The existence of a common gate has another fundamental implication: The two protochannels must be intimately associated in a proteic complex—the double-barreled shotgun model was born (Miller 1982). On the basis of stability reasons it was suggested that the two protopores would be arranged symmetrically around an axis constituted from the interface between the two subunits (Miller and White 1984).

These features, although solidly grounded on experiments that were elegant in their simplicity, were very original, not to say unfamiliar, for the "channel community," and therefore they stirred up considerable controversy. However, the progress made in the analysis of channel function and structure achieved throughout the last twenty years has spectacularly confirmed virtually all of them and provided deeper insights and new unexpected findings that we will try to summarize.

Cloning of the CLC family members

A critical turning point for the research on chloride channels was the cloning of the channel from *Torpedo marmorata*, called ClC-0, with an elegant but extremely labor-intensive expression cloning strategy (Jentsch et al. 1990). This exposed ClC-0 to the use of the powerful tools of molecular biology and allowed, by homology, the identification of several other CLC channels in organisms as diverse as animals, plants, yeast, archaebacteria, and eubacteria (Jentsch et al. 1999; Maduke et al. 2000).

Mammals possess nine different CLC genes, which, on the basis of sequence homology, can be grouped into three branches (Jentsch et al. 2002; Mindell and Maduke 2001). The first branch comprises plasma membrane channels, ClC-1, ClC-2, ClC-Ka, and ClC-Kb, whereas members of the two other branches (ClC-3, ClC-4, and ClC-5 in one branch and ClC-6 and ClC-7 in the other) function primarily in intracellular membranes.

The sequence, and structure, of CLC proteins bears no resemblance to any other class of membrane proteins. A very distinguishing element of all CLC channels and transporters, with respect to other Cl^--transporting membrane proteins, is their anion selectivity. First, members of the CLC family are practically completely impermeable to cations (except protons). Second, among the halides Cl^-, Br^-, and I^-, the selectivity and conductivity sequence for CLC proteins is generally $Cl^- > Br^- > I^-$. According to Wright and Diamond (Wright and Diamond 1977) this indicates a high-field-strength anion binding site in the transport pathway. In contrast, most other Cl^- channels (except CFTR) show an $I^- > Cl^-$ preference, suggestive of a larger pore in which ions are not completely dehydrated.

Crystal structure of the bacterial ClC-ec1

So far it has not been possible to obtain crystal structures from eukaryotic CLC members, and, therefore, all the structural information (for the transmembrane region) available to date has come from investigation of prokaryotic CLC counterparts, an approach that has been successful for a number of cation channels (Doyle et al. 1998; Zhou et al. 2001).

A projection structure of an *Escherichia coli* member of the CLC family, ClC-ec1, at 6.5-Å resolution, supported the dimeric nature of the channel but could not provide any molecular detail (Mindell et al. 2001). A much more thorough insight into the structure-function of CLC proteins was provided by two high-resolution structures of ClC-ec1 and StClC (from *Salmonella typhimurium*) obtained by Dutzler and coworkers (Dutzler et al. 2002, 2003).

The biology of prokaryotic CLC proteins is still largely unexplored. In particular, it was found that ClC-ec1 is actually a Cl^-/H^+ antiporter (Accardi and Miller 2004), a characteristic that conflicts with its proposed role as a shunt conductance relevant for acid resistance (Iyer et al. 2002). More importantly, this finding raises a number of issues regarding the possible extrapolation of features from the prokaryotic to the eukaryotic members of CLC family, some of which are discussed in later paragraphs. However, considering the sequence conservation between prokaryotic CLCs and eukaryotic members of the family, especially in the ion selectivity region (Maduke et al. 1999), there is confidence that the general structural elements apply to the entire family.

ClC-ec1 is a dimer composed of two identical subunits of triangular shape (Fig. 2a). The contact surface area between subunits is extensive (~2,300 $Å^2$), as expected because CLC channels are thought to exist and function only as dimers (Dutzler et al. 2002), even if it is not known at which stage of the biosynthesis dimerization occurs.

Each subunit contains Cl^- ions at its center, indicating a putative ion conduction pathway, with a mutual distance between the two pores of ~39 Å. The largest part of ClC-ec1 is embedded in the lipid bilayer, and only the N- and C-termini protrude into the cytoplasm (Fig. 2).

Each subunit consists of 18 α-helices (labeled A–R) organized in two topologically related domains that span the membrane in opposite directions in an arrangement called "antiparallel architecture" that has been found also in the structure of the aquaporins (Lee et al. 2005; Murata et al. 2000) and of a Na^+/H^+ antiporter from *E. coli* (Hunte et al. 2005).

The two domains are only weakly correlated in their sequence but show a significant similarity regarding the disposition of glycine residues (Dutzler et al. 2002). Some of the helices are long and tilted by about 45° with respect to the membrane; others are short and penetrate the membrane only halfway. The transmembrane structure is similar across the whole CLC family. One fundamental difference lies in the presence of large C-terminal intracellular domains in all eukaryotic and some prokaryotic CLC proteins that are absent in ClC-ec1 and StClC (Estévez and Jentsch 2002; Meyer and Dutzler 2006). Part of the isolated C-terminus of ClC-0 has been recently crystallized (Meyer and Dutzler 2006). Its structure is described below.

In agreement with the fact that ClC-ec1 is not an ion channel allowing the passive diffusive flow of ions but a stoichiometrically coupled ion transporter, ClC-ec1 lacks a real pore. In the structures of ClC-ec1, the central Cl^- ion is completely surrounded by protein and is not "visible" from either side of the membrane. The putative transport pathway is 15 Å long and contains three ion-binding sites named S_{int}, S_{cen}, and S_{ext}, starting from the one closer to the intracellular space. The S_{ext} site was found to be occupied by the negatively charged side chain of a critical glutamate residue (Glu-148) in the wild-type structure, but binds a Cl^- ion if Glu-148 is mutated to alanine or glutamine; no water molecules have been detected in the ion-binding region in the structures (Dutzler 2004; Dutzler et al. 2003) (Fig. 3).

Overall, the transport pathway across ClC-ec1 appears like a very narrow passage connecting intracellular and extracellular vestibules (Dutzler et al. 2002, 2003). The vestibules leading to the selectivity filter on both sides of the membrane contain basic (positively charged) amino acids, such as Arg-147 and Arg-451. The distribution of charges on the

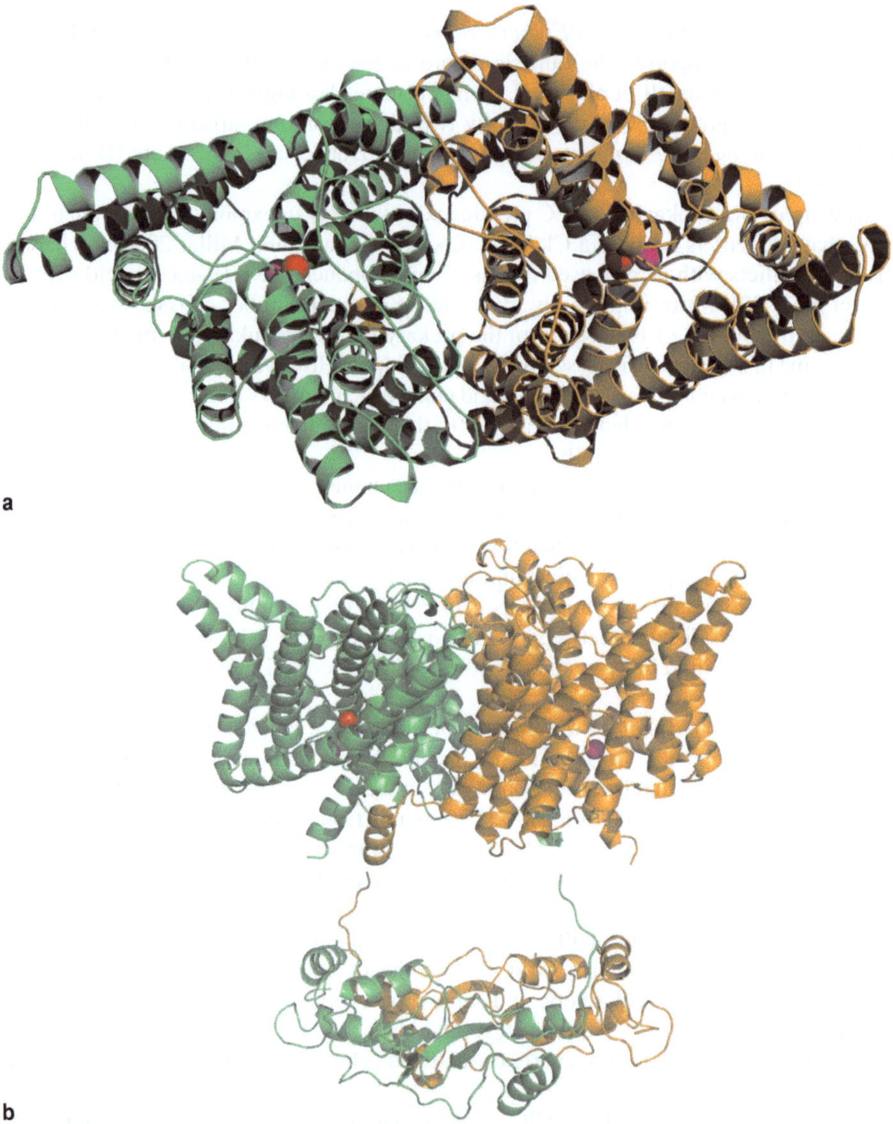

a

b

Fig. 2a, b Overall structure of ClC-ec1 and CBS domains. In **a**, ClC-ec1 (PDB accession no. 1KPK) is shown in a ribbon representation viewed from the extracellular side. The two subunits of the dimeric complex are shown in *green* and *orange*, respectively. The two Cl^- ions in the transport pathway of each subunit are shown in *red* (central chloride ion) and *magenta* (inner chloride ion). **b** Side view of ClC-ec1 assembled with the cytoplasmic C-terminal domains of ClC-0 from *Torpedo marmorata* (PDB accession no. 2D4Z). The relative orientation has been arbitrarily fixed, because the exact spatial arrangement of the C-terminal domain with respect to the membrane part is unknown

entire channel surface creates an electrostatic potential that probably funnels Cl^- ions into the pore entryways. The two pores of the dimer are separated by a large distance and by an electronegative region on the extracellular surface (Dutzler et al. 2002). These findings are consistent with the functional independence of the two pores in ClC-0 (Ludewig et al. 1996, 1997b; Middleton et al. 1996).

Amino acids from four separate protein regions are brought together near the membrane center to form the three ion-binding sites (Dutzler et al. 2002, 2003). These regions are highly conserved in CLC proteins; they include GSGIP in helix D (106–110), G(K/R)EGP in helix F (146–150), GXFXP in helix N (355–359), and Tyr-455 in helix R (Fig. 3a). These sequences occur at the N-termini of α-helices, where polypeptide loops precede α-helices D, F, and N. In agreement with this complex structural arrangement, several regions of CLC proteins influence pore properties like ion selectivity, single-channel conductance, and gating (Estévez and Jentsch 2002; Ludewig et al. 1997a, 1996; Pusch et al. 1995a,. 1995b; Wollnik et al. 1997).

Helices D, F, N, and R are oriented with their N-terminus pointing toward the central binding site. Because of the helix dipole, this arrangement of helices is expected to create a favorable environment for anion binding. This is, for example, the mechanism hypothesized to be at work in KcsA to favor ion binding to the pore (Roux and MacKinnon 1999). However, some authors have raised doubts against the generalization of such a mechanism to ClC-ec1. On the basis of electrostatic calculations, Faraldo-Gomez and Roux (Faraldo-Gomez and Roux 2004) proposed that in ClC-ec1 the energetic cost for desolvation of the anions on transfer into the protein is contributed only marginally by long-range interaction

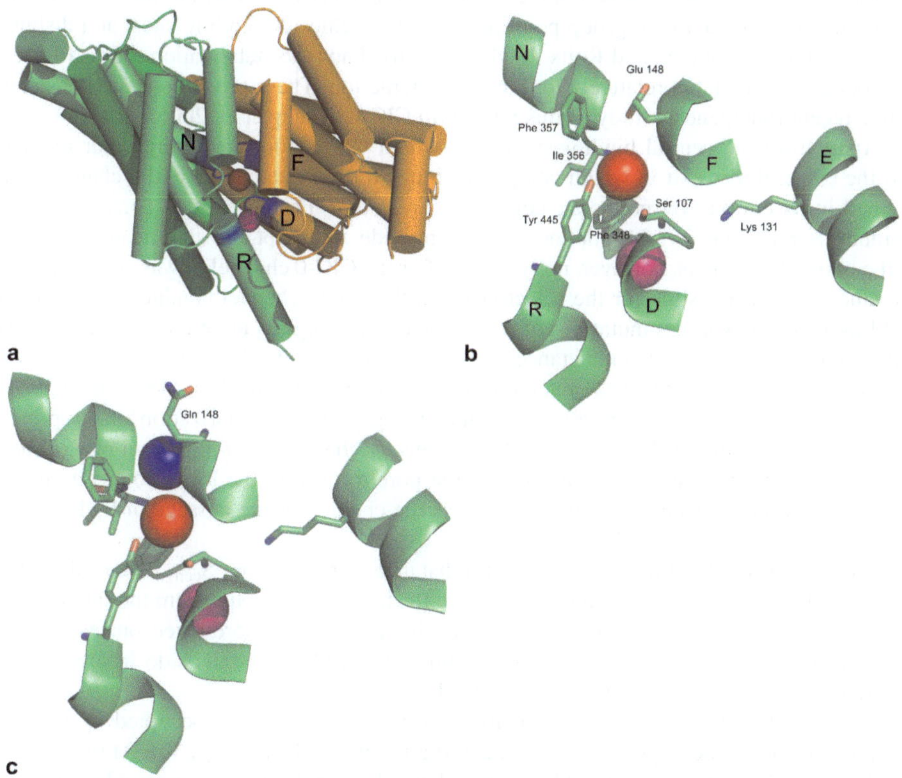

Fig. 3a–c The Cl⁻ transport pathway and Cl⁻ binding sites. **a** The position of the two Cl⁻ binding sites of ClC-ec1 (coloring of subunits and chloride ions as in Fig. 2) with the protein regions involved in coordination of the central Cl⁻ ion shown in *blue*. **b** Detail of the amino acids coordinating the central Cl⁻ ion in the wild-type ClC-ec1. **c** The central Cl⁻ binding site in the structure of the mutant Glu-148-Gln (PDB accession no. 1OTU). The side chain of Gln-148 is displaced from the permeation pathway, and a third Cl⁻ ion (shown in *blue*) is present at the position occupied by the side chain of Glu-148 in the wild-type structure

with the helix macrodipole and comes mainly from favorable electrostatic interactions with the backbone and side chains of residues that are not directly located in the permeation pathway.

This view is shared also by Cohen and Schulten (Cohen and Schulten 2004), who suggest, on the basis of molecular dynamics calculations, that the broken helix architecture does not constitute a prominent characteristic of the energy profile controlling Cl^- conduction and may possibly represent Nature's design evolved to expose backbone amide groups to the permeant anions.

In this respect, it is interesting to note that the bound Cl^- ions do not make direct contact with a full positive charge from lysine or arginine residues. It has been speculated that a full positive charge would create a deep energy well and cause Cl^- to bind too tightly, compromising the efficiency of transport (Dutzler 2004).

The Cl^- ion at the S_{cen} site is fully dehydrated and is coordinated by main chain amide nitrogen atoms from amino acids Ile-356 and Phe-357 and by side chain oxygen atoms from Ser-107 and Tyr-445 (Fig. 3b). On the basis of electrostatic calculations, however, it was hypothesized that the single most important favorable ion-side chain interaction in ClCec-1 originates not from Ser-107 or Tyr-445 but from Lys-131. The side chain of this residue is located in the transmembrane helix E, completely buried within the protein, with its positively charged amino group pointing toward the chloride-binding sites, at a distance of 7–9 Å (Faraldo-Gomez and Roux 2004) (Fig. 3b). Thus the stabilization seems to occur by a purely electrostatic, relatively long-range, interaction. These predictions are consistent with a recent mutagenesis study of this residue in ClC-0 (Zhang et al. 2006).

Apart from the central binding pocket in which Cl^- is coordinated by polar residues and the extracellular exit in which charged residues form a putative gate, the channel pore is lined in its entirety by nonpolar, noncharged residues. The pore's two conserved polar residues, Ser-107 and Tyr-445, define S_{cen} and provide an abrupt and significant narrowing of the pore. Their role is, however, not yet clear. For the ClC-0 channel, it was shown that the tyrosine is not responsible for the selectivity and the single-channel conductance (Accardi and Pusch 2003), whereas mutations of the serine residue slightly altered ion selectivity and reduced the single-channel conductance (Chen et al. 2003; Ludewig et al. 1996). Also, simulation studies suggested that the interaction energy of Ser-107 and Tyr-445 with Cl^- is not significant compared to the energy due to the strong electrical polarization of the protein (Cohen and Schulten 2004). It was therefore suggested that the most important role of these residues is to keep an anion permanently in the pore to prevent the formation of a proton-carrying continuous water file stretching across the channel or the passage of hydrophobic anions (Cohen and Schulten 2004).

The second ion binding-site, S_{int}, is at a distance of 6.5 Å from S_{cen}, toward the intracellular side. It is located at the interface where the aqueous vestibule from the intracellular solution meets the selectivity filter. The ion at this position is coordinated on one side by main chain amide nitrogen atoms from the end of helix D and on the side where it is exposed to the vestibule is probably still hydrated.

In the first structure of ClC-ec1 (Dutzler et al. 2002), S_{ext} was occupied by the side chain of the glutamate at position 148, occluding the ion pathway (Fig. 3b). At that time, it was believed that ClC-ec1 was a chloride ion channel, even if no direct electrophysiological data were available yet. It was therefore hypothesized that the crystal structure captured the channel in a state in which Cl^- was occluded, that is, did not have direct access to intracellular or extracellular space, and that Cl^- ions would activate conduction (gate the channel open) entering the pore from the extracellular side and inducing a conformational change that would displace the glutamate side chain.

This prediction was largely confirmed by a second structure of ClC-ec1 determined at 2.5-Å resolution in combination with parallel electrophysiological measurements performed on ClC-0 (Dutzler et al. 2003). When the corresponding Glu-148 of ClC-ec1 was mutated in ClC-0 into Ala (Glu-166-Ala), Gln (Glu-166-Gln), or Val (Glu-166-Val), it was found that fast gating transitions were practically abolished (Dutzler et al. 2003). Interestingly, lowering extracellular pH produced a similar open phenotype for wild-type ClC-0 (Chen and Chen 2001; Dutzler et al. 2003), suggesting that the protonation of the glutamate side chain from the extracellular side opens the wild-type channel (Fig. 4). The crystal structures of ClC-ec1 in which Glu-148 was mutated to Ala and Gln presented an anion at S_{ext} instead of the Glu side chain (Dutzler et al. 2003) (Fig. 3c). It was therefore suggested that when Glu-148 is mutated, the pore is open because it contains an uninterrupted queue of anions connecting the intracellular and the extracellular solutions.

In the structure of the Glu-148-Gln mutant of ClC-ec1, the side chain of Gln-148 is directed toward the extracellular solution rather than into the pore (Fig. 3c), and it was spec-

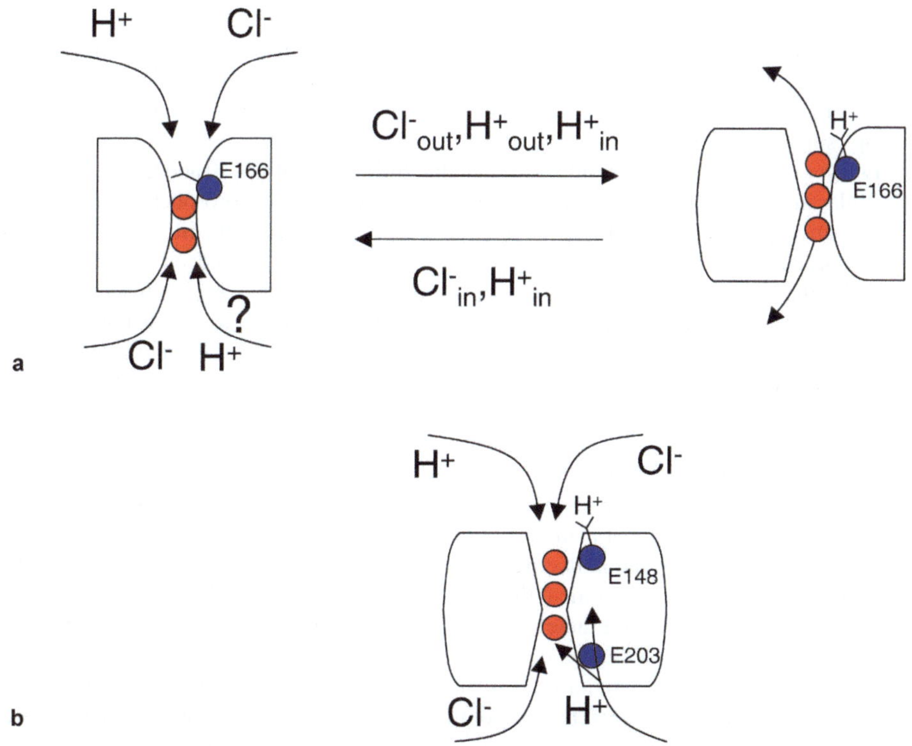

Fig. 4a, b Effect of Cl^- and H^+ on the operation of the protopore gate of CLC channels and transporters. Cl^- ions are indicated as *red spheres*. **a** Schematic representation of the transitions between the open and the closed state of CLC channels and of the physicochemical factors influencing forward and backward rates. Protonation of the E166 (numbering of ClC-0) side chain allows Cl^- flux. Possible additional rearrangements in the pore region involved in channel opening are also indicated. The pathway that intracellular protons have to follow to protonate E166 is not known, as indicated by *question mark*. **b** Schematic representation of the ClC-ec1 transporter. Protonation of E148 (E166 in ClC-0) and E203 are required for the coupled Cl^-/H^+ antiport activity, but the pathway that intracellular protons have to follow to reach E148 after protonation of E203 is not known. One possibility is that protons follow the Cl^- permeation pathway. Another possibility is that they reach E148 through a different route yet to be determined

ulated that this could be also the conformation assumed by the wild-type glutamate in the open—presumably protonated—state (Dutzler et al. 2003). However, this point is still under debate. For example, based on simulation studies, it was suggested that the side chain of the glutamate could swing out of the permeation pathway by a different type of movement (Bisset et al. 2005).

S_{ext} is located between the N-termini of helices F and N, where amide nitrogen atoms form a cage surrounding the ion, and is only 4 Å apart from S_{cen}. All three sites can simultaneously be occupied by Cl^- ions when the channel is open (Lobet and Dutzler 2006).

A very general point to be addressed is the extent to which the picture of the prokaryotic ClC-ec1 provides an accurate description of the eukaryotic counterparts. Sequence alignment exhibits a significant degree of conservation between bacterial and eukaryotic CLC channels; the similarity is especially strong in the selectivity filter region. Mutational studies on eukaryotic channels correlate well with the locations of key residues in the bacterial structures. Chen and Chen, using the cysteine accessibility method, were able to show that in ClC-0 the residues on the intracellular part of the putative helix R are arranged in an α-helical structure and line the wall of the ion permeation pathway as indicated by the crystal structure of the ClC-ec1 (Chen et al. 2003). The results of Engh and Maduke, based on the same approach, also suggest conservation of the overall architecture of the inner vestibule between ClC-0 and ClC-ec1 (Engh and Maduke 2005). Further support in this direction came from a recent biochemical evaluation of the membrane domain boundaries of ClC-2 (Ramjeesingh et al. 2006).

Estévez et al. showed that residues influencing the affinity of ClC-0 and ClC-1 for the intracellular inhibitors 9-anthracene carboxylic acid (9-AC) and *p*-chloro-phenoxy-acetic acid (CPA), partially overlapped with the Cl^- binding pocket identified in the StClC structure (Estévez et al. 2003). It seems, therefore, that the structure of ClC-ec1 indeed provides a good model for the description of other members of the CLC family. However, a potentially relevant difference between ClC-ec1 and CLC channels is the presence in the channels of more Arg and Lys residues near the pore (Corry et al. 2004). Moreover, the finding of Accardi and Miller that ClC-ec1 is not a chloride channel but a Cl^-/H^+ antiporter, with potentially a completely different mechanism of action, suggests caution in the extrapolation of structural features from ClC-ec1 to CLC channels (Accardi and Miller 2004). Subsequently, the eukaryotic ClC-4 and ClC-5 and the plant AtCLCa were also shown to be anion/proton antiporters and not chloride channels (De Angeli et al. 2006; Picollo and Pusch 2005; Scheel et al. 2005). It is surprising that members of the same protein family, sharing a fair degree of homology and high conservation in critical regions, behave in some cases as channels and in others as transporters. At the moment there is no evidence regarding the molecular determinants of such a difference, and therefore we also do not know whether ClC-ec1 represents a better model for ClC-4 and ClC-5 compared to the CLC channels.

The identification of the major molecular determinant of the fast gate, Glu-148 (166 in ClC-0), would explain two characteristics of the fast gate: (a) The fast gates of the two pores are independent because each pore contains its own glutamate residue and the conformational change associated with the swing of the glutamate side chain is local and probably does not influence the other pore; and (b) the fast gate is coupled to Cl^- permeation because Cl^- ions compete with the side chain of glutamate 166 for the occupancy of S_{ext} and once a Cl^- ion occupies this site there is no obstacle to the permeation process. This would also explain the relatively minor voltage dependence of gating of the kidney CLC channels, ClC-Ka and ClC-Kb, which carry a valine instead of a glutamate at the position equivalent to 166 of ClC-0 (Kieferle et al. 1994; Waldegger and Jentsch 2000).

As detailed below, the fast gate can be opened by a mechanism that is favored at low intracellular pH. Presumably, protonation of Glu-166 results in increased open probability due to neutralization of its side chain. For this second mechanism to occur, protons must access the Glu-166 side chain from the intracellular side (Fig. 4). Yin et al., on the basis of molecular simulations, suggested three proton pathways (Yin et al. 2004). One of these pathways involves glutamate residues at positions Glu-113, Glu-117, and Glu-203 that, interestingly, in ClC-0 are substituted by Lys, Leu, and Val, respectively. The residue Glu-203 in ClC-ec1 was in fact suggested by Accardi et al. (Accardi et al. 2005) to be an internally accessible acceptor for protons, as substitution of this residue with Gln completely abolished proton flux, underlining its importance for the mechanism of transport. Interestingly, all members of the CLC family known to be ion channels (ClC-0, ClC-1, ClC-2, ClC-Ka, ClC-Kb, and respective species homologs) present a Val in place of the Glu at position 203, suggesting a significant difference in the mechanism of transport between channel and antiporter members of the CLC family.

However, despite all the pieces of information gathered so far, our picture of the mechanism of gating is still incomplete; for example, some studies point to structural rearrangements of the pore associated with fast gate transitions, suggesting a larger conformational change than the one that would be produced by a simple swing of the Glu-148 side chain (Accardi and Pusch 2003; Traverso et al. 2003) (see "Use of CPA as a tool to explore the fast gate of ClC-0"). Moreover, a gating mechanism based solely on the movement of the Glu is unable to explain why the modulation of gating by Cl^-_{ext} is different from Cl^-_{int} (Chen 2003).

Use of CPA as a tool to explore the fast gate of ClC-0

Small ligand molecules have been very useful tools to explore gating mechanisms of voltage-dependent cation channels (Hille 2001). A classic example is the identification of the activation gate of K^+ channels by intracellularly applied tetraethylammonium (Armstrong 1966). In a similar manner, the small organic acid CPA and related compounds have been used as tools that interfere with the fast gate of ClC-0 (Accardi and Pusch 2003; Pusch et al. 2001; Traverso et al. 2003). CPA is the simplest derivative of 2-(*p*-chlorophenoxy)-3-phenylpropionic acid (CPP), a substance that is known to inhibit the macroscopic skeletal muscle conductance (Conte-Camerino et al. 1988). Later studies on heterologously expressed muscle ClC-1 revealed that CPP and analogs block ClC-1 exclusively from the intracellular side in a strongly voltage-dependent manner, leading to an apparent "shift" of the voltage dependence of opening (Aromataris et al. 1999; Liantonio et al. 2003; Pusch et al. 2000). The binding site of CPA and the unrelated 9-AC was mapped on ClC-1 with considerable detail (Estévez et al. 2003). CPA and 9-AC bind to the channel in a partially hydrophobic pocket adjacent to the central Cl^- binding site (when mapped onto the ClC-ec1 structure), even though the precise orientation of the drug molecule is unknown (Estévez et al. 2003). However, the small single-channel conductance (Pusch et al. 1994) and the relatively complex gating of ClC-1 (Accardi and Pusch 2000) made it difficult to understand the mechanism of CPP block in this channel. The prototype ClC-0 channel is more useful in this respect. Employing the point mutant Cys-212-Ser simplifies the system even more because this single amino acid substitution almost completely abolishes the common gating mechanism (Lin et al. 1999). CPA block of ClC-0 was extensively studied (Accardi and Pusch 2003; Pusch et al. 2001). It was found that CPA binds to closed channels with an about 20-fold higher affinity than to open channels. In this way, CPA stabilizes the closed state and leads to an apparent "shift" of the voltage dependence of opening. Open channel block is of low affinity and associated with rapid binding/unbinding kinetics (apparent K_D in the 20 mM range), whereas closed channel inhibition has much slower kinetics (Accardi and Pusch 2003). As discussed above, fast gating of ClC-0 has been proposed to reflect only the reorientation of the carboxylate side chain of the Glu-166 residue (Dutzler et al. 2003), without any further conformational change of the protein. In this model, the relatively large difference of the affinity and kinetics of open- and closed-channel binding of CPA is rather unexpected, but might be explained by different electrostatic repulsion between CPA and other anions in the pore. However, a recent crystallographic study by Lobet and Dutzler (Lobet and Dutzler 2006) suggested that, in both open and closed states of the fast gate, all three Cl^- ion binding sites are equally maximally occupied by Cl^- ions or by the carboxylate side chain of Glu-166. Thus the model advanced by Dutzler and colleagues appears unable to explain the characteristics of CPA block. Additional evidence in favor of a conformational change that accompanies opening of the fast gate was obtained by Accardi and Pusch from differential effects of pore mutants on closed- and open-channel block by CPA. For example, the mutant Thr-481-Ser exclusively altered the closed-channel affinity, whereas other mutations mostly altered the open-channel block (Accardi and Pusch 2003). Also, the data of Traverso et al. (Traverso et al. 2003), again using CPA as a tool, suggested that a conformational change, in addition to the glutamate swing-out, accompanies opening of ClC-0 protopores.

Thus several pieces of evidence argue against the simple gating model for the fast gate of ClC-0 in which the side chain of Glu-166 is the only moving part. Additional conformational changes, in particular on the intracellular side, would be more compatible with some of the data. However, a more precise definition of the mechanism of the fast gate will probably need direct structural information for a eukaryotic CLC homolog.

CBS domains

All eukaryotic CLC proteins have a long carboxy-terminal cytoplasmic region whose length ranges from 155 (ClC-Ks) to 398 amino acids (ClC-1) (Estévez et al. 2004). The C-terminal domain is essential for the functioning of the eukaryotic CLC proteins, as deletions and several point mutations in this region drastically affect transport activity and/or protein maturation and trafficking (see below). Indeed, several disease-causing mutations are found within the C-terminus (Estévez and Jentsch 2002; Jentsch et al. 2002; Pusch 2002), but, despite some recent progress, its precise functional and physiological role is unknown. The C-terminal region contains two so-called CBS domains (from cystathionine-β-synthase, the first protein in which these domains were identified). These structural domains normally occur in pairs and are found in several unrelated proteins from all organisms (Bateman 1997; Ponting 1997).

Recently, the crystal structure of the isolated cytoplasmic domain of ClC-0 from *Torpedo marmorata* was solved by Dutzler and coworkers (Meyer and Dutzler 2006) (see Fig. 2b). As previously described for a different protein (Sintchak et al. 1996; Zhang et al. 1999), the two CBS domains have a triangular shape and are made of three β-strands and two α-helices. Similar to other CBS-containing proteins, the two CBS domains (i.e., CBS1 and CBS2) were found to interact at the level of the β-strands, forming a typical CBS1-CBS2 complex. A portion of 95 residues of the linker between CBS1 and CBS2 was found to be disordered in the crystal structure, but it is not clear yet whether this reflects a crystallographic artifact or the intrinsic flexibility of the region. However, the residual C-terminal part of the linker, encompassing 25 residues, is well ordered. Interestingly, channel function was not affected by the removal of residues that were part of the disordered linker region, whereas no functional channels were obtained if the truncation was made in the structurally well-defined part of the linker region preceding CBS2 (Estévez et al. 2004).

Unfortunately, the protein did not associate in dimers in the crystallization conditions used by Meyer and Dutzler (Meyer and Dutzler 2006), and therefore critical information about the subunits' interaction had to be extrapolated from a modeling on the crystal structure of TM0935, a protein from *Thermotoga maritima* (Miller et al. 2004). However, even after this procedure, the surface of the domain in contact with the transmembrane region remained ambiguous, although CBS2 was suggested to be positioned closer to the pore than CBS1 (Meyer and Dutzler 2006). Moreover, the C-terminal part of the cytoplasmic domain, which is predicted to be relevant in the interaction between CBS1 and CBS2, was not included in the construct used for the crystallization.

Several functions have been proposed for CBS domains. Alanine scanning mutagenesis of the yeast Cl^- transporter ScClC (gef1p) suggested that CBS domains influenced the subcellular localization of the channel (Schwappach et al. 1998).

On truncation of ClC-0, ClC-1, and ClC-5 after the first CBS domain, the proteins did not give rise to current. However, their function could be restored by coexpression of the missing C-terminal CBS domain, suggesting that CBS2 may function as an independent structure (Maduke et al. 1998; Mo et al. 2004; Schmidt-Rose and Jentsch 1997). Estévez et al. showed that ClC-1 truncated after the CBS1 domain was not able to reach the plasma membrane by itself but that the expression could be restored to a normal level in the presence of the CBS2 domain in addition to a region of six amino acids at the N-terminal part of CBS2 (Estévez et al. 2004). It was also shown that CBS domains from different CLC members could be exchanged without abolishing channel function, demonstrating that the overall architectural conservation of the domain may suffice, despite the low sequence conservation, to preserve their role.

A first hint that the C-terminal region of the channel could be functionally linked to the slow gate came from Jentsch and coworkers (Fong et al. 1998), who made use of mutations in that region of the ClC-0 and of chimeric constructs and found that the C-terminal part is essential for functional expression of the channel and is involved in the operation of the slow gate. In particular, several point mutations in the CBS2 domain of ClC-0 and ClC-1 were found to influence the slow gate (Estévez et al. 2004).

Scott and Hawley found that a purified fragment comprising the last 260 C-terminal residues of ClC-2 was able to bind ATP and that mutations located in this region that are associated with genetic diseases lead to defects in ATP binding (Scott et al. 2004). It is interesting to correlate these findings with a study of Niemeyer et al. (Niemeyer et al. 2004). Analyzing the functional consequence of the mutation G715E in ClC-2 that was proposed to induce idiopathic generalized epilepsy (Haug et al. 2003), Niemeyer et al. could not find any gating alteration for the mutated channel but found that, in contrast to wild-type, it did not respond to the substitution of ATP with AMP with accelerated opening and closing kinetics, even though the effects were relatively minor. Recently, it was suggested that the isolated carboxy terminus of ClC-5 folds in a predominantly α-helical structure and it is able to bind ATP (Wellhauser et al. 2006). Interestingly, ATP modulates the activity of the common gate of ClC-1 channels such that increasing ATP concentration shifts the midpoint of the open probability distribution toward depolarized potentials and reduces the fraction of channels that remain open at strong hyperpolarized potentials (Bennetts et al. 2005). Bennetts et al.

suggested that the interaction with ATP is mediated by the CBS domains (Bennetts et al. 2005). Based on a homology model with the structure of a CBS dimer of IMPDH (inosine monophosphate dehydrogenase) and in silico docking, they identified a putative ATP binding pocket in a cleft between the two CBS domains of ClC-1 and confirmed their results, observing that mutations of residues that were predicted to interact with ATP reduced or ablated the ability of ATP to modulate channel function (Bennetts et al. 2005). However, no ATP binding could be detected in the CBS1-CBS2 complex of ClC-0, even at very high ATP concentrations (Meyer and Dutzler 2006). Physiologically, an increased ClC-1 activity due to ATP depletion during metabolic stress would stabilize the membrane potential and reduce muscle excitability, thereby preserving the viability of muscle fibers. Such a mechanism, however, has not been described *in vivo*. In fact, it is questionable that an increased chloride conductance, via a shift of the voltage dependence of the open probability, is able to suppress muscle excitation after nerve stimulation.

The fact that mutations in the CBS domains, per se or by affecting the ability to bind ATP, interfere with the operation of the common gate requires an interaction of the transmembrane part of the channel with the cytoplasmic terminus. An interesting possibility was suggested by Estévez and coworkers (Estévez and Jentsch 2002; Estévez et al. 2004) to explain this interaction: The last transmembrane helix R, whose N-terminal tyrosine coordinates a Cl^- ion in the middle of the pore and whose C-terminus extends into the cytosol, is directly connected to the CBS1-CBS2 complex. This helix may therefore be the structural link between the inner pore and CBS domains.

Additionally, CBS domains may be relevant in the interaction with other proteins. It has been found that deleting CBS1 and/or CBS2 impairs the interaction of ClC-5 with cofilin, an actin-associated protein that is crucial in the regulation of albumin uptake by the proximal tubule (Hryciw et al. 2003). Moreover, a PY motif is found between CBS1 and CBS2 of ClC-5 that probably interacts with HECT-ubiquitin ligases to modulate the retention of the channel in the plasma membrane (Schwake et al. 2001), and a splice variant of ClC-3 displays a PDZ-binding motif at its extreme carboxy terminus that can interact with the scaffolding proteins EBP50 (ERM-binding phosphoprotein 50), PDZK1, and GOPC (Golgi-associated PDZ and coiled-coil motif-containing protein) (Gentzsch et al. 2003; Ogura et al. 2002).

Gating of muscle-type ClC channels

According to the classic view, in voltage-dependent cation channels permeability and gating are considered, to a first approximation, as independent processes implying the presence of a permeable pore and of a separate structure that senses the transmembrane voltage and opens and closes the pore. This picture is completely inadequate for CLC channels. A first hint of the strong coupling of gating and permeation in ClC-0 came from the time asymmetry of the single-channel bursts implying that the gating transitions were not in thermodynamic equilibrium (Richard and Miller 1990) (see "Introduction"). Such a situation implies the existence of an external energy input into the system that was identified as the chloride electrochemical potential, anticipating one of the most eccentric features of CLC channels, a gating process that is mediated by the permeant ion.

A thorough investigation of the properties of ClC-0 expressed in oocytes and CHO cells allowed Pusch and coworkers to conclude that in ClC-0 permeation and gating are tightly linked (Pusch et al. 1995a). They found that only permeant anions affect gating, that the ion selectivity of conduction is reflected in the ion selectivity of gating, and that an anomalous mole fraction behavior in the conduction corresponds to a parallel behavior in the gating. Incidentally, the presence of such an anomalous mole fraction behavior showed for the first time that the channel pore contains more than one ion binding site, as was later confirmed by structural data (Dutzler et al. 2003).

As mentioned above, the conducting state of the ClC-0 channel is controlled by two different mechanisms defined as the slow gate and the fast gate.

The slow gate controls the opening (and closing) of both pores simultaneously (Miller and White 1980; White and Miller 1979).

There are different factors affecting the operation of the slow gate, such as potential, chloride concentration, pH, and temperature. Hyperpolarized potentials favor the opening of the slow gate (Miller and Richard 1990). The steady-state activation of the slow gate can be described by a Boltzmann function with a $V_{1/2}$ of approximately –80 mV and an apparent gating valence of ~2 (Pusch et al. 1997). Moreover, the slow gate apparently does not deactivate completely at depolarized voltages, leading to an offset of the open probability of the slow gate at positive voltages. Interestingly, this offset seems to correlate with the expression level of ClC-0 in oocytes (Pusch et al. 1997).

Chen and Miller (Chen and Miller 1996), found in single-channel recordings, that increasing $[Cl^-]_{ext}$ shortened the mean closed time and increased the mean open time of the slow gate. Also, $[Cl^-]_{int}$ influences the operation of the slow gate. Decreasing intracellular Cl^- shifted the p_{open} of the slow gate to more negative potentials and reduced the maximal activation at the most negative voltages (Pusch et al. 1999). Temperature is another variable that markedly influences the operation of the slow gate (Pusch et al. 1997). In particular, the kinetics of closing of the slow gate showed a Q_{10} of ~40 at 20°C, suggesting that the transition between the open and the closed state requires a complex rearrangement of the protein. The effect of an increase of temperature is on one hand to inactivate the channels in a more complete fashion at positive voltages and on the other hand to decrease the fraction of channels that can be activated by the slow gate at negative voltages (Pusch et al. 1997). However, the voltage of half-maximal activation is relatively independent of temperature. This complex behavior cannot be correctly described by a simple two-state model (open-closed states) but requires at least two open and two closed states for its description. The effect of temperature was assessed also on the single-channel level, with the finding that increasing the temperature increases the frequency of closure of the slow gate. As expected, single-channel currents increase with temperature, but the dependence is shallow, consistent with a diffusion-regulated process (Pusch et al. 1997).

In ClC-1, which normally lacks the typical slow gate activation at negative voltages (Steinmeyer et al. 1991b), a hyperpolarization-activated component of the current becomes apparent at low pH_{ext} (5.5), which is reminiscent of the activation of the slow gate in ClC-0 (Rychkov et al. 1996).

The mechanism responsible for the slow gating has not yet been identified. The fact that the slow gate acts on both pores simultaneously suggests, on the structural level, that it relies on subunit interactions (Estévez and Jentsch 2002), in agreement with the finding that concatemers comprising subunits of different CLC members led invariably to loss of slow gating transitions (Lorenz et al. 1996; Weinreich and Jentsch 2001).

The interaction of the subunits in the dimeric architecture of CLC proteins can involve the interface between the transmembrane segments or the cytosolic portions that are of substantial length in eukaryotic channels, or both.

Most ClC-1 mutations leading to dominant myotonia change the voltage dependence of the channel and most likely involve the slow gate (Pusch et al. 1995b; Saviane et al. 1999).

These mutations are scattered along the channel amino acid sequence (Pusch 2002) and therefore prove that different regions of the channel probably interact to determine slow gate transitions. However, several mutations cluster in helices at the dimer interface that probably are important for subunit contacts: Mutations Val-286-Ala and Ile-290-Met change residues in helix H, whereas mutations Phe-307-Ser, Ala-313-Thr, and Arg-317-Gln change residues in helix I (Duffield et al. 2003; Pusch 2002). Moreover, several point mutations in ClC-0 that are distant from the dimer interface have also been shown to eliminate slow gate transitions (Lin et al. 1999; Ludewig et al. 1996; Traverso et al. 2006). As explained in the section on the CBS domains, the C-terminus also appears to be a major determinant of the slow gate (Estévez et al. 2004; Fong et al. 1998).

The fast protopore gate of ClC-0

The fast gate acts individually on the single pores of the dimer (Miller 1982). In single-channel recordings of ClC-0 incorporated into planar lipid bilayers, it was found that the fast gate operates in the milliseconds time range and the open probability of the single protopore increases with voltage with an apparent gating charge of ~1 (Miller 1982) and follows a Boltzmann distribution as predicted for a two-state channel model (Hanke and Miller 1983).

$[Cl^-]_{ext}$ influences the open probability of the fast gate (Pusch et al. 1995a), with high extracellular Cl^- favoring the opening of the channel, shifting the voltage dependence of the open probability toward negative potentials (Fig. 4). Using single-channel recordings, Chen and Miller (Chen and Miller 1996) showed that the open probability approaches a nonzero asymptote at very negative potentials, an effect that can be described as incomplete closure of the channel. The basis of this phenomenon is that the opening rate does not depend in a monotonic manner on voltage. At depolarized potentials the opening rate increases exponentially with voltage; at hyperpolarized voltages, however, the opening rate decreases at intermediate potentials but increases again at highly hyperpolarized potentials. The result is that the opening rate has a minimum at negative voltages. On increase in $[Cl^-]_{ext}$, the voltage activation curve shifts to the left along the voltage axis without significant change in the apparent gating charge.

The closing rate of the fast gate depends on voltage, decreasing exponentially with depolarization. Importantly, the closing rate is only slightly affected by $[Cl^-]_{ext}$. Therefore, whereas the voltage dependence of the open probability is determined by both the opening and the closing rate, the external Cl^- dependence derives almost completely from an effect on opening.

The operation of the fast gate depends also on $[Cl^-]_{int}$. In particular, the effect on the opening rate is very small, whereas lowering $[Cl^-]_{int}$ substantially increases the closing rate (Chen and Miller 1996) (Fig. 4). As a result, increasing $[Cl^-]_{int}$ shifts the steady-state activation curve to the left, as with high $[Cl^-]_{ext}$. However, $[Cl^-]_{int}$ exerts a more prominent effect on the degree of incomplete closure at hyperpolarized potentials, which was not observed changing $[Cl^-]_{ext}$. In particular, as $[Cl^-]_{int}$ increases, the asymptote of the open probability at negative voltages also increases (Chen and Miller 1996; Ludewig et al. 1997a).

These observations were rationalized by a model in which the fast gate of ClC-0 may open through two different routes with opposite voltage dependence (Chen and Chen 2001; Chen and Miller 1996). In one mode, opening is favored by membrane depolarization and is sensitive to $[Cl^-]_{ext}$. A plausible mechanism for this gate would be that Cl^- first binds to the channel and then travels through the pore to reach an inner binding site, spanning some distance in the membrane electric field, as already suggested by Pusch (Pusch 1996; Pusch et al. 1995a). The other mode does not depend on $[Cl^-]_{ext}$ and is favored by hyperpolarized potentials (Chen and Chen 2001). A more quantitative analysis of the $[Cl^-]_{int}$ dependence of the fast gate was performed by Chen et al. (Chen et al. 2003). Their results confirmed that $[Cl^-]_{int}$ almost exclusively affects the closing rate (increasing $[Cl^-]_{int}$ decreased the closing rate). The effect of $[Cl^-]_{int}$ on the closing rate was saturable, suggesting that it is mediated by a Cl^--binding site. This was confirmed by experiments in which Cl^- was substituted with Br^- and SO_4^{2-}, showing how the impermeant ion SO_4^{2-} did not have any such effect, whereas Br^-, which binds to the pore more tightly than Cl^-, had a stronger effect (Chen et al. 2003).

The fast gate is also affected by alterations of the intrinsic electrostatic potential of the pore (Chen and Chen 2003; Zhang et al. 2006). In particular, mutating several residues known to line the pore or located close to it affected the closing rate, with very little effect on the opening rate. Introducing positively charged residues (or removing negatively charged residues) in the pore consistently increased the closing rate; vice versa, introducing negatively charged residues decreased the closing rate. It seems therefore that increasing $[Cl^-]_{int}$ and introducing more negative charges in the pore lead to a similar effect (Chen and Chen 2003; Chen et al. 2003). Chen and coworkers proposed two mechanisms to explain these results, both based on the assumption that Glu-166 is the fast gate in ClC-0. The negative charge of the glutamate side chain could directly interact with charged residues in the pore region. In this scenario, negative charges in the inner pore would repel the negative charge on the glutamate so that the gate would be more difficult to close, that is, it would be more difficult for the carboxylate side chain of Glu-166 to occupy the S_{ext} position. However, as judged from the structure of ClC-ec1, some mutations tested in the study would be more than 20 Å away from Glu-166. More importantly, the behavior of the double mutant E127K/K519E is not in agreement with this model (Chen and Chen 2003). The alternative possibility is that the effect of the electrostatic potential of the pore on gating is mediated by the permeant anion. For example, a more positive charge at the amino acid positions 127, 515, and 519 that are located near S_{int} would decrease the ability of Cl^- present at this site to displace Cl^- at S_{cen} and at S_{ext}. This, in turn, would decrease the ability of Cl^- to compete with Glu-166 for S_{ext}, leading to faster closing of the protopore gate.

This hypothesis is especially appealing because it would explain the behavior of the mutant Glu-127-Gln, for which the effect on the fast gate mirrors the effect on channel conductance. However, not all the mutants affect both fast gate and conductance. Chen and colleagues therefore suggested that the charge of residues in the pore and the charge carried by the permeant ion both can contribute to the overall gating process and that the location of the charge in the pore determines their relative contribution (Chen and Chen 2003; Chen et al. 2003).

Very recently it has been found that the residue K149 in ClC-0 (corresponding to K131 in ClC-ec1), although not directly lining the pore, plays a very important role in the electrostatics of the channel, as mutations of this residue reduce the opening rate of the fast gate (Zhang et al. 2006). Interestingly, the mutation K131M in ClC-ec1 results in a perturbation of Cl^-/H^+ antiporter function (Accardi et al. 2005).

The electrostatics of the pore is also a major determinant of the single-channel conductance of ClC-0 (Chen and Chen 2003). For example, it was found that mutations changing the charge in the inner pore (e.g., Lys-519-Glu) reduce the conductance at "physiological" Cl^- concentrations, but not at saturating $[Cl^-]_{int}$ (Chen and Chen 2003). In contrast, for the mutation Ser-123-Thr, which changed the highly conserved serine in the selectivity filter, the decrease in conductance could not be rescued by manipulation of the internal Cl^-. Notably, the mutant Tyr-512-Phe, located in the selectivity filter, produced an increase in conductance of 30% compared to wild-type. This suggests that the regulation of channel conductance by mutations in the selectivity filter and in the channel inner mouth is different. The hydroxyl groups of Ser and Tyr are clearly shown in the ClC-ec1 structure to coordinate a Cl^- ion at S_{cen} and are conserved in the CLC family. The fact that mutations in the corresponding residues in ClC-0 have such a different influence on conductance is still difficult to explain and may suggest a complex effect of these mutations on channel conductance and some difference between the transporter and the channel members of CLC proteins.

The modulation of the fast gate by external protons was first studied by Chen and Chen (Chen and Chen 2001), who showed that reducing pH_{ext} increases the open probability, mostly at hyperpolarized potentials, almost exclusively increasing the opening rate (Fig. 4). The macroscopic effect of a decrease in pH_{ext} is therefore mostly an increase in the minimal open probability (P_{min}) at hyperpolarized potentials and not a shift of the $p_{open}(V)$ curve, which is instead seen on changing $[Cl^-]_{ext}$. Chen and Chen (Chen and Chen 2001) proposed that the effect of pH_{ext} on the fast gate is not mediated by a change in the affinity of the Cl^- binding site that regulates channel opening (Chen and Miller 1996; Pusch 1996) and that therefore the mechanism of pH_{ext} regulation must be intrinsically different from the $[Cl^-]_{ext}$-dependent channel opening. The regulation by external protons, Cl^- independent and mostly effective at hyperpolarized potentials, is similar to one of the mechanisms of opening described by Chen (Chen et al. 2003), potentially indicating that the two processes are linked (Chen and Chen 2001).

Moreover, the fact that the modulation by pH_{ext} is stronger at negative voltages is reminiscent of the action of $[Cl^-]_{int}$ on the fast gate. Chen and Chen (Chen and Chen 2001) indeed suggested that the action of external protons is more pronounced at higher $[Cl^-]_{int}$.

The ClC-0 mutant Glu-166-Asp has a drastically reduced open probability compared to wild-type (Traverso et al. 2006) and is thus expected to display an even stronger response to the external pH. Traverso et al. (Traverso et al. 2006) found instead that decreasing pH_{ext} did not increase outward currents. In particular, low pH_{ext} increased a persistent inward current that was characterized by a smaller single-channel conductance. These results suggested that Asp-166 can be protonated from the intracellular side in a voltage-dependent manner or from the extracellular side in a voltage-independent manner, resulting in open states of different conductance (Traverso et al. 2006). In ClC-1 it was found that decreasing pH_{ext} affected the macroscopic current, mostly by increasing the steady-state component at the expense of the deactivating portion. At variance with the behavior of external Cl^- at low pH_{int}, it was found that at low pH_{ext}, external $Cl^-$was not able to influence channel gating (Rychkov et al. 1996).

The influence of the internal pH on the fast gate transitions was investigated in the reconstituted *Torpedo* channel (Hanke and Miller 1983) (Fig. 4). Low pH_{int} drives the protochannel open without changing its conductance. The effect was interpreted in terms of a shift of the voltage dependence of the open probability toward negative potentials. Hanke and Miller suggested that on opening of the channel a titratable group exposed to the intracellular solution changes its pK from 6 to 9 and that this change in pK underlies the ability of protons to drive the channel into its open state (Hanke and Miller 1983). In ClC-1, internal pH had a very similar effect (Rychkov et al. 1996). Hanke and Miller also investigated the pH dependence of the opening and closing rate constants. They found that at all pH values tested, those rates vary exponentially with voltage and at all voltages both opening and closing rate constants vary with proton concentration. However, with an increase in the proton concentration, the closing rate constant decreases whereas the opening rate increases. Therefore, the effect of pH_{int} changes mainly translates into a shift of the p_{open} along the voltage axis. The pH dependence implies that a simple two-state model is insufficient to describe the channel behavior and that a protonation reaction must be added to the scheme. Hanke and Miller (Hanke and Miller 1983) suggested, however, that the protonation step does not contribute to the voltage dependence of gating, which in their model is brought about only by the transition between open and closed states. Such an interpretation was recently challenged by Pusch and coworkers, who investigated the pH dependence of the Glu-166-Asp ClC-0 mutant (Traverso et al. 2006). This mutant strongly affects the operation of the fast gate, dramatically reducing the open probability of the channel. This drastic effect of the conservative Glu→Asp mutation (Traverso et al. 2006) probably reflects the sensitivity of ClC-0 gating on the protonation state and flexibility of this key acidic residue. Lowering pH_{int} increased current of the Glu→Asp mutant, in agreement with the behavior of the wild-type channel (Hanke and Miller 1983; Traverso et al. 2006). However, the pH_{int} dependence of this mutant is not consistent with a model in which the protonation step is voltage independent, but could be better described by a model in which the protonation/deprotonation reactions carry most of the voltage dependence. This suggestion also opens up new questions. It is reasonably well established that Glu-166 is the proton acceptor responsible for the regulation of the fast gate by pH_{ext}. On the other hand, we still do not know which residue(s) is involved in the control of the fast gate by intracellular protons. An interesting hypothesis is that opening of the fast gate requires the protonation of Glu-166. Protonation may occur, in a relatively voltage-independent manner, from the extracellular solution or, in a voltage-dependent manner, from the intracellular side. A protonation of Glu-166 (or Asp-166) from the intracellular side was also proposed recently by Miller as the possible major source of voltage dependence of the fast gate of ClC-0 (Miller 2006) (see Fig. 4).

Zinc and cadmium—inhibitors of ClC-0, ClC-1, and ClC-2

ClC-0, ClC-1, and ClC-2 are inhibited by Zn^{2+} and Cd^{2+} ions (Chen 1998; Clark et al. 1998; Kürz et al. 1997; Rychkov et al. 1997), confirming results obtained for the Cl^- conductance of frog skeletal muscle (Hutter and Warner 1967). A first mechanistic insight into the interaction between Zn^{2+} ions and CLC channels came, however, from an analysis of the Zn^{2+} block of ClC-0 (Chen 1998). For ClC-0 the inhibition is reversible with an IC_{50} of 1–3 μM. The effect of Zn^{2+} did not seem to be mediated by an interaction with the fast gate, whose voltage dependence of the open probability and of the kinetics remained unaltered in the presence of Zn^{2+}. The apparent on- and off-rates of Zn^{2+} inhibition were slow and showed pronounced temperature dependence, from which it was suggested that the inhibition was unlikely to stem from a simple open channel block and probably involved a more complicated process (Chen 1998). In particular, the temperature dependence of the effect directly suggested a possible link of the inhibition with the operation of the slow gate (Chen 1998; Pusch et al. 1997). It was found that indeed increasing Zn^{2+} concentration facilitated the slow gating process (Chen 1998). Specifically, the effect of Zn^{2+} on slow gating equilibrium appears to come mostly from an increase in the forward rate of inactivation. Interestingly, the mutation Cys-212-Ser in ClC-0, which was shown to eliminate the slow gating process, also drastically reduces the channel's sensitivity to Zn^{2+} (Lin et al. 1999), further supporting the association between the slow gate and the mechanism of Zn^{2+} inhibition.

As described below, the common gate of ClC-1 has quite different features from that in ClC-0, such as, for example, an opposite voltage dependence, and vastly different kinetics and temperature sensitivity. The IC_{50} for Zn^{2+} inhibition of ClC-1 has been found to be 0.35 mM (Rychkov et al. 1997). In contrast to ClC-0 and ClC-2 (Chen 1998; Clark et al. 1998), Zn^{2+} and Cd^{2+} block appear to be irreversible for ClC-1 (Kürz et al. 1997; Rychkov et al. 1997). Interestingly, also in ClC-1 the mutation Cys-277-Ser, corresponding to the mutation Cys-212-Ser of ClC-0, drastically reduces the closure of the slow gate (Accardi et al. 2001) and virtually eliminates Zn^{2+} block, suggesting a similarity in the mechanism of Zn^{2+} block on the two channels (Duffield et al. 2005). At variance with ClC-0, however, in ClC-1 the block by Zn^{2+} is too slow to be a simple function of the open probability of either the fast or the putative slow gate. Moreover, the temperature dependence of Zn^{2+} inhibition (Q_{10} ~13°) is much higher than the Q_{10} of the putative slow gate, which is ~4° (Bennetts et al. 2001). Both elements indicate that in ClC-1 the mechanism of Zn^{2+} inhibition, although founded on the interaction with the slow gate as in ClC-0, may present significant differences, and Duffield et al. (Duffield et al. 2005) proposed that in ClC-1 Zn^{2+} acts by binding to a closed substate of the common gate that has very low probability in the wild-type channel and was therefore not previously identified.

Extracellular Cd^{2+} produces a concentration-dependent block of ClC-1 expressed in the Sf-9 cell line, with an IC_{50} of 1 mM (Rychkov et al. 1997). It was suggested that ClC-1 has at least two binding sites for Cd^{2+} in which His residues may play a prominent role (Rychkov et al. 1997).

Zúñiga et al. found that Cd^{2+} block of ClC-2 is mediated by an acceleration of the rate of deactivation (Zúñiga et al. 2004). Mutation of Cys-256 in ClC-2, corresponding to a cysteine residue known to affect the operation of the slow gate in ClC-0 (Cys-212-Ser) (Lin et al. 1999) and ClC-1 (Cys-277-Ser) (Accardi et al. 2001) and to drastically reduce Zn^{2+} block, also reduced the effect of Cd^{2+} compared to wild-type, indicating that Cd^{2+} would exert its action through an interaction with the gating machinery of the channel (Zúñiga et al. 2004). However, at variance with the action of Zn^{2+} on ClC-0 and ClC-1, Cd^{2+} affected both the fast and the slow gating process of ClC-2 (Yusef et al. 2006), indicating a strong coupling between fast and slow gating, similar to what was proposed for ClC-1 (Accardi et al. 2001). Moreover, the mutation His-811-Ala in ClC-2, corresponding to a mutation that completely and selectively abolishes slow gating in ClC-0 (Estévez et al. 2004) and that is located in the highly conserved CBS2 domain, affected both fast and slow gating of ClC-2. Interestingly, combining this mutation with Glu-217-Val ablates all gating transitions (Yusef et al. 2006).

ClC-1—the skeletal muscle chloride channel

ClC-1 was cloned from rat skeletal muscle by homology screening with a probe derived form the *Torpedo* ClC-0, with which it shares 54% sequence identity (Steinmeyer et al. 1991b). It is predominantly expressed in skeletal muscle, where it accounts for the large Cl^- conductance responsible for the resting membrane potential (Bretag 1987; Steinmeyer et al. 1991b). Low transcript levels could also be detected in kidney, heart, and smooth muscle (Steinmeyer et al. 1991b).

Analysis of dominant-negative mutations suggested that ClC-1 has a multimeric architecture (Pusch et al. 1995b; Steinmeyer et al. 1994). This view was supported by Lorenz et al. (Lorenz et al. 1996), who showed that ClC-1 and ClC-2, on coexpression in *Xenopus* oocytes, form heterooligomers.

Even though a quantitative single-channel analysis of ClC-1 is difficult because of its small single-channel conductance (Pusch et al. 1994), an extension of the double-barreled architecture from ClC-0 to ClC-1 was strongly supported by inspection of the single-channel behavior displaying two equidistant conductance levels of 1.2 and 2.4 pS (Saviane et al. 1999). Incidentally, the small single-channel conductance explains why many previous attempts to detect its activity in intact muscle preparations failed.

Despite the similarities with ClC-0, there are a number of functional characteristics that differentiate ClC-1 from ClC-0. In contrast to ClC-0, gating and permeation apparently do not seem to be so closely linked in ClC-1 (Rychkov et al. 1998) as anions like cyclamate and methanesulfonate can have a substantial effect on gating without being permeant. However, these results can probably be explained by an external anion binding site with relatively high affinity for organic anions in ClC-1 but not in ClC-0 (Rychkov et al. 2001). Occupation of this site by organic anions indirectly influences the occupation by chloride of deeper anion binding sites. Gating of ClC-1 is similar to the fast gating of ClC-0 in that it also activates with depolarization and can be described by a Boltzmann function with an apparent gating charge of ~0.9 (Pusch et al. 1994; Rychkov et al. 1996; Steinmeyer et al. 1991b).

Even if under normal conditions ClC-1 lacks a slow hyperpolarization-activated gate, such a gate becomes visible al low pH_{ext} and positive holding potentials (Rychkov et al. 1996).

Under physiological pH conditions, ClC-1-mediated currents display a deactivation comprising two exponential components (Rychkov et al. 1996). Accardi and Pusch (Accardi and Pusch 2000) showed that these components have time constants that are quite similar at negative voltages but grow apart as the voltage is increased. At a voltage of 200 mV they differ almost by a factor of 100, mimicking the difference between fast and slow gates in ClC-0. In particular, investigation of the dependence of the two components on $[Cl^-]_{ext}$ and pH_{int} suggested that the faster gating components found for ClC-1 behaved very similarly to the fast gate of ClC-0 and the slower component of ClC-1 was similar to the slow gate of ClC-0. It was therefore proposed that also for ClC-1 the two gating components correlated with the operation of fast and slow gates. However, it was shown that the voltage dependence of the slow gate in ClC-1 is reversed compared to ClC-0 (Accardi and Pusch 2000; Saviane et al. 1999).

The physiological role of ClC-1 is discussed below in the context of its involvement in congenital myotonia.

Myotonia

Myotonia—or muscle stiffness—is a symptom that is associated with various genetic diseases. In "chloride-channel" recessive (Becker) (Becker 1957) and dominant (Thomsen) (Thomsen 1876) myotonia congenita, myotonia is practically the only symptom. Muscle diseases caused by mutations in the SCN4A sodium channel have overlapping but not identical symptoms (Lehmann-Horn and Jurkat-Rott 1999).

Myotonia is caused by hyperexcitability of the muscle plasma membrane, such that normal nerve stimulation produces an exaggerated and possibly repetitive firing of muscle action potentials (myotonic runs) (Adrian and Bryant 1974). About 80% of the resting conductance of skeletal muscle consists of a chloride conductance, g_{Cl}, the majority of which is carried by ClC-1 (Steinmeyer et al. 1991a, 1991b). A marked reduction of g_{Cl} thus decreases the depolarizing and stabilizing conductance, causing hyperexcitability. In most neurons, equivalently stabilizing and repolarizing conductances are mostly carried by K^+ channels. It is thought that in skeletal muscle a K^+ conductance is not adequate for such a role because of an expected buildup of K^+ ions in the restricted space inside the t-tubules (Cannon 2000; Pusch 2001). In fact, detubulation of rat skeletal muscle reduces g_{Cl} but not g_K (Palade and Barchi 1977), However, the two studies that specifically investigated the subcellular localization of ClC-1 with immunofluorescence found the protein in the sarcolemma and not in the t-tubules (Gurnett et al. 1995; Papponen et al. 2005).

In myotonic dystrophy (DM), for which myotonia is only one of many symptoms, it has recently been shown that the RNA coding for the ClC-1 protein is strongly reduced by an alteration of its correct splicing (Berg et al. 2004; Charlet et al. 2002; Mankodi et al. 2002).

Dominant and recessive myotonia congenita are instead caused by mutations in *CLCN1*, the gene coding for ClC-1. A mouse model for recessive myotonia, the *adr* mouse (Mehrke et al. 1988), helped to identify ClC-1 as the major skeletal muscle Cl^- channel (Steinmeyer et al. 1991a). In the *adr* mouse, no ClC-1 protein is made because both alleles are practically destroyed by a homozygous transposon insertion. Similarly, most mutations that lead to recessive myotonia in humans either completely abolish channel function (like, e.g., early stop codons) or drastically reduce channel function (see Pusch 2002 for an overview of possible effects of recessive mutations). There may be several reasons for the fact that heterozygous carriers of such recessive mutations (50% gene dosage) are generally asymptomatic. A 50% gene dosage could be functionally compensated at the RNA level (transcription, splicing, processing, turnover) or at the protein level (translation, processing, sorting, targeting, turnover). In fact, heterozygous *adr* mice show an almost unaltered muscle chloride conductance (Chen et al. 1997). It remains, however, as an interesting problem if and how much the Cl^- conductance is reduced in human heterozygous carriers of recessive mutations. Pharmacological experiments indicate that more than 50% of the Cl^- conductance must be inhibited in order to cause myotonia (Furman and Barchi 1978). This observation and the fact that heterozygous carriers of recessive mutations are generally asymptomatic demonstrate that in order for a *CLCN1* mutation to be inherited in a dominant manner it must produce a dominant-negative effect. That is, it must reduce g_{Cl} more than a heterozygous loss of function, beyond the threshold that is necessary to precipitate myotonia. These considerations are in agreement with the fact that far more recessive than dominant *CLCN1* mutations have been described (Pusch 2002): Channel function is easily destroyed, for example, by early stop codons, but a dominant-negative effect requires a specific association with a wild-type subunit. The first dominant mutation, P480L, was identified in descendants of Thomsen, who himself suffered from the disease (Steinmeyer et al. 1994). When coexpressed with wild-type subunits in *Xenopus* oocytes, the mutation exerted a strong dominant-negative effect, and this was the first indication that CLC channels are homomultimers (Steinmeyer et al. 1994). The mechanism of action remained unclear, however, and the initial estimate of the number of subunits (4) turned out to be wrong. Later, it was found that several dominant mutations, including P480L, exert a dominant-negative effect by "shifting" the voltage dependence of channel activation to more positive voltages, such that channels are less active at the skeletal muscle resting membrane potential (Pusch et al. 1995b). In the context of the double-barreled structure of CLC channels with two separate gates (fast, protopore gate and slow, common gate) it was later found that most dominant mutations act primarily on the common gate of ClC-1 (Aromataris et al. 2001; Saviane et al. 1999). It also must be said, however, that the distinction between dominant and recessive forms of the disease is not very clear-cut. The same mutation may appear as dominant in some pedigrees and as recessive in others (Plassart-Schiess et al. 1998). Thus other factors, independent of ClC-1, seem to contribute to the severity of myotonia.

The shift of the voltage dependence is not the only dominant-negative mechanism. For example, the C-terminal truncation R894X has a quite strong dominant-negative effect, without an apparent change of the voltage dependence (Meyer-Kleine et al. 1995). This mechanism remains to be identified. As we hope to have illustrated above, understanding the pathophysiology of myotonia provides a valuable insight into the general function of ClC-1.

ClC-2—a complex chloride channel of epithelial and nonepithelial cells

ClC-2 has been cloned from rat heart and brain. The 907-amino acid protein shares 49% identity with ClC-0 and 55% with ClC-1. It is broadly expressed in several tissues and in cell lines of different origin such as epithelial, fibroblast, and neuronal (Thiemann et al. 1992).

On expression in *Xenopus* oocytes, ClC-2 gave rise to currents that were slowly activated and inwardly rectifying, unlike ClC-0 and ClC-1 (Gründer et al. 1992; Thiemann et al. 1992). Moreover, these currents were activated only at unphysiological, hyperpolarized potentials. The instantaneous *I-V* curve observed after activation of the hyperpolarized current revealed a linear current-voltage relationship. Similar to ClC-0, iodide is less permeant than chloride. Extracellular 9-anthracene carboxylic acid (1 mM) and diphenylaminecarboxylate (1 mM) inhibited the conductance by 50%, whereas 1 mM DIDS was almost ineffective (Thiemann et al. 1992).

Single-channel analysis applied on concatemeric constructs of ClC-0 and ClC-2 demonstrated a functional dimeric architecture of ClC-2 (Weinreich and Jentsch 2001) in analogy with ClC-0 (Bauer et al. 1991; Ludewig et al. 1996; Middleton et al. 1996; Miller and White 1984) and ClC-1 (Saviane et al. 1999). Unfortunately, the single-channel conductance of ClC-2 is only 2.6 pS, a factor that has so far hampered attempts at a thorough characterization of the channel properties at the single-channel level.

Superfusion of oocytes expressing ClC-2 with hypotonic solution produced currents with faster kinetics that were activated at less hyperpolarized potentials and therefore in the physiological voltage range, suggesting that ClC-2 is involved in volume regulation (Furukawa et al. 1998; Gründer et al. 1992; Jentsch et al. 2002). In particular, the overall current amplitude significantly increased on superfusion with hypotonic solution (Gründer et al. 1992). The activation was fully reversible and needed around 10 min to set in, suggesting that the effect was probably due to slow intracellular changes rather than to a direct effect on the channel. Hypertonicity did not have any effect on wild-type ClC-2. A chimeric approach allowed the identification of the N-terminal domain as determinant for the volume sensitivity of the channel (Gründer et al. 1992).

In particular, deletions in the first 31 amino acids led to constitutively open channels that were also unresponsive to hyper- or hypotonicity (when analyzed with the two-electrode voltage-clamp technique), whereas upstream from this essential domain, deletions produced channels with an intermediate phenotype. The effect of these domains was independent from their position, as the N-terminal region could be transplanted to the C-terminus, retaining its effect (Gründer et al. 1992).

Extracellular pH significantly affects the operation of ClC-2, with moderate acidification leading to channel activation already at ~ –30 mV and to increased steady-state currents (Jordt and Jentsch 1997). It was suggested that the mechanism of action is a shift of the voltage dependence of the common gating mechanism, and Jordt and Jentsch proposed that, in analogy with the effect of hyperpolarization and cell swelling, the response to extracellular pH depends on the N-terminal domain (Jordt and Jentsch 1997). However, the role of the N-terminal domain of the channel is still not very clear. In fact, in contrast with the voltage-independent phenotype of the N-terminal deletion described previously with two-electrode voltage-clamp recordings (Gründer et al. 1992), in inside-out patches the same construct gave rise to channels that conserved the characteristic activation at hyperpolarized potentials of the wild-type, albeit with a faster kinetics (Pusch et al. 1999). Similarly, deletions of amino acids 16–61 of rClC-2 expressed in HEK cells, although producing faster opening and closing kinetics compared to wild-type, did not produce significant changes in voltage and pH dependence (Varela et al. 2002). However, it was observed that with nystatin-perforated patches, which allow the selective exchange of cations between the cytoplasm and the pipette solution, currents of the amino terminal-deleted mutant lost their voltage dependence (Varela et al. 2002), suggesting that the differential effect of the deletion in different expression systems and measuring conditions may depend on factors such as osmotic state of the cells, cytoskeleton structure integrity, or diffusible cytoplasmic components, as already discussed by Pusch et al. (Pusch et al. 1999).

In a study of currents in mouse parotid acinar cells that were probably mediated by ClC-2, Arreola et al. found a bimodal pH_{ext} effect with a conductance maximum around pH 6.5 (Arreola et al. 2002). Interestingly, acidification to pH 5.5 applied during opening by hyperpolarization led first to a transient activation followed by inhibition, suggesting the existence of two different proton-binding sites. Occupation of one of these can exert a stimulatory effect, but the site becomes accessible to extracellular protons only in the open state of the channel (Arreola et al. 2002). Very similar results were found for guinea pig ClC-2 expressed in HEK cells (Niemeyer et al. 2003). In particular, because the transient activation by external protons was ablated in the mutant Glu-217-Val, it was suggested that the residue Glu-217 is the acceptor site for protons responsible for the stimulatory effect of low pH on ClC-2 (Niemeyer et al. 2003).

Mutating Lys-566, located at the end of the transmembrane-spanning domain, to glutamate was found to shift the voltage dependence of gating and to change the inward rectification of the open channel *I-V* relationship of wild-type ClC-2 to outward rectifying, in analogy with the effect of the corresponding mutant (Lys-519-Glu) on the fast gate of ClC-0 (Pusch et al.1999, 1995a). However, mutation of this lysine did not modify activation by hyperpolarization, cell swelling, and acidification. In contrast, mutations in helix I and the preceding loop abolished all three modes of activation by constitutively opening the channel without changing its pore properties (Jordt and Jentsch 1997).

It has been clearly established that gating of ClC-2 depends on intracellular $[Cl^-]$, whose increase shifts the open probability of the channel toward positive potentials in a N-terminal-deleted construct of rat ClC-2 expressed in oocytes (Pusch et al. 1999), the full-length rat ClC-2 expressed in HEK cells (Niemeyer et al. 2003), and the human ClC-2 (Haug et al. 2003).

The role of extracellular chloride is still controversial, as Pusch et al. (Pusch et al. 1999) showed that, surprisingly, decreasing $[Cl^-]_{ext}$ increases the open probability of ClC-2, whereas Niemeyer et al. (Niemeyer et al. 2003) were not able to observe any effect of extracellular Cl^- on gating.

Activation of macroscopic ClC-2 currents follows a double exponential time course, with time constants differing roughly by one order of magnitude (de Santiago et al. 2005; Pusch et al. 1999; Zúñiga et al. 2004). Both time constants are voltage dependent, becoming faster on hyperpolarization (Zúñiga et al. 2004). The opposite behavior was found for ClC-1, in which both gates are opened by depolarization (Accardi and Pusch 2000), while the slow gate of ClC-0 also opens in response to hyperpolarization (Pusch et al. 1997). Interestingly, the amplitudes of the two components in ClC-2 have an opposite voltage dependence (de Santiago et al. 2005). The voltage dependence of macroscopic currents could be described by a Boltzmann function with half-maximal activation and slope factor of, respectively, −117 and 22 mV at 22°C (Zúñiga et al. 2004). Activation and deactivation time constants were reduced on temperature increase without major changes in the steady-state activation curve. The Q_{10} values calculated for both fast and slow time constants are between 4 and 5, suggesting a significant conformational change associated with those processes (Zúñiga et al. 2004). This value is larger than the Q_{10} factor related to the fast gate of ClC-0 but much smaller than that for the slow gate (Pusch et al. 1997). On the other hand, it is in the same range of the Q_{10} factor measured for the common gate of ClC-1 (Bennetts et al. 2001).

An attempt at quantitative dissection of the properties of the fast and slow gating processes in ClC-2 was made by de Santiago et al. (de Santiago et al. 2005). These authors found that the open probability of the protopore gate can be described by a Boltzmann distribution with half-maximal activation at −63 mV and an apparent gating charge of −1.22, whereas the common gate remains about 55% open at positive voltages and is associated with an apparent gating charge of −0.99 with half-maximal activation at −134 mV. The mutation Cys-258-Ser affected the voltage dependence of both gates. This observation is in contrast with the fact that the double mutant Glu-217-Ala/Cys-258-Ser produced currents very similar to that of the single mutant Glu-217-Ala, probably dominated by the slow component (Niemeyer et al. 2003), suggesting that the mutation Cys-258-Ser mostly affects the operation of the fast gate.

Moreover, the mutation His-811-Ala in ClC-2, corresponding to a mutation that completely and selectively abolishes slow gating in ClC-0 (Estévez et al. 2004) and that is located in the highly conserved CBS2 domain, affected both fast and slow gating of ClC-2. Interestingly, combining this mutation with Glu-217-Val ablates all gating transitions (Yusef et al. 2006).

Collectively, these pieces of evidence point to a gating mechanism composed of a fast and slow component that bear some resemblance to the protopore and common gate identified in ClC-0 and ClC-1 but that appear also to have very distinctive features whose nature is only poorly understood. In particular, a very specific element of ClC-2 is the stronger correlation of the two gating modes compared to the other channel members of the CLC family (Zúñiga et al. 2004).

Mutation of the conserved Glu at position 217 into Val (Glu-148 in ClC-ec1) produces a loss of sensitivity of the channel to Cl^{-}_{int} and almost abolishes its voltage dependence and the characteristic inward rectification of the current (de Santiago et al. 2005; Zúñiga et al. 2004). It has been conjectured that in ClC-2 fast gating is due to the movement of the side chain of Glu-217 in a mechanism similar to that proposed for the fast gate of ClC-0 and ClC-1 (Niemeyer et al. 2003; Yusef et al. 2006; Zúñiga et al. 2004). In this scenario, the residual voltage dependence observed for ClC-2 at strongly hyperpolarized potentials could be explained in terms of transitions of the common gate (de Santiago et al. 2005; Niemeyer et al. 2003). The analogy between ClC-2 and other channel members of ClC family can be extended also to the extracellular pH dependence as the transient activation of ClC-2 at acidic extracellular pH is reminiscent of the increase in the open probability of the fast gate of ClC-0 at low external pH and is mediated by the corresponding glutamate residue (Niemeyer et al. 2003).

Even more speculatively, slow gating transitions in ClC-2 have been proposed to arise from conformational changes in the pore, which are known to accompany protopore gating in ClC-0 (Accardi and Pusch 2003), possibly mediated through rearrangements of the helix R, which can be influenced by movements of the C-terminus of the protein (Yusef et al. 2006). However, in the absence of single-channel measurements, the separation between fast and slow gating transitions is only tentative and requires a more solid experimental basis.

Physiological role of ClC-2

In contrast to the tissue distribution, the subcellular localization of ClC-2 channels is still controversial. On one hand, antibodies against ClC-2 detected signals in apical membranes of intestinal tissues in humans (Murray et al. 1996) and mice (Gyömörey et al. 2000) and in rat lung epithelium (Blaisdell et al. 2000). On the other hand, with different antibodies, a basolateral localization was suggested in rat small intestine and colon (Lipecka et al. 2002) and guinea pig colonocytes (Catalán et al. 2002).

As pointed out by Zdebik et al. (Zdebik et al. 2004), none of these studies included controls with knockout tissues, raising the possibility that some of these results may reflect unspecific binding. Such an approach was recently undertaken by Sepúlveda and coworkers (Peña-Münzenmayer et al. 2005) who, by an immunohistochemical approach using ClC-2-null mice as a negative control, showed that ClC-2 has a basolateral localization in intestinal epithelial cells of wild-type mice. Moreover, heterologous expression in epithelial cell lines of a ClC-2 construct with a C-terminally fused GFP in combination with confocal fluorescence imaging confirmed a basolateral expression. The polarized expression seems to depend on the AP-1AB clathrin adaptor protein, which is known to be an epithelium-specific complex involved in basolateral sorting (Fölsch et al. 1999). In particular, a dileucine motif, which is normally recognized by AP-1AB (Nakatsu and Ohno 2003), encoded in the CBS2 domain of ClC-2 was found to be critical for the basolateral localization as its disruption produced apical localization (Peña-Münzenmayer et al. 2005). This motif is conserved in ClC-2 from different organisms and in the other plasma membrane CLC members but not in ClC-3, -4, -5, -6, and -7, all mainly expressed intracellularly.

A basolateral localization of ClC-2, implying that it does not contribute to chloride secretion, is compatible with the observation that homozygous mice harboring the ΔF508 mutation in the CFTR gene, leading to cystic fibrosis through impaired Cl^- secretion, survived better when ClC-2 was additionally disrupted (Zdebik et al. 2004). This issue is of significant medical importance because the pathological changes in cystic fibrosis are predominantly due to defective Cl^- conductance on the apical side of the cells, and therefore pharmacological intervention on Cl^- channels, in particular ClC-2, to be beneficial, will have to take into account its subcellular localization. For example, since ClC-2 is expressed basolaterally, maneuvers that decrease the channel conductance might have a positive effect on the pathology (Zdebik et al. 2004). A possible physiological mechanism that potentially modulates Cl^- fluxes through the plasma membrane of epithelial cells depending on their absorption-secretion activity is provided by the regulation of the open probability of ClC-2 operated by $[Cl^-]_{int}$ (Catalán et al. 2004).

Several recent papers reported on the interaction of ClC-2 with other cellular proteins. Hinzpeter et al. (Hinzpeter et al. 2006) presented evidence that ClC-2 can interact with Hsp70 and Hsp90 in HEK cells and with Hsp90 in mouse brain, producing a reduction of channel expression at the plasma membrane. However the molecular basis for the interaction remains to be determined. Bali et al. (Bali et al. 2001) suggested a role of vesicular trafficking in the regulation of ClC-2 plasma membrane expression. It has been indicated that such a role could be mediated by the dynein motor complex that coimmunoprecipitates with ClC-2 in rat hippocampal slices and controls retrograde trafficking of the channel between plasma membrane and endosomes in COS-7 cells (Dhani et al. 2003).

The expression level of ClC-2 in rat renal proximal tubules is influenced by thyroid hormones (Santos Ornellas et al. 2003) and estrogens (Nascimento et al. 2003), suggesting the relevance of this channel for Cl^- transport in the kidney, even though no renal phenotype was reported in ClC-2 knockout mice (Bösl et al. 2001). It was found that the transcription factors SP1 and SP3 influence the expression level of ClC-2 in lung epithelial cells (Holmes et al. 2003) binding to the ClC-2 promoter (Chu et al. 1999). In particular, for SP1, it has been shown that such regulation is exerted by the glycosylated isoform (Vij and Zeitlin 2006).

In contrast with what could be expected from its ubiquitous expression, ClC-2-deficient mice only manifest severe degeneration of the retina and the testes leading to male infertility (Bösl et al. 2001). Both effects have been attributed to a defective transport by epithelia that would normally control the ionic environment of sensitive germ cells and photoreceptors (Bösl et al. 2001). Although the mouse system does not always represent an accurate model for humans, these findings suggest close reconsideration of the proposed role of ClC-2 in lung development (Blaisdell et al. 2000; Murray et al. 1995), nephrogenesis (Huber et al. 1998), gastric acid secretion (Malinowska et al. 1995), and modulation of postsynaptic response to GABA and glycine (Smith et al. 1995; Staley et al. 1996).

An interesting, although controversial, insight into the physiological role of ClC-2 in humans is provided by the identification of three mutations in the *CLCN2* gene causing idiopathic generalized epilepsy (Haug et al. 2003). One mutation produced a truncation of the channel just after the beginning of helix F. Heterologous expression of this construct alone or of a concatameric construct with wild-type ClC-2 in tsA201 cells led to a complete loss of channel function. Coexpression experiments resulted in a significantly smaller Cl^- current compared to wild-type. A second mutation, which has also been found in healthy controls, produces a splice variant with a 33-amino acid deletion involving helix B and had the same effect as the previous mutation. As channel constructs harboring the mutations and tagged with yellow fluorescent protein were expressed at the plasma membrane, it has been claimed that both M200fsX231 and Δ74–117 mutants of ClC-2 reach the membrane to exert dominant-negative effects that markedly inhibit the activity of wild-type ClC-2 (Haug et al. 2003). It was conjectured that these mutations would decrease Cl^- efflux from neurons, resulting in Cl^- accumulation with consequent impairment of the inhibitory GABA response (Staley et al. 1996), which in turn may lead to hyperexcitability.

The third mutation results in the amino acid substitution Gly-715-Glu, located between the two CBS domains, and gives rise to functional channels with altered voltage dependence (channels opened at less negative potentials compared to wild-type). In contrast to the other mutations, this effect represents a gain of function. It was speculated that the pathophysiological effect of this mutation is to sustain a significant Cl^- efflux during the repolarization phase of the action potential that tends to depolarize neurons, generating hyperexcitability (Haug et al. 2003). These experimental results were therefore interpreted as being compatible with a role of ClC-2 in Cl^- efflux as an essential element for normal neuronal excitation.

However, an analysis of the functional consequences of these mutations performed in HEK cells gave drastically different results (Niemeyer et al. 2004). The GFP-labeled mutations M200fsX231 and Δ74–117 were localized only intracellularly, and they did not affect the maximal cellular conductance, severely questioning a dominant-negative effect (Niemeyer et al. 2004). It is interesting to note that most truncating mutations of ClC-1 have been found to be associated with recessive and not dominant myotonia. It was therefore proposed that haploinsufficiency would be the mechanism leading to the epilepsy in patients with these mutations (Niemeyer et al. 2004).

On the other hand, the only functional consequence of the mutation Gly-715-Glu was to affect the AMP sensitivity of the channel, pointing to a completely different pathophysiological consequence from that previously proposed (Niemeyer et al. 2004). These conflicting *in vitro* results, and the lack of signs of epilepsy in ClC-2 knockout mice, call for additional human genetic evidence before *CLCN2* can be firmly classified as an epilepsy susceptibility gene.

Interestingly, ClC-2 has been shown to be inhibited by venom from the scorpion *L. quinquestriatus hebraeus*, which appears to shift the voltage dependence of activation toward hyperpolarizing potentials (Thompson et al. 2005). However, unspecific effects cannot be fully excluded, requiring the purification of the putative peptide.

The renal and inner ear ClC-K channels

ClC-K channels were identified by homology cloning in rat and human kidney (Kieferle et al. 1994; Uchida et al. 1993). In particular, the two channels from rat (rClC-K1 and rClC-K2) are 80% identical, whereas the human channels (hClC-Ka and hClC-Kb) show 90% identity with each other, indicating a comparatively recent evolutionary divergence (Kieferle et al. 1994). Besides the kidney, these channels are also expressed in the inner ear (Uchida et al. 1995; Vandewalle et al. 1997).

Expression of ClC-K1 in *Xenopus* oocytes gave rise to small, slightly outwardly rectifying currents that showed some time-dependent gating at voltages more positive than +40 mV or more negative than −100 mV, similar to chloride currents of the thin ascending limb observed in *in vitro* perfusion experiments (Uchida et al. 1993; Waldegger and Jentsch 2000). It is interesting to correlate the lack of gating with the observation that ClC-K channels are the only CLC members that have a valine in place of the critical glutamate (Glu-166 in ClC-0) that was shown to be a major determinant of the fast gate in ClC-0, ClC-1, and ClC-2. In fact, in mutating this valine into glutamate, a significant gating was introduced in the channel behavior (Waldegger and Jentsch 2000). The permeability sequence is $Cl^- > Br^- > NO_3^- > I^-$. The current amplitude increases on extracellular alkalinization to pH 8.0 and strongly decreases when the extracellular pH is reduced from pH 7.5 to 6.5 (Uchida et al. 1995; Waldegger and Jentsch 2000). Moreover, ClC-K1 activity is modulated by extracellular Ca^{2+} (Uchida et al. 1995) with an increase of $[Ca^{2+}]_{ext}$ from 1 to 5 mM producing a fourfold increase in channel currents (Waldegger and Jentsch 2000). Sensitivity to pH and Ca^{2+} also correlates with *in vitro* microperfusion experiments on the thin ascending limb (Uchida et al. 1995).

The fact that the other ClC-K channels (i.e., rat ClC-K2, human ClC-Ka and ClC-Kb), on expression in oocytes, did not give rise to current despite correct protein synthesis and a very high sequence identity with ClC-K1 was puzzling and led to the hypothesis that an auxiliary β-subunit would be necessary for their functional expression (Waldegger and Jentsch 2000). This speculation was later confirmed by Estévez et al. (Estévez et al. 2001), who showed that a gene mutated in a specific form of Bartter syndrome encodes a β-subunit (called barttin) of ClC-K channels. Barttin consists of two putative transmembrane domains and strictly colocalizes with both ClC-Ks in kidney and cochlea (Estévez et al. 2001). When coexpressed with ClC-Ka and ClC-Kb in heterologous systems it induces detectable currents, and coexpression with ClC-K1 dramatically increases the currents that are elicited by ClC-K1 expression alone (Estévez et al. 2001). Interestingly, heteromeric ClC-K1/barttin channels appear to have a modified Ca^{2+} sensitivity compared to ClC-K1 alone (Waldegger et al. 2002), even though the relative influence of unspecific background currents was clearly larger for the pure ClC-K1 currents, whereas series resistance problems may be large for barttin-increased currents.

Immunoprecipitation and immunofluorescence experiments indicated that barttin is physically associated with ClC-K proteins and increases their surface expression (Estévez et al. 2001; Waldegger et al. 2002). ClC-Ks seem to be localized mainly in the Golgi apparatus without barttin coexpression (Uchida and Sasaki 2005), but it is not clear whether barttin binds and recruits ClC-Ks to the appropriate location in the plasma membrane or whether barttin, on binding, masks Golgi-localizing signals of ClC-Ks, thus releasing it from the Golgi apparatus (Hayama et al. 2003). Also, the stoichiometry of barttin-ClC-K complexes and the respective interacting regions are unknown.

Barttin contains a putative PY motif that is a potential site for binding of WW domain-containing ubiquitin ligases or may serve as a tyrosine-based endocytosis signal (Estévez et al. 2001). When the tyrosine residue of the PY motif was mutated (Tyr-98-Ala), stimulation of ClC-Ka and ClC-Kb currents by barttin was enhanced, but macroscopic currents did not differ qualitatively from those of wild-type heteromers (Estévez et al. 2001). An interaction with ubiquitin ligases was suggested on the basis of the reduction of ClC-K/barttin currents on overexpression of the ubiquitin ligase Nedd4-2 (Embark et al. 2004). However, at variance with ClC-5 (Schwake et al. 2001) and ENaC (Abriel et al. 1999), also containing a PY motif, the expression of an inactive form of Nedd4-2 did not increase ClC-K/barttin currents (Embark et al. 2004).

The two ClC-K isoforms are differentially distributed in nephrons (Kieferle et al. 1994; Vandewalle et al. 1997). ClC-Ka (in rat: ClC-K1) is exclusively expressed in a particular nephron segment, the thin limb of Henle's loop (Fig. 5), whereas ClC-Kb (in rat: ClC-K2) has a broader expression in kidney but is especially abundant in the thick ascending limb, a nephron segment specialized in NaCl reabsorption (Fig. 5) (Jentsch et al. 2005a). In particular, it was shown by immunohistochemistry that ClC-K1 and ClC-K2 are expressed exclusively in basolateral membranes of renal (Vandewalle et al. 1997) and cochlear (Estévez et al. 2001) epithelia, although another group proposed that ClC-K1 is present in both apical and basolateral membranes of the thin limb of Henle's loop (Uchida et al. 1995). It should be noted that all localization studies have been performed in rodents. It is not clear whether the same tissue distribution applies to humans, especially because the functional/physiological equivalence of ClC-K1/ClC-Ka and ClC-K2/ClC-Kb, respectively, is based on relatively vague arguments.

Impairment of Cl^- transport as the underlying cause of renal salt-wasting diseases was suggested already about thirty years ago (Gill and Bartter 1978). ClC-Kb mutations are associated with Bartter syndrome type III (Simon et al. 1997), an autosomal recessive salt-wasting disorder characterized by reduced sodium chloride reabsorption underlying the fundamental role of the channel in this physiological process. During reabsorption, Na^+, K^+, and Cl^- ions enter the tubule cells that line the nephrons through apical Na-K-2Cl cotransporters. The subsequent extrusion of Cl^- through the basolateral side prevents accumulation of Cl^- that would oppose inward transport of Na^+, which is of paramount importance for water and salt homeostasis (Hunter 2001; Jeck et al. 2004a). The particularly high expression of ClC-Kb in a compartment specialized in salt reabsorption (the thick ascending limb), its basolateral localization, and its relevance for normal reabsorption activity support the role of ClC-Kb as the basolateral Cl^- pathway in this schematic model.

Mutations affecting only ClC-Ka have not been implicated in human diseases so far, but simultaneous mutations in ClC-Ka and ClC-Kb lead to Bartter syndrome with deafness (see below). Even if the physiological role of ClC-Ka in humans is not completely clear,

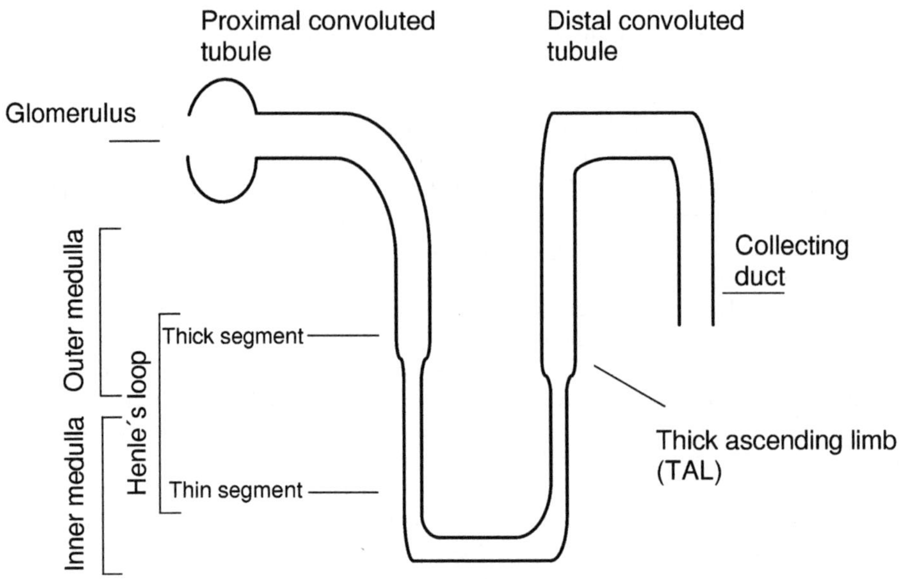

Fig. 5 Schematic representation of the nephron

an interesting insight into its function is provided by disruption of the presumed mouse ortholog ClC-K1 that produces apparent nephrogenic diabetes insipidus, a defect of urinary concentration (Matsumura et al. 1999).

Urinary concentration is determined by water reabsorption in the collecting duct (Fig. 5) that is stimulated when the kidney medulla is hypertonic. In humans, the establishment of such a situation requires the exquisite coordination of many ionic transport systems along the different segments of the nephron (the so-called countercurrent system), as exemplified by the fact that mutations in almost all of these produce a pathogenic phenotype (Sands and Bichet 2006). The study of Matsumara et al. (Matsumura et al. 1999) suggested that ClC-K1 has a critical role in the urine concentrating mechanism as already speculated by Uchida et al. (Uchida et al. 1995). This is in functional agreement with the expression of ClC-K1 in the thin ascending limb, whose Cl^- permeability was found to be impaired in the ClC-K1 knockout in *in vitro* microperfusion experiments (Matsumura et al. 1999). Interestingly, such a role for ClC-K1 correlates with the finding that maximum urine concentrating ability in mice and rats is observed 2–3 weeks after birth and parallels a gradual increase in ClC-K1 expression within the ascending limb of Henle's loop (Kobayashi et al. 2001; Liu et al. 2001) (Fig. 5).

In the course of genetic analysis of patients suffering from salt-losing tubular disorders, several mutations and molecular variants of ClC-Kb have been identified (Konrad et al. 2000). One in particular leads to the amino acid substitution Thr-481-Ser, which is also found at a frequency of 20%–40% in nonaffected individuals and produces a 20-fold increase in current induced by heterologously expressed CLC-Kb, probably due to increased open probability of the channel (Jeck et al. 2004a). This mutation has been associated with high blood pressure (Jeck et al. 2004b), but recent publications contradict this hypothesis (Kokubo et al. 2005; Speirs et al. 2005).

Human mutations in the β-subunit barttin lead to Bartter syndrome type IV characterized by both severe renal salt loss and congenital deafness (Birkenhäger et al. 2001). According to the mechanism proposed to explain the physiopathology of the deafness, ClC-Ka and -Kb represent essential basolateral exit pathways to keep internal Cl^- concentration at a level that is compatible with the efficient accumulation of K^+ into marginal cells of the cochlear stria vascularis. K^+ is then secreted into the endolymph, where its high concentration (150 mM) is required for the stimulation of sensory hair cells (Jentsch 2000). Mutations involving ClC-Ka or ClC-Kb alone have not been implicated in deafness. This probably reflects the fact that coexpression of ClC-Ka and -Kb in the cochlea preserves a sufficient level of Cl^- extrusion even in the case of mutations that impair one of them. This functional rescue cannot occur in case of barttin mutations that indeed invariably cause the renal and inner ear phenotypes. This scenario is confirmed by the finding that simultaneous ClC-Ka and -Kb mutations result in a phenotype that mimics type IV Bartter syndrome (Schlingmann et al. 2004).

Considering the physiological relevance of ClC-K and their involvement in pathogenic state in human and mouse, the identification of pharmacological tools to modify their properties can have important medical consequences and may represent a tool to better understand their biophysical properties (Pusch et al. 2006).

In contrast to the behavior of other ClC channels (Conte-Camerino et al. 1988; Liantonio et al. 2002; Pusch et al. 2000), ClC-K channels are inhibited by derivatives of CPP and DIDS from the extracellular side (Liantonio et al. 2002; Picollo et al. 2004). In particular, for ClC-Ka and ClC-K1, it was found that the block by 3-phenyl-CPP was quickly reversible and competitive with extracellular Cl^-, suggesting that the binding site for the compound is exposed to the extracellular side and is located close to the ion-conducting pore (Liantonio et al. 2004; Picollo et al. 2004).

Surprisingly, the apparent affinity of ClC-Kb for the compounds was found to be five- to six fold lower than for ClC-Ka (K_D of ~80 and 90 μM for 3-phenyl-CPP and DIDS, respectively) despite the very high sequence identity between them. An elegant approach allowed Picollo et al. (Picollo et al. 2004) to identify a critical residue at position 68 as the major molecular determinant for the differential behavior, as ClC-Ka has a neutral asparagine at this position whereas ClC-Kb has a charged aspartate.

Very recently, Liantonio et al. (Liantonio et al. 2006) showed that niflumic acid (NFA) and flufenamic acid (FFA), drugs belonging to a class of nonsteroid anti-inflammatory fenamates, modulate ClC-K channel activity in a singular manner. NFA applied extracellularly at concentrations up to 1 mM increased ClC-Ka current amplitudes by a factor of two in a voltage-independent manner, whereas higher concentrations blocked the current. Such a biphasic behavior was tentatively explained by the presence of two different binding sites. In contrast with this behavior, NFA produced only activation of ClC-Kb (Liantonio et al. 2006). On the other hand, FFA blocked ClC-Ka but activated ClC-Kb (Liantonio et al. 2006).

Although the molecular basis for the effect of these molecules is still poorly understood, they provide a promising starting point for identification of diuretics and for the treatment of Bartter syndrome (Liantonio et al. 2006).

ClC-3—a transporter with special importance in the brain

ClC-3 was first cloned by Kawasaki et al. (Kawasaki et al. 1994). It has an ubiquitous expression pattern, but is predominantly found in brain, most notably in the olfactory bulb, hippocampus, and cerebellum, and in kidney and colon (Kawasaki et al. 1994). Ogura et al. (Ogura et al. 2002) described a splice variant of ClC-3, called ClC-3B, that has a different and slightly longer C-terminal end and is expressed mostly in epithelial cells. The ClC-3 protein is found predominantly in late endosomes, lysosomes, and synaptic vesicles and is important for their acidification (Hara-Chikuma et al. 2005b; Stobrawa et al. 2001). No human disease caused by mutations in ClC-3 has been reported so far, but its disruption in mice leads to a progressive degeneration of the hippocampus and a complete loss of photoreceptors (Stobrawa et al. 2001). In addition to these findings, an independently generated ClC-3 KO mouse (Yoshikawa et al. 2002) also showed markers of lysosomal storage disease that partially overlapped with neuronal ceroid lipofuscinosis (NCL), but no association of ClC-3 mutations with NCL could be detected in several dog lineages that suffered from late-onset NCL (Wohlke et al. 2006). Importantly, there was no significant difference in swelling-activated currents between wild-type and knockout mice (Stobrawa et al. 2001; Wang et al. 2006), proving that ClC-3 does not underlie the swelling-activated chloride current as previously suggested (Duan et al. 1997; see also discussion in Jentsch et al. 2002).

The mechanism underlying the ClC-3 knockout phenotypes is still unknown, but, based on an increased glutamate uptake in synaptic vesicles of knockout mice, it was speculated that the neurodegeneration might be caused by a neurotoxic effect of glutamate due to the altered intracellular vesicle pH (Stobrawa et al. 2001). However, a different mechanism based on trafficking defects of other membrane proteins brought about by altered acidification of intracellular compartments produced by dysfunction of ClC-3 could not be excluded. In this respect, it is interesting to note that Salazar et al. (Salazar et al. 2004) revealed that a mouse deficient in AP-3, an adaptor protein responsible for the correct sorting of membrane proteins in synaptic vesicles, also manifested a marked decrease in the expression of ClC-3. Moreover ClC-3 co-localized with a zinc transporter (ZnT3) and modulated Zn^{2+} level in a specific subpopulation of synaptic vesicles (Salazar et al. 2004), a finding of possible physiological relevance considering the inhibitory effect of Zn^{2+} on NMDA-mediated response in the hippocampus (Vogt et al. 2000).

Robinson et al. (Robinson et al. 2004) suggested that calcium-calmodulin-protein kinase II (CaMKII) is able to activate ClC-3 in different cell types and proposed Ser-109 as the phosphorylation site. In particular in transfected tsA cells and HT29 cells (human colonic tumor cell line), ClC-3 was reported to have a substantial plasma membrane expression. These studies, however, seem to need further confirmation, as no other group has reported similar results yet.

Recently, an interesting involvement of ClC-3 activity in the oxidative function of neutrophils has been discovered (Moreland et al. 2006). Starting from the initial observation that ClC-3 KO mice, but not wild-type mice, died frequently from sepsis following intravascular catheter placement, Moreland et al. (Moreland et al. 2006) found that neutrophils from knockout mice showed an impaired NADPH oxidase activity. ClC-3 was found in particular in secretory vesicles and secondary granule compartments. The precise role of ClC-3 in neutrophil oxidative function remains, however, to be elucidated.

The biophysical properties of ClC-3 have been notoriously difficult to analyze, and different groups have reported conflicting results (see Jentsch et al. 2002). We consider the studies of the Weinman group (Li et al. 2000, 2002) as the most reliable. Weinman and colleagues expressed ClC-3 in mammalian cell lines and could detect relatively small membrane currents in highly overexpressing cells, in which most of the expressed protein remained intracellular (Li et al. 2002). Interestingly, these cells showed enlarged and acidic intracellular structures (Li et al. 2002). Importantly, the biophysical properties of these currents were very similar to those of ClC-4 and ClC-5 (Li et al. 2000), CLC proteins that can be reliably expressed in *Xenopus* oocytes (Friedrich et al. 1999; Steinmeyer et al. 1995) and are highly homologous to ClC-3. Based on the functional and structural similarity of ClC-3 with ClC-4 and ClC-5, it has been suggested that ClC-3, like these other two proteins, is actually a Cl^- / H^+ antiporter, and not a Cl^- channel. However, lacking direct experimental evidence, this conclusion must still be considered as tentative.

ClC-4—a transporter whose physiological role is poorly understood

ClC-4 has been identified by van Slegtenhorst et al. (van Slegtenhorst et al. 1994); it shares 78% sequence identity with ClC-5 and shows very similar biophysical properties (Friedrich et al. 1999). ClC-4 is mainly found in brain, liver, and kidney, where its subcellular localization closely resembles that of ClC-5, that is, it colocalizes mainly with endosomal markers (Mohammad-Panah et al. 2003). It was suggested that ClC-4 facilitates endosomal acidification and is important for endocytosis (Mohammad-Panah et al. 2003). Biochemical and functional lines of evidence suggested that ClC-4 and ClC-5 can form heterodimers (Mohammad-Panah et al. 2003; Suzuki et al. 2006). However, unlike ClC-5, ClC-4 is not crucial for renal endocytosis because ClC-4 knockout mice do not display proteinuria (Jentsch et al. 2005b). ClC-4 was proposed to facilitate incorporation of copper into ceruloplasmin by shunting currents of Cu^{2+}-ATPases in the secretory pathway (Wang and Weinman 2004). In rodent and human intestinal epithelia ClC-4 has been reported to colocalize with CFTR in apical membrane and subapical vesicles, and it has been suggested to mediate chloride current across the plasma membrane of Caco-2 cells (which represent a model for human enterocytes). This would support a role of ClC-4 in intestinal chloride secretion, suggesting that it might be capable of functionally complementing CFTR *in vivo* (Wang and Weinman 2004). These results, however, are difficult to reconcile with the Cl^-/H^+ antiporter function of ClC-4 (Picollo and Pusch 2005; Scheel et al. 2005) and its biophysical properties (Friedrich et al. 1999) and need further experimental verification.

ClC-5—a transporter involved in endocytosis

ClC-5 is the most extensively studied member of the CLC branch also comprising ClC-3 and ClC-4, being the only one for which an involvement in a genetic disease has been described. In fact, the identification of ClC-5 is due to its link to Dent disease, an X-linked hereditary disorder that is always associated with low-molecular-weight proteinuria and less frequently with hypercalciuria, which in turn causes nephrocalcinosis, nephrolithiasis (kidney stones), and eventual renal failure (Dent and Friedman 1964; Günther et al. 1998; Wrong et al. 1994). Fisher et al. (Fisher et al. 1994) identified a gene whose transcripts were heavily expressed in kidney and which was partially deleted in individuals affected by Dent disease. The predicted amino acid sequence of the gene product showed a high degree of homology to previously isolated members of the CLC family.

Steinmeyer et al. (Steinmeyer et al. 1995) independently cloned full-length ClC-5 from rat brain (rClC-5) by homology to other CLC proteins. The open reading frame of 2,238 bp predicts a protein of 746 amino acids with a molecular mass of 83 kDa. It is highly expressed in kidney, but mRNA is also detectable in brain and liver and to a lesser extent in lung and testis (Steinmeyer et al. 1995). A splice variant with an additional 70 amino acids at the intracellular amino terminus has been detected at the mRNA level but not at the protein level, and no mutations have been identified so far in the exons encoding the 70 additional amino acids (Ludwig et al. 2003).

ClC-5 is predominantly expressed in kidney but is also found in other tissues, such as intestinal epithelia. In these tissues it is present in vesicles of the endosomal pathway (Devuyst et al. 1999; Günther et al. 1998; Jentsch et al. 2005c; Vandewalle et al. 2001). In particular, the expression of ClC-5 is very high in the proximal tubule (PT), which is responsible for the endocytotic uptake of low-molecular-weight proteins (Jentsch et al. 2005a) (Fig. 5). In the PT and in intercalated cells of the collecting duct (Günther et al. 1998), in rat fetal lung (Edmonds et al. 2002), and in human retinal pigment epithelium (Weng et al. 2002), ClC-5 colocalizes with a V-type H^+-ATPase in vesicles below the apical membrane. In particular, transmission electron microscopy of PT cells shows the presence of ClC-5 in vesicles that are concentrated below the microvilli of the brush border (Günther et al. 1998). This specialized region contains an extensive endocytotic apparatus necessary for the pronounced endocytotic activity of proximal tubule cells (Günther et al. 1998).

Moreover, it was found that ClC-5 colocalizes with endosomal markers (Günther et al. 1998) and endocytosed proteins early after uptake (Devuyst et al. 1999; Günther et al. 1998; Piwon et al. 2000), arguing for its presence in early endosomes (Jentsch et al. 2005a).

Expression of rat ClC-5 in *Xenopus* oocytes produced strongly outwardly rectifying chloride currents for which it was not possible to detect any gating relaxations (Steinmeyer et al. 1995). The rectification also prevented the determination of a true reversal potential and consequently the assessment of relative permeability ratios. From the current magnitude, however, the conductivity sequence was determined as $NO_3^- > Cl^- > Br^- > I^-$ > glutamate, in agreement with the behavior of other CLC proteins (Steinmeyer et al. 1995). Several classic Cl^- channel inhibitors (DIDS, 9-AC, CPA) had no effect on rClC-5. In *Xenopus* oocytes, rClC-5 elicited chloride currents only at potentials more positive than 20 mV, a value that is not reached across the plasma membrane of most cells. It was therefore speculated that rClC-5 may be localized to intracellular compartments characterized by a different transmembrane voltage and that the currents observed in oocytes may result from "spillover expression," whereby vesicles normally targeted to an intracellular compartment reach the plasma membrane because of overexpression (Steinmeyer et al. 1995). This was also confirmed by immunofluorescence performed on COS-7 cells transfected with ClC-5 (Günther et al. 1998). Whether this occurs *in vivo* is not yet clear (Jentsch 2005). However, Wang et al. proposed from biotinylation studies that in proximal tubule cells, about 8% of the total cellular pool of ClC-5 is located at the plasma membrane (Wang et al. 2005).

Recently, Suzuki et al. (Suzuki et al. 2006), using immunofluorescence and immunoprecipitation, reported that ClC-3, ClC-4, and ClC-5 show a high degree of colocalization in intracellular organelles on expression in HEK cells and potentially form heteromultimers.

Dent disease can be caused by nonsense mutations, deletions, and also missense mutations in the *CLCN5* gene (reviewed in Jentsch et al. 2005c; Uchida and Sasaki 2005). The missense mutations are interspersed along the whole secondary structure of the protein. Nevertheless, on heterologous expression in oocytes, most of them produce a similar phenotype, namely impaired trafficking to the plasma membrane. This seems to imply that several distinct protein regions are essential for correct targeting and/or protein stability. Interestingly, the mutation Arg-516-Trp, located only 5 amino acids away from the mutant Leu-521-Arg that abolishes channel expression at the plasma membrane, shows a normal level of expression but leads nonetheless to a drastic reduction of Cl^- current, implying that this mutation does not affect targeting but severely impairs transport activity (Ludwig et al. 2005). In contrast, truncation of ClC-5 at position 648, located just after the CBS1 domain, although not functional, does not impair trafficking but instead produces a paradoxical increase in plasma membrane expression (Ludwig et al. 2005). It was hypothesized that the effect could be due to the deletion of a PY motif located between the two CBS domains. In fact, Schwake et al. (Schwake et al. 2001) showed that mutating the PY motif of ClC-5 almost doubled surface expression and channel activity and suggested that the effect could be due to impaired internalization of the protein. Such a motif had been also implicated in internalization and ubiquitination of the amiloride-sensitive sodium channel (ENaC), and mutations in it lead to Liddle syndrome, another human inherited kidney disorder associated with hypertension (Hansson et al. 1995a, 1995b).

Low-molecular-weight proteinuria is a hallmark of Dent disease. Proteins of low molecular weight are filtered at the glomerulus and are normally reabsorbed in the proximal tubule (Fig. 5) by fluid-phase and receptor-mediated endocytosis (Mellman 1996). After being endocytosed, the proteins are subsequently degraded in lysosomal compartments (Maack and Park 1990). Acidification of the endosomes is essential for the progression along the endocytic apparatus to lysosomes (Mellman et al. 1986) (but see Günther et al. 1998 and references therein), and it is mediated by a V-type H^+-ATPase (Gluck et al. 1996). Interestingly, immunohistochemistry studies of biopsies of Dent disease patients revealed a consistent inversion of H^+-ATPase polarity in PT cells, showing a basolateral distribution contrasting with its apical location in the normal kidney (Moulin et al. 2003). These modifications in polarity and/or expression of the H^+-ATPase are compatible with an interaction between ClC-5 and the H^+-ATPase that would be essential for the proper targeting or stability of the latter and may explain the deficit in urinary acidification observed in some patients with Dent disease (Moulin et al. 2003). The colocalization of ClC-5 with the H^+-ATPase suggested that ClC-5 might be important for endocytosis, and it was speculated that its role was to provide an electrical shunt for the efficient accumulation of protons by the H^+-ATPase (Günther et al. 1998). Indirect support for a role of CLC-5 in the acidification of intracellular compartments comes also from yeast: Disruption of either the yeast CLC (GEF1) (Greene et al. 1993), which resides in intracellular vesicles (Hechenberger et al. 1996; Schwappach et al. 1998), or of GEF2, a subunit of the vacuolar H^+-ATPase, caused an increased sensitivity to more alkaline pH (Gaxiola et al. 1998; Schwappach et al. 1998).

A knockout mouse approach provided a powerful insight into the physiological role of ClC-5 (Piwon et al. 2000; Wang et al. 2000). In very elegant experiments, Piwon et al. exploited the fact that ClC-5 is encoded on the X chromosome, which is subject to random inactivation in females, leading to a mosaic expression of ClC-5 in heterozygous (+/–) females. In this way, cell-autonomous phenotypes could be distinguished from non-cell-autonomous effects. In particular, cells lacking ClC-5 endocytosed much less protein than ClC-5-expressing cells, explaining the low-molecular-weight proteinurea. ClC-5 disruption affected both receptor-mediated and fluid-phase endocytosis (Günther et al. 2003; Piwon et al. 2000). In this respect, it is important to note that in the ClC-5 knockout mouse the amount of megalin at the plasma membrane was also reduced (Christensen et al. 2003; Piwon et al. 2000), probably because of impaired endosome recycling (Piwon et al. 2000) (see below). Moreover, the *in vitro* acidification of cortical renal endosomes prepared from ClC-5 knockout animals was reduced, supporting the proposed role of ClC-5 in endosomal acidification (Günther et al. 2003; Hara-Chikuma et al. 2005a; Piwon et al. 2000).

It is generally accepted that altered endosomal acidification impairs endocytosis (see above), and this might be due to a pH-dependent association of endosomes with regulatory proteins such as the GTPase Arf6 (Maranda et al. 2001). However, the details of the progressive acidification in the maturing endosomes are not yet very clear. For example, it has been suggested that primary endocytic vesicles are not acidified (Fuchs et al. 1994), and pharmacological inhibition of endosomal acidification does not affect the primary endocytic rate (Cupers et al. 1997; Tyteca et al. 2002).

In this scenario, the role of ClC-5 also does not appear to be completely clear. The fact that Günther et al. (Günther et al. 2003) found a significant degree of acidification also in endosomes of ClC-5 knockout mice that depended on the amount of Cl^- in the medium (Günther et al. 2003) could be explained by some contamination in the preparation but also by the presence, in endosomes, of Cl^- conductances that are not mediated by ClC-5. This is in agreement with the observation of Hara-Chikuma et al. on primary culture of proximal tubule cells from wild-type and ClC-5 KO mice (Hara-Chikuma et al. 2005a). They found that early endosomes lacking ClC-5 showed slightly impaired acidification and Cl^- accumulation compared to wild-type. Importantly, the acidification and Cl^- accumulation was almost completely abolished by the nonspecific Cl^- channel inhibitor NPPB in both wild-type and KO endosomes (Hara-Chikuma et al. 2005a). Altogether these lines of evidence raise the possibility that the central function of ClC-5 might not be the acidification of these compartments.

Another important observation concerning the impaired endocytosis in Dent disease is that in ClC-5 KO mice megalin and cubilin expression at the plasma membrane was reduced and these proteins were redistributed in intracellular organelles (Christensen et al. 2003; Piwon et al. 2000). These proteins belong to the family of multiligand tandem receptors involved in endocytosis, and their decrease at the plasma membrane is also compatible with the proteinuria phenotype of patients with Dent disease (Devuyst et al. 2005). Moreover, it is interesting to correlate this finding with the presumed preferential role of the subapical endosomes expressing ClC-5 in the recycling endosomal activity (Hara-Chikuma et al. 2005a).

Additional information about the role of ClC-5 in receptor-mediated endocytosis has been provided from analysis of albumin reabsorption in the PT of opossum kidney (OK) cells, which occurs through the megalin/cubulin receptor complex (Hryciw et al. 2005). Poronnik and coworkers observed that the level of ClC-5 expression at the plasma membrane of OK cells is influenced by the amount of albumin present extracellularly (Hryciw et al. 2004). The authors speculated that the effect is mediated by ubiquitination of ClC-5 operated by ubiquitin-protein ligase Nedd4-2 and that ClC-5 mediates the formation of an endocytic complex at the plasma membrane that contains the albumin-binding receptor megalin/cubilin (Hryciw et al. 2005). In the light of this model, the observed interaction between the C-terminus of ClC-5 and the actin-depolymerizing protein cofilin (Hryciw et al. 2003) was proposed to be required for the localized disruption of the actin cytoskeleton (Hryciw et al. 2005) that allows the endosomes to pass into the cytoplasm (Qualmann et al. 2000). Recently, it was shown that ClC-5 coimmunoprecipitates with the Na^+/H^+ exchanger regulatory factor NHERF2, a PDZ scaffold protein that may be relevant for the assembly of macromolecular complexes at the plasma membrane comprising the Na^+/H^+ exchanger (Hryciw et al. 2006). In particular, silencing NHERF produced a decrease of albumin uptake that was paralleled by a decreased surface expression of ClC-5 (Hryciw et al. 2006).

Disruption of the *clcn5* gene also produced defective internalization of the apical NaP_i-2 (sodium-phosphate cotransporter) and the apical Na^+/H^+ exchanger NHE3 (involved in reabsorption of Na^+, HCO_3^-, and fluid) (Piwon et al. 2000). The effect is mediated by parathyroid hormone (PTH) whose endocytosis is also defective in *clcn5* KO mice, leading to a progressive increase in luminal PTH levels that in turn stimulates endocytosis of those transporters (Jentsch et al. 2005b; Piwon et al. 2000). The decreased plasma membrane level of NaP_i-2 possibly explains the hyperphosphaturia phenotype in Dent disease patients.

It has been more difficult to explain the pathophysiological progression of Dent disease patients to nephrocalcinosis and kidney stones. In particular, the ClC-5 knockout mouse strain established by Jentsch and coworkers (Piwon et al. 2000) did not show hypercalciuria, while a strain obtained by Guggino and coworkers (Wang et al. 2000) showed hypercalciuria, which may then potentially lead to renal stones. This difference has been explained by Günther et al. (Günther et al. 2003) with the fact that hormones involved in Ca^{2+} homeostasis are subject in the PT to tight regulation that could be altered by slight genetic differences and/or diet.

Recently, it has been found that the disruption of ClC-5 in a collecting duct cell model brings about an increase in plasma membrane level of annexin A2 (Carr et al. 2006), which has been characterized as a crystal-binding molecule in renal epithelial cells (Kumar et al. 2003). This, in turn, may produce agglomeration and retention of calcium crystals, which potentially leads to nephrocalcinosis and renal stones. It has been hypothesized that ablation of ClC-5 impaired endosomal acidification rerouting annexin to the recycling pathway, resulting in an increase in plasma membrane expression (Carr et al. 2006). In this respect, it is important to note that with a ClC-5 knockout mouse model that displays hypercalciuria Silva et al. (Silva et al. 2003) suggested that the hypercalciuria is of bone and renal origin and is not caused by elevated intestinal calcium absorption. Interestingly, an interaction between ClC-5 and other proteins was also found by Mo and Wills (Mo and Wills 2004). These authors presented evidence that overexpression of ClC-5 in oocytes can alter the normal translation or trafficking of ENaC, CFTR, and NaDC-1 (sodium dicarboxylate cotransporter) to the plasma membrane by a mechanism that is independent from ClC-5-mediated chloride conductance, because a ClC-5 fragment comprising only amino acids 347–647 was sufficient to produce the same results. The mechanism behind such an effect, however, remains obscure.

Another interesting physiological role of ClC-5 has been recently suggested by van den Hove et al. (van den Hove et al. 2006), who showed that ClC-5 is significantly expressed in the thyroid (in particular in plasma membrane and late endosomes of thyrocytes) and that ClC-5 knockout mice develop a goiter with accumulation of iodide and thyroglobulin that seemed not to be caused by a defective endocytosis. It was therefore speculated that ClC-5 is involved in the regulation of plasma membrane expression of pendrin, an I^-/Cl^- exchanger responsible for iodide efflux or that ClC-5 can function as an additional iodide conductance in thyrocytes.

All these observations have vastly improved our understanding of the physiological role of ClC-5, but they still do not allow the unambiguous identification of the molecular mechanism that links ClC-5 dysfunction to the impaired endocytosis observed in Dent disease, and this reflects also the complexity of the underlying cellular processes.

We can schematically summarize the state of our present knowledge about the function of ClC-5 by formulating three possibilities.

It may be that ClC-5 is central for the acidification of endosomes. This could be rate limiting for their capacity to progress either in the degradation pathway or in the control of their redistribution to the recycling pathway, thereby determining the plasma membrane expression of proteins responsible for receptor-mediated endocytosis. A third possibility is that ClC-5 has only a marginal role in the acidification of endosomes but is an essential factor in very early endocytic processes, like endocytic vesicle budding. None of these alternatives necessarily excludes the others, and more experiments are needed to verify these possibilities or to suggest new ones.

Another element in this scenario is provided by the finding that ClC-5 (and ClC-4) are actually not chloride ion channels as it has been assumed (Jentsch et al. 2002) but rather

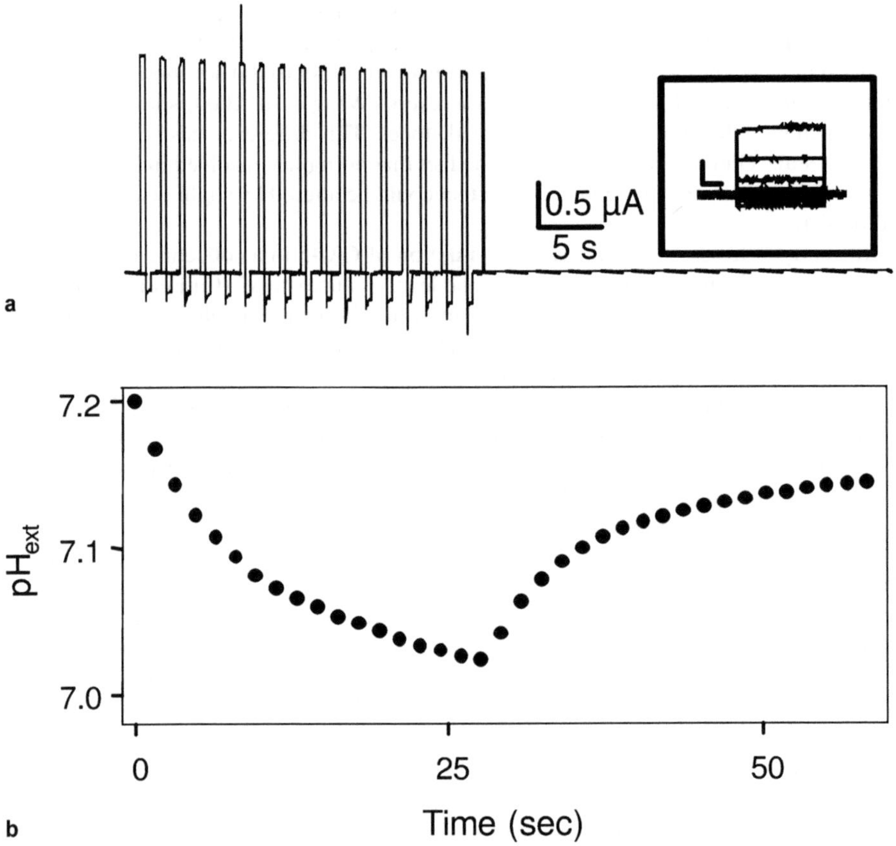

Fig. 6a, b Proton transport activity of ClC-5. **a** Current response of a voltage-clamped ClC-5-expressing oocyte stimulated by a train of pulses to 60 mV. After about 28 s the voltage clamp was switched off. **b** Simultaneously recorded pH close to the oocyte surface is plotted versus time. The *inset* in **a** shows a family of currents traces elicited by voltage pulses from −140 to 80 mV in 20-mV increments, recorded from the same oocyte. The marked outward rectification of ClC-5 is evident

transporters in which the inward movement of Cl^- is stochiometrically coupled to the outward movement of H^+ (Picollo and Pusch 2005; Scheel et al. 2005) (see Figs. 4 and 6). Such a transport mechanism seems, at first sight, to conflict with the accepted view of these proteins as passive Cl^- conductance allowing efficient acidification of vesicles by the proton pump because the Cl^-/H^+ antiporter activity of ClC-5 would actually lead to a partial dissipation of the proton gradient and ultimately to a waste of energy (Pusch et al. 2006).

However, the physiological implications of the transport activity of ClC-5 have not been explored yet and might be more complex than previously outlined.

The mechanism of transport couples the luminal pH to the Cl^- gradient across the vesicular membrane. However, the degree of coupling between the Cl^- and H^+ fluxes mediated by the ClC-5 depends on the stoichiometry of the transport, for which, at the moment, we only have a rough guess (Picollo and Pusch 2005; Scheel et al. 2005).

Our understanding of the biophysical and physiological function of ClC-5 is at a very early stage. The antiporter activity emerged recently, and a consistent corpus of new experimental evidence regarding the interaction of ClC-5 with other proteins suggests a more diversified and articulated function than just a passive Cl^- efflux to allow acidification of intracellular vesicles, but our knowledge is still too limited to formulate a more specific description of what this function could actually be.

The intracellular ClC-6 and ClC-7 proteins

On the basis of sequence conservation ClC-6 and ClC-7 form a separate branch of the CLC family with a sequence homology between them of 45%. Both have a very broad tissue distribution (Brandt and Jentsch 1995). These two proteins have so far escaped any attempt at biophysical characterization, as it has not been possible to express them in heterologous systems yet.

ClC-7 has been found to be a late endosomal/lysosomal Cl^- channel with a very broad tissue distribution (Kasper et al. 2005; Kornak et al. 2001). In particular, it is highly expressed in osteoclasts, the cells involved in bone degradation.

Its physiological relevance is highlighted by the finding that the *CLCN7* gene, encoding the human ClC-7 protein, is mutated in the disease osteopetrosis and that knockout mice for the corresponding gene develop severe osteopetrosis and retinal degeneration as also reported for some patients affected by malignant infantile ostepetrosis (Kornak et al. 2001). Also less severe dominant osteopetrosis can be caused by mutations in the *CLCN7* gene (Cleiren et al. 2001; Frattini et al. 2003; Sobacchi et al. 2001). In ClC-7 knockout mice, skeletal abnormalities include loss of bone marrow cavities that are instead filled with bone material and failure of teeth to erupt, but the mice also display neurodegeneration in the CNS (Kasper et al. 2005; Kornak et al. 2001).

Bone degradation is carried out by a specialized osteoclast plasma membrane domain, the ruffled border, through acidification of the resorption lacuna. In fact, the ruffled border is formed by the exocytotic insertion of vesicles of late endosomal/lysosomal origin, containing the H^+-ATPase. ClC-7 colocalizes with the proton pump in this membrane and was suggested to function as a shunt for the efficient acidification (Jentsch et al. 2005a). This hypothesis is in agreement with the finding that ClC-7 knockout osteoclasts still attach to ivory but fail to acidify the resorption lacuna and are unable to degrade the bone surrogate (Kornak et al. 2001). Moreover, polymorphisms in the gene coding for ClC-7 have been associated with alterations in bone mineral density and bone resorption markers in post-menopausal women and have been found to modulate the phenotypes of patients affected by autosomal dominant osteopetrosis type II (Kornak et al. 2006).

Very recently, Lange et al. (Lange et al. 2006) found that the ClC-7 protein is associated with the β-subunit Ostm1, which was known to produce osteopetrosis when mutated but whose function was unclear. The interaction of ClC-7 with Ostm1 is important for the stability of ClC-7, as ClC-7 protein levels are greatly reduced in mice lacking Ostm1. It was speculated that the role of Ostm1 is to shield ClC-7 from lysosomal degradation, as ClC-7 is the only mammalian CLC protein lacking N-linked glycosylation sites.

Given the essential role of ClC-7 for proper bone resorption, the protein has been suggested as a target for the treatment of osteoporosis that is characterized by excessive bone resorption (or too little bone formation). The compound NS3736 belongs to the group of acidic diphenylureas that has been shown to block Cl^- conductance in human erythrocytes (Bennekou et al. 2001). Schaller et al. (Schaller et al. 2004) found that this compound blocks acidification in resorption compartments and inhibits osteoclastic resorption *in vitro*. The ability of NS3736 to prevent bone loss *in vivo* was tested in aged ovariectomized rats taken as a model of osteoporosis, and it could be shown that daily treatment with 30 mg/kg protected bone strength dose-dependently, leaving bone formation unaffected. In a recent study Karsdal et al. (Karsdal et al. 2005) found that the compounds NS5818 and NS3696, close analogs of NS3637, have a very similar effect. Taken together these results suggest that chloride channel inhibitors might be useful in the treatment of osteoporosis.

Outlook

In the past 16 years (since the cloning of ClC-0) our knowledge about CLC chloride channels and transporters has increased enormously. It is no overstatement that the discovery of the CLC family has opened new horizons in fields as diverse as biophysics of membrane transport, physiology, pharmacology, and molecular medicine. Nevertheless, there are still many unsettled questions. Among the most pertinent questions are those related to a full understanding of the physiological roles of the intracellular CLC proteins: Are they really shunts? Are we still missing essential β-subunits? What are the functional properties of ClC-6 and ClC-7? Also, we are still lacking really high-affinity blockers (or activators) for any CLC protein, and there are no pharmacological tools available for ClC-3, ClC-4, and ClC-5. From a biophysical point of view, it would be interesting to decipher the rules that render a CLC protein a passive channel or, alternatively, an active transporter. It seems that CLCs will keep scientists in different areas busy for quite some time.

Acknowledgements. We thank Elena Babini for critical review of the manuscript. The financial support by Telethon Italy (grant GGP04018) is gratefully acknowledged.

References

Abramson J, Smirnova I, Kasho V, Verner G, Kaback HR, Iwata S (2003) Structure and mechanism of the lactose permease of *Escherichia coli*. Science 301:610–615

Abriel H, Loffing J, Rebhun JF, Pratt JH, Schild L, Horisberger JD, Rotin D, Staub O (1999) Defective regulation of the epithelial Na^+ channel by Nedd4 in Liddle's syndrome. J Clin Invest 103:667–673

Accardi A, Ferrera L, Pusch M (2001) Drastic reduction of the slow gate of human muscle chloride channel (ClC-1) by mutation C277S. J Physiol 534:745–752

Accardi A, Miller C (2004) Secondary active transport mediated by a prokaryotic homologue of ClC Cl^- channels. Nature 427:803–807

Accardi A, Pusch M (2000) Fast and slow gating relaxations in the muscle chloride channel CLC-1. J Gen Physiol 116:433–444

Accardi A, Pusch M (2003) Conformational changes in the pore of CLC-0. J Gen Physiol 122:277–293

Accardi A, Walden M, Nguitragool W, Jayaram H, Williams C, Miller C (2005) Separate ion pathways in a Cl^-/H^+ exchanger. J Gen Physiol 126:563–570.

Adrian RH, Bryant SH (1974) On the repetitive discharge in myotonic muscle fibres. J Physiol 240:505–515

Armstrong CM (1966) Time course of TEA^+-induced anomalous rectification in squid giant axons. J Gen Physiol 50:491–503

Aromataris EC, Astill DS, Rychkov GY, Bryant SH, Bretag AH, Roberts ML (1999) Modulation of the gating of ClC-1 by S-(–) 2-(4-chlorophenoxy)propionic acid. Br J Pharmacol 126:1375–1382

Aromataris EC, Rychkov GY, Bennetts B, Hughes BP, Bretag AH, Roberts ML (2001) Fast and slow gating of CLC-1:differential effects of 2-(4-chlorophenoxy)propionic acid and dominant negative mutations. Mol Pharmacol 60:200–208

Arreola J, Begenisich T, Melvin JE (2002) Conformation-dependent regulation of inward rectifier chloride channel gating by extracellular protons. J Physiol 541:103–112

Arreola J, Melvin JE (2003) A novel chloride conductance activated by extracellular ATP in mouse parotid acinar cells. J Physiol 547:197–208

Bali M, Lipecka J, Edelman A, Fritsch J (2001) Regulation of ClC-2 chloride channels in T84 cells by TGF-α. Am J Physiol Cell Physiol 280:C1588-C1598

Barbier-Brygoo H, Vinauger M, Colcombet J, Ephritikhine G, Frachisse J, Maurel C (2000) Anion channels in higher plants:functional characterization, molecular structure and physiological role. Biochim Biophys Acta 1465:199–218

Bateman A (1997) The structure of a domain common to archaebacteria and the homocystinuria disease protein. Trends Biochem Sci 22:12–13

Bauer CK, Steinmeyer K, Schwarz JR, Jentsch TJ (1991) Completely functional double-barreled chloride channel expressed from a single *Torpedo* cDNA. Proc Natl Acad Sci USA 88:11052–11056

Becker PE (1957) Zur Frage der Heterogenie der erblichen Myotonien. Nervenarzt 28:455–460

Bennekou P, de Franceschi L, Pedersen O, Lian L, Asakura T, Evans G, Brugnara C, Christophersen P (2001) Treatment with NS3623, a novel Cl– conductance blocker, ameliorates erythrocyte dehydration in transgenic SAD mice:a possible new therapeutic approach for sickle cell disease. Blood 97:1451–1457

Bennetts B, Roberts ML, Bretag AH, Rychkov GY (2001) Temperature dependence of human muscle ClC-1 chloride channel. J Physiol 535:83–93

Bennetts B, Rychkov GY, Ng H-L, Morton CJ, Stapleton D, Parker MW, Cromer BA (2005) Cytoplasmic ATP-sensing domains regulate gating of skeletal muscle ClC-1 chloride channels. J Biol Chem 280:32452–32458

Berg J, Jiang H, Thornton CA, Cannon SC (2004) Truncated ClC-1 mRNA in myotonic dystrophy exerts a dominant-negative effect on the Cl current. Neurology 63:2371–2375

Birkenhäger R, Otto E, Schurmann MJ, Vollmer M, Ruf EM, Maier-Lutz I, Beekmann F, Fekete A, Omran H, Feldmann D, Milford DV, Jeck N, Konrad M, Landau D, Knoers NV, Antignac C, Sudbrak R, Kispert A, Hildebrandt F (2001) Mutation of BSND causes Bartter syndrome with sensorineural deafness and kidney failure. Nat Genet 29:310–314

Bisset D, Corry B, Chung SH (2005) The fast gating mechanism in ClC-0 channels. Biophys J 89:179–186

Blaisdell CJ, Edmonds RD, Wang XT, Guggino S, Zeitlin PL (2000) pH-regulated chloride secretion in fetal lung epithelia. Am J Physiol Lung Cell Mol Physiol 278:L1248–L1255

Bösl MR, Stein V, Hübner C, Zdebik AA, Jordt SE, Mukhopadhyay AK, Davidoff MS, Holstein AF, Jentsch TJ (2001) Male germ cells and photoreceptors, both dependent on close cell-cell interactions, degenerate upon ClC-2 Cl^- channel disruption. EMBO J 20:1289–1299

Brandt S, Jentsch TJ (1995) ClC-6 and ClC-7 are two novel broadly expressed members of the CLC chloride channel family. FEBS Lett 377:15–20

Bretag AH (1987) Muscle chloride channels. Physiol Rev 67:618–724

Cannon SC (2000) Spectrum of sodium channel disturbances in the nondystrophic myotonias and periodic paralyses. Kidney Int 57:772–779

Carr G, Simmons NL, Sayer JA (2006) Disruption of clc-5 leads to a redistribution of annexin A2 and promotes calcium crystal agglomeration in collecting duct epithelial cells. Cell Mol Life Sci 63:367–377

Catalán M, Cornejo I, Figueroa CD, Niemeyer MI, Sepúlveda FV, Cid LP (2002) ClC-2 in guinea pig colon:mRNA, immunolabeling, and functional evidence for surface epithelium localization. Am J Physiol Gastrointest Liver Physiol 283:G1004-G1013

Catalán M, Niemeyer MI, Cid LP, Sepúlveda FV (2004) Basolateral ClC-2 chloride channels in surface colon epithelium:regulation by a direct effect of intracellular chloride. Gastroenterology 126:1104–1114

Charlet BN, Savkur RS, Singh G, Philips AV, Grice EA, Cooper TA (2002) Loss of the muscle-specific chloride channel in type 1 myotonic dystrophy due to misregulated alternative splicing. Mol Cell 10:45–53

Chen MF, Chen TY (2001) Different fast-gate regulation by external Cl^- and H^+ of the muscle- type ClC chloride channels. J Gen Physiol 118:23–32

Chen MF, Chen TY (2003) Side-chain charge effects and conductance determinants in the pore of ClC-0 chloride channels. J Gen Physiol 122:133–145

Chen MF, Niggeweg R, Iaizzo PA, Lehmann-Horn F, Jockusch H (1997) Chloride conductance in mouse muscle is subject to post-transcriptional compensation of the functional Cl^- channel 1 gene dosage. J Physiol 504:75–81

Chen TY (1998) Extracellular zinc ion inhibits ClC-0 chloride channels by facilitating slow gating. J Gen Physiol 112:715–726

Chen TY (2003) Coupling gating with ion permeation in ClC channels. Sci STKE 2003:pe23

Chen TY, Chen MF, Lin CW (2003) Electrostatic control and chloride regulation of the fast gating of ClC-0 chloride channels. J Gen Physiol 122:641–651

Chen TY, Miller C (1996) Nonequilibrium gating and voltage dependence of the ClC-0 Cl^- channel. J Gen Physiol 108:237–250

Christensen EI, Devuyst O, Dom G, Nielsen R, Van der Smissen P, Verroust P, Leruth M, Guggino WB, Courtoy PJ (2003) Loss of chloride channel ClC-5 impairs endocytosis by defective trafficking of megalin and cubilin in kidney proximal tubules. Proc Natl Acad Sci USA 100:8472–8477

Chu S, Blaisdell CJ, Liu M-ZM, Zeitlin PL (1999) Perinatal regulation of the ClC-2 chloride channel in lung is mediated by Sp1 and Sp3. Am J Physiol Lung Cell Mol Physiol 276:L614–L624

Clark S, Jordt SE, Jentsch TJ, Mathie A (1998) Characterization of the hyperpolarization-activated chloride current in dissociated rat sympathetic neurons. J Physiol 506:665–678

Cleiren E, Benichou O, Van Hul E, Gram J, Bollerslev J, Singer FR, Beaverson K, Aledo A, Whyte MP, Yoneyama T, deVernejoul MC, Van Hul W (2001) Albers-Schonberg disease (autosomal dominant osteopetrosis, type II) results from mutations in the ClCN7 chloride channel gene. Hum Mol Genet 10:2861–2867

Cohen J, Schulten K (2004) Mechanism of anionic conduction across ClC. Biophys J 86:836–845

Conte-Camerino D, Mambrini M, DeLuca A, Tricarico D, Bryant SH, Tortorella V, Bettoni G (1988) Enantiomers of clofibric acid analogs have opposite actions on rat skeletal muscle chloride channels. Pflügers Arch 413:105–107

Corry B, O'Mara M, Chung SH (2004) Conduction mechanisms of chloride ions in ClC-type channels. Biophys J 86:846–860

Cupers P, Veithen A, Hoekstra D, Baudhuin P, Courtoy PJ (1997) Three unrelated perturbations similarly uncouple fluid, bulk-membrane, and receptor endosomal flow in rat fetal fibroblasts. Biochem Biophys Res Commun 236:661–664

De Angeli A, Monachello D, Ephritikhine G, Frachisse JM, Thomine S, Gambale F, Barbier-Brygoo H (2006) AtCLCa, a proton/nitrate antiporter, mediates nitrate accumulation in plant vacuoles. Nature. In press.

de Santiago JA, Nehrke K, Arreola J (2005) Quantitative analysis of the voltage-dependent gating of mouse parotid ClC-2 chloride channel. J Gen Physiol 126:591–603

Dent CE, Friedman M (1964) Hypercalcuric rickets associated with renal tubular damage. Arch Dis Child 39:240–249

Devuyst O, Christie PT, Courtoy PJ, Beauwens R, Thakker RV (1999) Intra-renal and subcellular distribution of the human chloride channel, CLC-5, reveals a pathophysiological basis for Dent's disease. Hum Mol Genet 8:247–257

Devuyst O, Jouret F, Auzanneau C, Courtoy PJ (2005) Chloride channels and endocytosis:new insights from Dent's disease and ClC-5 knockout mice. Nephron Physiol 99:69–73

Dhani SU, Bear CE (2006) Role of intramolecular and intermolecular interactions in ClC channel and transporter function. Pflügers Arch 451:708–715

Dhani SU, Mohammad-Panah R, Ahmed N, Ackerley C, Ramjeesingh M, Bear CE (2003) Evidence for a functional interaction between the ClC-2 chloride channel and the retrograde motor dynein complex. J Biol Chem 278:16262–16270

Doyle DA, Morais Cabral J, Pfuetzner RA, Kuo A, Gulbis JM, Cohen SL, Chait BT, MacKinnon R (1998) The structure of the potassium channel:molecular basis of K^+ conduction and selectivity. Science 280:69–77

Duan D, Winter C, Cowley S, Hume JR, Horowitz B (1997) Molecular identification of a volume-regulated chloride channel. Nature 390:417–421

Duffield M, Rychkov G, Bretag A, Roberts M (2003) Involvement of helices at the dimer interface in ClC-1 common gating. J Gen Physiol 121:149–161

Duffield MD, Rychkov GY, Bretag AH, Roberts ML (2005) Zinc inhibits human ClC-1 muscle chloride channel by interacting with its common gating mechanism. J Physiol 568:5–12

Dutzler R (2004) The structural basis of ClC chloride channel function. Trends Neurosci 27:315–320

Dutzler R, Campbell EB, Cadene M, Chait BT, MacKinnon R (2002) X-ray structure of a ClC chloride channel at 3.0 Å reveals the molecular basis of anion selectivity. Nature 415:287–294

Dutzler R, Campbell EB, MacKinnon R (2003) Gating the selectivity filter in ClC chloride channels. Science 300:108–112

Edmonds RD, Silva IV, Guggino WB, Butler RB, Zeitlin PL, Blaisdell CJ (2002) ClC-5:ontogeny of an alternative chloride channel in respiratory epithelia. Am J Physiol Lung Cell Mol Physiol 282:L501–L507

Eggermont J, Trouet D, Carton I, Nilius B (2001) Cellular function and control of volume-regulated anion channels. Cell Biochem Biophys 35:263–274

Embark HM, Bohmer C, Palmada M, Rajamanickam J, Wyatt AW, Wallisch S, Capasso G, Waldegger P, Seyberth HW, Waldegger S, Lang F (2004) Regulation of CLC-Ka/barttin by the ubiquitin ligase Nedd4-2 and the serum- and glucocorticoid-dependent kinases. Kidney Int 66:1918–1925

Engh AM, Maduke M (2005) Cysteine accessibility in ClC-0 supports conservation of the ClC intracellular vestibule. J Gen Physiol 125:601–617

Estévez R, Boettger T, Stein V, Birkenhäger R, Otto E, Hildebrandt F, Jentsch TJ (2001) Barttin is a Cl^- channel beta-subunit crucial for renal Cl^- reabsorption and inner ear K^+ secretion. Nature 414:558–561

Estévez R, Jentsch TJ (2002) CLC chloride channels:correlating structure with function. Curr Opin Struct Biol 12:531–539

Estévez R, Pusch M, Ferrer-Costa C, Orozco M, Jentsch TJ (2004) Functional and structural conservation of CBS domains from CLC channels. J Physiol 557:363–378

Estévez R, Schroeder BC, Accardi A, Jentsch TJ, Pusch M (2003) Conservation of chloride channel structure revealed by an inhibitor binding site in ClC-1. Neuron 38:47–59

Faraldo-Gomez JD, Roux B (2004) Electrostatics of ion stabilization in a ClC chloride channel homologue from *Escherichia coli.* J Mol Biol 339:981–1000

Fisher SE, Black GC, Lloyd SE, Hatchwell E, Wrong O, Thakker RV, Craig IW (1994) Isolation and partial characterization of a chloride channel gene which is expressed in kidney and is a candidate for Dent's disease (an X-linked hereditary nephrolithiasis). Hum Mol Genet 3:2053–2059

Fölsch H, Ohno H, Bonifacino JS, Mellman I (1999) A novel clathrin adaptor complex mediates basolateral targeting in polarized epithelial cells. Cell 99:189–198

Fong P, Rehfeldt A, Jentsch TJ (1998) Determinants of slow gating in ClC-0, the voltage-gated chloride channel of *Torpedo marmorata.* Am J Physiol Cell Physiol 274:C966–C973

Frattini A, Pangrazio A, Susani L, Sobacchi C, Mirolo M, Abinun M, Andolina M, Flanagan A, Horwitz EM, Mihci E, Notarangelo LD, Ramenghi U, Teti A, Van Hove J, Vujic D, Young T, Albertini A, Orchard PJ, Vezzoni P, Villa A (2003) Chloride channel ClCN7 mutations are responsible for severe recessive, dominant, and intermediate osteopetrosis. J Bone Miner Res 18:1740–1747

Friedrich T, Breiderhoff T, Jentsch TJ (1999) Mutational analysis demonstrates that ClC-4 and ClC-5 directly mediate plasma membrane currents. J Biol Chem 274:896–902

Fuchs R, Ellinger A, Pavelka M, Mellman I, Klapper H (1994) Rat liver endocytic coated vesicles do not exhibit ATP-dependent acidification in vitro. Proc Natl Acad Sci USA 91:4811–4815

Furman RE, Barchi RL (1978) The pathophysiology of myotonia produced by aromatic carboxylic acids. Ann Neurol 4:357–365

Furukawa T, Ogura T, Katayama Y, Hiraoka M (1998) Characteristics of rabbit ClC-2 current expressed in *Xenopus* oocytes and its contribution to volume regulation. Am J Physiol Cell Physiol 274:C500–C512

Gaxiola RA, Yuan DS, Klausner RD, Fink GR (1998) The yeast CLC chloride channel functions in cation homeostasis. Proc Natl Acad Sci USA 95:4046–4050

Gentzsch M, Cui L, Mengos A, Chang XB, Chen JH, Riordan JR (2003) The PDZ-binding chloride channel ClC-3B localizes to the Golgi and associates with cystic fibrosis transmembrane conductance regulator-interacting PDZ proteins. J Biol Chem 278:6440–6449

Gill JR Jr, Bartter FC (1978) Evidence for a prostaglandin-independent defect in chloride reabsorption in the loop of Henle as a proximal cause of Bartter's syndrome. Am J Med 65:766–772

Gluck SL, Underhill DM, Iyori M, Holliday LS, Kostrominova TY, Lee BS (1996) Physiology and biochemistry of the kidney vacuolar H^+-ATPase. Annu Rev Physiol 58:427–445

Greene JR, Brown NH, DiDomenico BJ, Kaplan J, Eide DJ (1993) The GEF1 gene of *Saccharomyces cerevisiae* encodes an integral membrane protein; mutations in which have effects on respiration and iron-limited growth. Mol Gen Genet 241:542–553

Gründer S, Thiemann A, Pusch M, Jentsch TJ (1992) Regions involved in the opening of CIC-2 chloride channel by voltage and cell volume. Nature 360:759–762

Guggino WB (2004) The cystic fibrosis transmembrane regulator forms macromolecular complexes with PDZ domain scaffold proteins. Proc Am Thorac Soc 1:28–32

Günther W, Luchow A, Cluzeaud F, Vandewalle A, Jentsch TJ (1998) ClC-5, the chloride channel mutated in Dent's disease, colocalizes with the proton pump in endocytotically active kidney cells. Proc Natl Acad Sci USA 95:8075–8080

Günther W, Piwon N, Jentsch TJ (2003) The ClC-5 chloride channel knock-out mouse—an animal model for Dent's disease. Pflügers Arch 445:456–462

Gurnett CA, Kahl SD, Anderson RD, Campbell KP (1995) Absence of the skeletal muscle sarcolemma chloride channel ClC-1 in myotonic mice. J Biol Chem 270:9035–9038

Gyömörey K, Yeger H, Ackerley C, Garami E, Bear CE (2000) Expression of the chloride channel ClC-2 in the murine small intestine epithelium. Am J Physiol Cell Physiol 279:C1787–C1794

Hanke W, Miller C (1983) Single chloride channels from *Torpedo* electroplax. Activation by protons. J Gen Physiol 82:25–45

Hanrahan JW, Wioland MA (2004) Revisiting cystic fibrosis transmembrane conductance regulator structure and function. Proc Am Thorac Soc 1:17–21

Hansson JH, Nelson-Williams C, Suzuki H, Schild L, Shimkets R, Lu Y, Canessa C, Iwasaki T, Rossier B, Lifton RP (1995a) Hypertension caused by a truncated epithelial sodium channel gamma subunit:genetic heterogeneity of Liddle syndrome. Nat Genet 11:76–82

Hansson JH, Schild L, Lu Y, Wilson TA, Gautschi I, Shimkets R, Nelson-Williams C, Rossier BC, Lifton RP (1995b) A de novo missense mutation of the beta subunit of the epithelial sodium channel causes hypertension and Liddle syndrome, identifying a proline-rich segment critical for regulation of channel activity. Proc Natl Acad Sci USA 92:11495–11499

Hara-Chikuma M, Wang Y, Guggino SE, Guggino WB, Verkman AS (2005a) Impaired acidification in early endosomes of ClC-5 deficient proximal tubule. Biochem Biophys Res Commun 329:941–946

Hara-Chikuma M, Yang B, Sonawane ND, Sasaki S, Uchida S, Verkman AS (2005b) ClC-3 chloride channels facilitate endosomal acidification and chloride accumulation. J Biol Chem 280:1241–1247

Hartzell C, Putzier I, Arreola J (2005) Calcium-activated chloride channels. Annu Rev Physiol 67:719–758

Haug K, Warnstedt M, Alekov AK, Sander T, Ramirez A, Poser B, Maljevic S, Hebeisen S, Kubisch C, Rebstock J, Horvath S, Hallmann K, Dullinger JS, Rau B, Haverkamp F, Beyenburg S, Schulz H, Janz D, Giese B, Muller-Newen G, Propping P, Elger CE, Fahlke C, Lerche H, Heils A (2003) Mutations in CLCN2 encoding a voltage-gated chloride channel are associated with idiopathic generalized epilepsies. Nat Genet 33:527–532

Hayama A, Rai T, Sasaki S, Uchida S (2003) Molecular mechanisms of Bartter syndrome caused by mutations in the BSND gene. Histochem Cell Biol 119:485–493

Hechenberger M, Schwappach B, Fischer WN, Frommer WB, Jentsch TJ, Steinmeyer K (1996) A family of putative chloride channels from *Arabidopsis* and functional complementation of a yeast strain with a CLC gene disruption. J Biol Chem 271:33632–33638

Hille B (2001) Ion channels of excitable membranes. Sinauer, Sunderland, MA

Hinzpeter A, Lipecka J, Brouillard F, Baudoin-Legros M, Dadlez M, Edelman A, Fritsch J (2006) Association between Hsp90 and the ClC-2 chloride channel upregulates channel function. Am J Physiol Cell Physiol 290:C45–C56.

Holmes KW, Hales R, Chu S, Maxwell MJ, Mogayzel PJ Jr, Zeitlin PL (2003) Modulation of Sp1 and Sp3 in lung epithelial cells regulates ClC-2 chloride channel expression. Am J Respir Cell Mol Biol 29:499–505

Hryciw DH, Ekberg J, Ferguson C, Lee A, Wang D, Parton RG, Pollock CA, Yun CC, Poronnik P (2006) Regulation of albumin endocytosis by PSD95/Dlg/ZO-1 (PDZ) scaffolds:interaction of Na^+-H^+ exchange regulatory factor-2 with ClC-5. J Biol Chem 281:16068–16077

Hryciw DH, Ekberg J, Lee A, Lensink IL, Kumar S, Guggino WB, Cook DI, Pollock CA, Poronnik P (2004) Nedd4-2 functionally interacts with ClC-5:involvement in constitutive albumin endocytosis in proximal tubule cells. J Biol Chem 279:54996–55007

Hryciw DH, Ekberg J, Pollock CA, Poronnik P (2005) ClC-5:A chloride channel with multiple roles in renal tubular albumin uptake. Int J Biochem Cell Biol 3:3

Hryciw DH, Wang Y, Devuyst O, Pollock CA, Poronnik P, Guggino WB (2003) Cofilin interacts with ClC-5 and regulates albumin uptake in proximal tubule cell lines. J Biol Chem 278:40169–40176

Huber S, Braun G, Schroppel B, Horster M (1998) Chloride channels ClC-2 and ICln mRNA expression differs in renal epithelial ontogeny. Kidney Int Suppl 67:S149–S151

Hübner CA, Stein V, Hermans-Borgmeyer I, Meyer T, Ballanyi K, Jentsch TJ (2001) Disruption of KCC2 reveals an essential role of K-Cl cotransport already in early synaptic inhibition. Neuron 30:515–524

Hunte C, Screpanti E, Venturi M, Rimon A, Padan E, Michel H (2005) Structure of a Na^+/H^+ antiporter and insights into mechanism of action and regulation by pH. Nature 435:1197–1202

Hunter M (2001) Accessory to kidney disease. Nature 414:502–503

Hutter OF, Warner AE (1967) Action of some foreign cations and anions on the chloride permeability of frog muscle. J Physiol 189:445–460

Iyer R, Iverson TM, Accardi A, Miller C (2002) A biological role for prokaryotic ClC chloride channels. Nature 419:715–718

Jeck N, Waldegger P, Doroszewicz J, Seyberth H, Waldegger S (2004a) A common sequence variation of the CLCNKB gene strongly activates ClC-Kb chloride channel activity. Kidney Int 65:190–197

Jeck N, Waldegger S, Lampert A, Boehmer C, Waldegger P, Lang PA, Wissinger B, Friedrich B, Risler T, Moehle R, Lang UE, Zill P, Bondy B, Schaeffeler E, Asante-Poku S, Seyberth H, Schwab M, Lang F (2004b) Activating mutation of the renal epithelial chloride channel ClC-Kb predisposing to hypertension. Hypertension 43:1175–1181

Jentsch TJ (2000) Neuronal KCNQ potassium channels:physiology and role in disease. Nat Rev Neurosci 1:21–30

Jentsch TJ (2005) Chloride transport in the kidney:lessons from human disease and knockout mice. J Am Soc Nephrol 16:1549–1561

Jentsch TJ, Friedrich T, Schriever A, Yamada H (1999) The CLC chloride channel family. Pflügers Arch 437:783–795

Jentsch TJ, Maritzen T, Zdebik AA (2005a) Chloride channel diseases resulting from impaired transepithelial transport or vesicular function. J Clin Invest 115:2039–2046

Jentsch TJ, Neagoe I, Scheel O (2005b) CLC chloride channels and transporters. Curr Opin Neurobiol 15:319–325

Jentsch TJ, Poet M, Fuhrmann JC, Zdebik AA (2005c) Physiological functions of CLC Cl channels gleaned from human genetic disease and mouse models. Annu Rev Physiol 67:779–807

Jentsch TJ, Stein V, Weinreich F, Zdebik AA (2002) Molecular structure and physiological function of chloride channels. Physiol Rev 82:503–568

Jentsch TJ, Steinmeyer K, Schwarz G (1990) Primary structure of *Torpedo marmorata* chloride channel isolated by expression cloning in *Xenopus* oocytes. Nature 348:510–514

Jordt SE, Jentsch TJ (1997) Molecular dissection of gating in the ClC-2 chloride channel. EMBO J 16:1582–1592

Karsdal MA, Henriksen K, Sørensen MG, Gram J, Schaller S, Dziegiel MH, Heegaard AM, Christophersen P, Martin TJ, Christiansen C, Bollerslev J (2005) Acidification of the osteoclastic resorption compartment provides insight into the coupling of bone formation to bone resorption. Am J Pathol 166:467–476

Kasper D, Planells-Cases R, Fuhrmann JC, Scheel O, Zeitz O, Ruether K, Schmitt A, Poet M, Steinfeld R, Schweizer M, Kornak U, Jentsch TJ (2005) Loss of the chloride channel ClC-7 leads to lysosomal storage disease and neurodegeneration. EMBO J 24:1079–1091

Kawasaki M, Uchida S, Monkawa T, Miyawaki A, Mikoshiba K, Marumo F, Sasaki S (1994) Cloning and expression of a protein kinase C-regulated chloride channel abundantly expressed in rat brain neuronal cells. Neuron 12:597–604

Kibble JD, Trezise AE, Brown PD (1996) Properties of the cAMP-activated Cl^- current in choroid plexus epithelial cells isolated from the rat. J Physiol 496:69–80

Kieferle S, Fong P, Bens M, Vandewalle A, Jentsch TJ (1994) Two highly homologous members of the ClC chloride channel family in both rat and human kidney. Proc Natl Acad Sci USA 91:6943–6947

Kobayashi K, Uchida S, Mizutani S, Sasaki S, Marumo F (2001) Developmental expression of CLC-K1 in the postnatal rat kidney. Histochem Cell Biol 116:49–56

Kokubo Y, Iwai N, Tago N, Inamoto N, Okayama A, Yamawaki H, Naraba H, Tomoike H (2005) Association analysis between hypertension and CYBA, CLCNKB, and KCNMB1 functional polymorphisms in the Japanese population—the Suita Study. Circ J 69:138–142

Konrad M, Vollmer M, Lemmink HH, van den Heuvel LP, Jeck N, Vargas-Poussou R, Lakings A, Ruf R, Deschenes G, Antignac C, Guay-Woodford L, Knoers NV, Seyberth HW, Feldmann D, Hildebrandt F (2000) Mutations in the chloride channel gene CLCNKB as a cause of classic Bartter syndrome. J Am Soc Nephrol 11:1449–1459

Kornak U, Kasper D, Bösl MR, Kaiser E, Schweizer M, Schulz A, Friedrich W, Delling G, Jentsch TJ (2001) Loss of the ClC-7 chloride channel leads to osteopetrosis in mice and man. Cell 104:205–215

Kornak U, Ostertag A, Branger S, Benichou O, de Vernejoul MC (2006) Polymorphisms in the CLCN7 gene modulate bone density in postmenopausal women and in patients with autosomal dominant osteopetrosis type II. J Clin Endocrinol Metab 91:995–1000

Kumar V, Farell G, Deganello S, Lieske JC (2003) Annexin II is present on renal epithelial cells and binds calcium oxalate monohydrate crystals. J Am Soc Nephrol 14:289–297

Kürz L, Wagner S, George AL Jr, Rüdel R (1997) Probing the major skeletal muscle chloride channel with Zn^{2+} and other sulfhydryl-reactive compounds. Pflügers Arch 433:357–363

Lange PF, Wartosch L, Jentsch TJ, Fuhrmann JC (2006) ClC-7 requires Ostm1 as a beta-subunit to support bone resorption and lysosomal function. Nature 440:220–223

Lee JK, Kozono D, Remis J, Kitagawa Y, Agre P, Stroud RM (2005) Structural basis for conductance by the archaeal aquaporin AqpM at 1.68 Å. Proc Natl Acad Sci USA 102:18932–18937

Lehmann-Horn F, Jurkat-Rott K (1999) Voltage-gated ion channels and hereditary disease. Physiol Rev 79:1317–1372

Li X, Shimada K, Showalter LA, Weinman SA (2000) Biophysical properties of ClC-3 differentiate it from swelling-activated chloride channels in Chinese hamster ovary-K1 cells. J Biol Chem 275:35994–35998

Li X, Wang T, Zhao Z, Weinman SA (2002) The ClC-3 chloride channel promotes acidification of lysosomes in CHO-K1 and Huh-7 cells. Am J Physiol Cell Physiol 282:C1483–C1491

Liantonio A, Accardi A, Carbonara G, Fracchiolla G, Loiodice F, Tortorella P, Traverso S, Guida P, Pierno S, De Luca A, Camerino DC, Pusch M (2002) Molecular requisites for drug binding to muscle CLC-1 and renal CLC-K channel revealed by the use of phenoxy-alkyl derivatives of 2-(*p*-chlorophenoxy)propionic acid. Mol Pharmacol 62:265–271

Liantonio A, De Luca A, Pierno S, Didonna MP, Loiodice F, Fracchiolla G, Tortorella P, Antonio L, Bonerba E, Traverso S, Elia L, Picollo A, Pusch M, Conte Camerino D (2003) Structural requisites of 2-(*p*-chlorophenoxy)propionic acid analogues for activity on native rat skeletal muscle chloride conductance and on heterologously expressed CLC-1. Br J Pharmacol 139:1255–1264

Liantonio A, Picollo A, Babini E, Carbonara G, Fracchiolla G, Loiodice F, Tortorella V, Pusch M, Camerino DC (2006) Activation and inhibition of kidney CLC-K chloride channels by fenamates. Mol Pharmacol 69:165–173

Liantonio A, Pusch M, Picollo A, Guida P, De Luca A, Pierno S, Fracchiolla G, Loiodice F, Tortorella P, Conte Camerino D (2004) Investigations of pharmacologic properties of the renal CLC-K1 chloride channel co-expressed with barttin by the use of 2-(*p*-chlorophenoxy)propionic acid derivatives and other structurally unrelated chloride channel blockers. J Am Soc Nephrol 15:13–20

Lin YW, Lin CW, Chen TY (1999) Elimination of the slow gating of ClC-0 chloride channel by a point mutation. J Gen Physiol 114:1–12

Lipecka J, Bali M, Thomas A, Fanen P, Edelman A, Fritsch J (2002) Distribution of ClC-2 chloride channel in rat and human epithelial tissues. Am J Physiol Cell Physiol 282:C805–C816

Liu W, Morimoto T, Kondo Y, Iinuma K, Uchida S, Imai M (2001) “Avian-type” renal medullary tubule organization causes immaturity of urine-concentrating ability in neonates. Kidney Int 60:680–693

Lobet S, Dutzler R (2006) Ion-binding properties of the ClC chloride selectivity filter. EMBO J 25:24–33

Lorenz C, Pusch M, Jentsch TJ (1996) Heteromultimeric CLC chloride channels with novel properties. Proc Natl Acad Sci USA 93:13362–13366

Ludewig U, Jentsch TJ, Pusch M (1997a) Analysis of a protein region involved in permeation and gating of the voltage-gated *Torpedo* chloride channel ClC-0. J Physiol 498:691–702

Ludewig U, Pusch M, Jentsch TJ (1996) Two physically distinct pores in the dimeric ClC-0 chloride channel. Nature 383:340–343

Ludewig U, Pusch M, Jentsch TJ (1997b) Independent gating of single pores in CLC-0 chloride channels. Biophys J 73:789–797

Ludwig M, Doroszewicz J, Seyberth HW, Bokenkamp A, Balluch B, Nuutinen M, Utsch B, Waldegger S (2005) Functional evaluation of Dent's disease-causing mutations:implications for ClC-5 channel trafficking and internalization. Hum Genet 117:228–237

Ludwig M, Waldegger S, Nuutinen M, Bokenkamp A, Reissinger A, Steckelbroeck S, Utsch B (2003) Four additional CLCN5 exons encode a widely expressed novel long CLC-5 isoform but fail to explain Dent's phenotype in patients without mutations in the short variant. Kidney Blood Press Res 26:176–184

Maack T, Park CH (1990) Endocytosis and lysosomal hydrolysis of proteins in proximal tubules. Methods Enzymol 191:340–354

Maduke M, Miller C, Mindell JA (2000) A decade of CLC chloride channels:structure, mechanism, and many unsettled questions. Annu Rev Biophys Biomol Struct 29:411–438

Maduke M, Pheasant DJ, Miller C (1999) High-level expression, functional reconstitution, and quaternary structure of a prokaryotic ClC-type chloride channel. J Gen Physiol 114:713–722

Maduke M, Williams C, Miller C (1998) Formation of CLC-0 chloride channels from separated transmembrane and cytoplasmic domains. Biochemistry 37:1315–1321

Malinowska DH, Kupert EY, Bahinski A, Sherry AM, Cuppoletti J (1995) Cloning, functional expression, and characterization of a PKA-activated gastric Cl^- channel. Am J Physiol Cell Physiol 268:C191–C200

Mankodi A, Takahashi MP, Jiang H, Beck CL, Bowers WJ, Moxley RT, Cannon SC, Thornton CA (2002) Expanded CUG repeats trigger aberrant splicing of ClC-1 chloride channel pre-mRNA and hyperexcitability of skeletal muscle in myotonic dystrophy. Mol Cell 10:35–44

Maranda B, Brown D, Bourgoin S, Casanova JE, Vinay P, Ausiello DA, Marshansky V (2001) Intra-endosomal pH-sensitive recruitment of the Arf-nucleotide exchange factor ARNO and Arf6 from cytoplasm to proximal tubule endosomes. J Biol Chem 276:18540–18550

Marty A, Llano I (2005) Excitatory effects of GABA in established brain networks. Trends Neurosci 28:284–289

Matsumura Y, Uchida S, Kondo Y, Miyazaki H, Ko SB, Hayama A, Morimoto T, Liu W, Arisawa M, Sasaki S, Marumo F (1999) Overt nephrogenic diabetes insipidus in mice lacking the CLC-K1 chloride channel. Nat Genet 21:95–98

Mehrke G, Brinkmeier H, Jockusch H (1988) The myotonic mouse mutant ADR:electrophysiology of the muscle fiber. Muscle Nerve 11:440–446

Mellman I (1996) Endocytosis and molecular sorting. Annu Rev Dev Biol 12:575–625

Mellman I, Fuchs R, Helenius A (1986) Acidification of the endocytic and exocytic pathways. Annu Rev Biochem 55:663–700

Meyer S, Dutzler R (2006) Crystal structure of the cytoplasmic domain of the chloride channel ClC-0. Structure 14:299–307

Meyer-Kleine C, Steinmeyer K, Ricker K, Jentsch TJ, Koch MC (1995) Spectrum of mutations in the major human skeletal muscle chloride channel gene (CLCN1) leading to myotonia. Am J Hum Genet 57:1325–1334

Middleton RE, Pheasant DJ, Miller C (1996) Homodimeric architecture of a ClC-type chloride ion channel. Nature 383:337–340

Miller C (1982) Open-state substructure of single chloride channels from *Torpedo* electroplax. Philos Trans R Soc Lond B Biol Sci 299:401–411

Miller C (2006) ClC chloride channels viewed through a transporter lens. Nature 440:484–489

Miller C, Richard EA (1990) The voltage-dependent chloride channel of *Torpedo* electroplax. Intimations of molecular structure from quirks of single-channel function. In:Chloride Channels and Carriers in Nerve, Muscle and Glial Cells. F.J. Alvarez-Leefmans and J.M. Russell, editors. pp. 383–405. Plenum, New York

Miller C, White MM (1980) A voltage-dependent chloride conductance channel from *Torpedo* electroplax membrane. Ann NY Acad Sci 341:534–551

Miller C, White MM (1984) Dimeric structure of single chloride channels from *Torpedo* electroplax. Proc Natl Acad Sci USA 81:2772–2775

Miller MD, Schwarzenbacher R, von Delft F, Abdubek P, Ambing E, Biorac T, Brinen LS, Canaves JM, Cambell J, Chiu HJ, Dai X, Deacon AM, DiDonato M, Elsliger MA, Eshagi S, Floyd R, Godzik A, Grittini C, Grzechnik SK, Hampton E, Jaroszewski L, Karlak C, Klock HE, Koesema E, Kovarik JS, Kreusch A, Kuhn P, Lesley SA, Levin I, McMullan D, McPhillips TM, Morse A, Moy K, Ouyang J, Page R, Quijano K, Robb A, Spraggon G, Stevens RC, van den Bedem H, Velasquez J, Vincent J, Wang X, West B, Wolf G, Xu Q, Hodgson KO, Wooley J, Wilson IA (2004) Crystal structure of a tandem cystathionine-beta-synthase (CBS) domain protein (TM0935) from *Thermotoga maritima* at 1.87 Å resolution. Proteins 57:213–217

Mindell JA, Maduke M (2001) ClC chloride channels. Genome Biol 2:REVIEWS3003

Mindell JA, Maduke M, Miller C, Grigorieff N (2001) Projection structure of a ClC-type chloride channel at 6.5 Å resolution. Nature 409:219–223

Misgeld U, Deisz RA, Dodt HU, Lux HD (1986) The role of chloride transport in postsynaptic inhibition of hippocampal neurons. Science 232:1413–1415

Miyazawa A, Fujiyoshi Y, Unwin N (2003) Structure and gating mechanism of the acetylcholine receptor pore. Nature 423:949–955

Mo L, Wills NK (2004) ClC-5 chloride channel alters expression of the epithelial sodium channel (ENaC). J Membr Biol 202:21–37

Mo L, Xiong W, Qian T, Sun H, Wills NK (2004) Coexpression of complementary fragments of ClC-5 and restoration of chloride channel function in a Dent's disease mutation. Am J Physiol Cell Physiol 286:C79–C89

Mohammad-Panah R, Harrison R, Dhani S, Ackerley C, Huan LJ, Wang Y, Bear CE (2003) The chloride channel ClC-4 contributes to endosomal acidification and trafficking. J Biol Chem 278:29267–29277

Moreland JG, Davis AP, Bailey G, Nauseef WM, Lamb FS (2006) Anion channels including CLC-3 are required for normal neutrophil oxidative function, phagocytosis, and transendothelial migration. J Biol Chem 7:7

Moulin P, Igarashi T, Van der Smissen P, Cosyns JP, Verroust P, Thakker RV, Scheinman SJ, Courtoy PJ, Devuyst O (2003) Altered polarity and expression of H^+-ATPase without ultrastructural changes in kidneys of Dent's disease patients. Kidney Int 63:1285–1295

Murata K, Mitsuoka K, Hirai T, Walz T, Agre P, Heymann JB, Engel A, Fujiyoshi Y (2000) Structural determinants of water permeation through aquaporin-1. Nature 407:599–605

Murray CB, Chu S, Zeitlin PL (1996) Gestational and tissue-specific regulation of C1C-2 chloride channel expression. Am J Physiol Lung Cell Mol Physiol 271:L829–L837

Murray CB, Morales MM, Flotte TR, McGrath-Morrow SA, Guggino WB, Zeitlin PL (1995) CIC-2:a developmentally dependent chloride channel expressed in the fetal lung and downregulated after birth. Am J Respir Cell Mol Biol 12:597–604

Nakatsu F, Ohno H (2003) Adaptor protein complexes as the key regulators of protein sorting in the post-Golgi network. Cell Struct Funct 28:419–429

Nascimento DS, Reis CU, Goldenberg RC, Ortiga-Carvalho TM, Pazos-Moura CC, Guggino SE, Guggino WB, Morales MM (2003) Estrogen modulates ClC-2 chloride channel gene expression in rat kidney. Pflügers Arch 446:593–599

Niemeyer MI, Cid LP, Zúñiga L, Catalán M, Sepúlveda FV (2003) A conserved pore-lining glutamate as a voltage- and chloride-dependent gate in the ClC-2 chloride channel. J Physiol 553:873–879

Niemeyer MI, Yusef YR, Cornejo I, Flores CA, Sepúlveda FV, Cid LP (2004) Functional evaluation of human ClC-2 chloride channel mutations associated with idiopathic generalized epilepsies. Physiol Genomics 19:74–83

Nobles M, Higgins CF, Sardini A (2004) Extracellular acidification elicits a chloride current that shares characteristics with $I_{Cl(swell)}$. Am J Physiol Cell Physiol 287:C1426–C1435

Ogura T, Furukawa T, Toyozaki T, Yamada K, Zheng YJ, Katayama Y, Nakaya H, Inagaki N (2002) ClC-3B, a novel ClC-3 splicing variant that interacts with EBP50 and facilitates expression of CFTR-regulated ORCC. FASEB J 16:863–865

Palade PT, Barchi RL (1977) Characteristics of the chloride conductance in muscle fibers of the rat diaphragm. J Gen Physiol 69:325–342

Papponen H, Kaisto T, Myllyla VV, Myllyla R, Metsikko K (2005) Regulated sarcolemmal localization of the muscle-specific ClC-1 chloride channel. Exp Neurol 191:163–173

Peña-Münzenmayer G, Catalán M, Cornejo I, Figueroa CD, Melvin JE, Niemeyer MI, Cid LP, Sepúlveda FV (2005) Basolateral localization of native ClC-2 chloride channels in absorptive intestinal epithelial cells and basolateral sorting encoded by a CBS-2 domain di-leucine motif. J Cell Sci 118:4243–4252

Picollo A, Liantonio A, Didonna MP, Elia L, Camerino DC, Pusch M (2004) Molecular determinants of differential pore blocking of kidney CLC-K chloride channels. EMBO Rep 5:584–589

Picollo A, Pusch M (2005) Chloride/proton antiporter activity of mammalian CLC proteins ClC-4 and ClC-5. Nature 436:420–423

Piwon N, Günther W, Schwake M, Bösl MR, Jentsch TJ (2000) ClC-5 Cl^- -channel disruption impairs endocytosis in a mouse model for Dent's disease. Nature 408:369–373

Plassart-Schiess E, Gervais A, Eymard B, Lagueny A, Pouget J, Warter JM, Fardeau M, Jentsch TJ, Fontaine B (1998) Novel muscle chloride channel (CLCN1) mutations in myotonia congenita with various modes of inheritance including incomplete dominance and penetrance. Neurology 50:1176–1179

Ponting CP (1997) CBS domains in ClC chloride channels implicated in myotonia and nephrolithiasis (kidney stones). J Mol Med 75:160–163

Pusch M (1996) Knocking on channel's door. The permeating chloride ion acts as the gating charge in ClC-0. J Gen Physiol 108:233–236

Pusch M (2001) Chloride channelopathies. Pharmaceutical News 8:45–51

Pusch M (2002) Myotonia caused by mutations in the muscle chloride channel gene CLCN1. Hum Mutat 19:423–434

Pusch M, Accardi A, Liantonio A, Ferrera L, De Luca A, Camerino DC, Conti F (2001) Mechanism of block of single protopores of the *Torpedo* chloride channel ClC-0 by 2-(*p*-chlorophenoxy)butyric acid (CPB). J Gen Physiol 118:45–62

Pusch M, Jordt SE, Stein V, Jentsch TJ (1999) Chloride dependence of hyperpolarization-activated chloride channel gates. J Physiol 515:341–353

Pusch M, Liantonio A, Bertorello L, Accardi A, De Luca A, Pierno S, Tortorella V, Camerino DC (2000) Pharmacological characterization of chloride channels belonging to the ClC family by the use of chiral clofibric acid derivatives. Mol Pharmacol 58:498–507

Pusch M, Ludewig U, Jentsch TJ (1997) Temperature dependence of fast and slow gating relaxations of ClC-0 chloride channels. J Gen Physiol 109:105–116

Pusch M, Ludewig U, Rehfeldt A, Jentsch TJ (1995a) Gating of the voltage-dependent chloride channel ClC-0 by the permeant anion. Nature 373:527–531

Pusch M, Steinmeyer K, Jentsch TJ (1994) Low single channel conductance of the major skeletal muscle chloride channel, ClC-1. Biophys J 66:149–152

Pusch M, Steinmeyer K, Koch MC, Jentsch TJ (1995b) Mutations in dominant human myotonia congenita drastically alter the voltage dependence of the ClC-1 chloride channel. Neuron 15:1455–1463

Pusch M, Zifarelli G, Murgia AR, Picollo A, Babini E (2006) Channel or transporter? The CLC saga continues. Exp Physiol 91:149–152.

Qualmann B, Kessels MM, Kelly RB (2000) Molecular links between endocytosis and the actin cytoskeleton. J Cell Biol 150:F111–F116.

Ramjeesingh M, Li C, She YM, Bear CE (2006) Evaluation of the membrane domain of CLC-2. Biochem J 9:9

Richard EA, Miller C (1990) Steady-state coupling of ion-channel conformations to a transmembrane ion gradient. Science 247:1208–1210

Riordan JR (2005) Assembly of functional CFTR chloride channels. Annu Rev Physiol 67:701–718

Riordan JR, Rommens JM, Kerem B, Alon N, Rozmahel R, Grzelczak Z, Zielenski J, Lok S, Plavsic N, Chou JL, Drumm ML, Iannuzzi MC, Collins FS, Tsui L-C (1989) Identification of the cystic fibrosis gene:cloning and characterization of complementary DNA. Science 245:1066–1073

Robinson NC, Huang P, Kaetzel MA, Lamb FS, Nelson DJ (2004) Identification of an N-terminal amino acid of the CLC-3 chloride channel critical in phosphorylation-dependent activation of a CaMKII-activated chloride current. J Physiol 556:353–368

Rosenthal R, Bakall B, Kinnick T, Peachey N, Wimmers S, Wadelius C, Marmorstein A, Strauss O (2006) Expression of bestrophin-1, the product of the VMD2 gene, modulates voltage-dependent Ca^{2+} channels in retinal pigment epithelial cells. FASEB J 20:178–180.

Roux B, MacKinnon R (1999) The cavity and pore helices in the KcsA K^{+} channel:electrostatic stabilization of monovalent cations. Science 285:100–102

Rychkov G, Pusch M, Roberts M, Bretag A (2001) Interaction of hydrophobic anions with the rat skeletal muscle chloride channel ClC-1:effects on permeation and gating. J Physiol 530:379–393

Rychkov GY, Astill DS, Bennetts B, Hughes BP, Bretag AH, Roberts ML (1997) pH-dependent interactions of Cd^{2+} and a carboxylate blocker with the rat ClC-1 chloride channel and its R304E mutant in the Sf-9 insect cell line. J Physiol 501:355–362

Rychkov GY, Pusch M, Astill DS, Roberts ML, Jentsch TJ, Bretag AH (1996) Concentration and pH dependence of skeletal muscle chloride channel ClC- 1. J Physiol 497:423–435

Rychkov GY, Pusch M, Roberts ML, Jentsch TJ, Bretag AH (1998) Permeation and block of the skeletal muscle chloride channel, ClC-1, by foreign anions. J Gen Physiol 111:653–665

Salas-Casas A, Ponce-Balderas A, Garcia-Perez RM, Cortes-Reynosa P, Gamba G, Orozco E, Rodriguez MA (2006) Identification and functional characterization of EhClC-A, an *Entamoeba histolytica* ClC chloride channel located at plasma membrane. Mol Microbiol 59:1249–1261

Salazar G, Love R, Styers ML, Werner E, Peden A, Rodriguez S, Gearing M, Wainer BH, Faundez V (2004) AP-3-dependent mechanisms control the targeting of a chloride channel (ClC-3) in neuronal and non-neuronal cells. J Biol Chem 279:25430–25439

Sands JM, Bichet DG (2006) Nephrogenic diabetes insipidus. Ann Intern Med 144:186–194

Santos Ornellas D, Grozovsky R, Goldenberg RC, Carvalho DP, Fong P, Guggino WB, Morales M (2003) Thyroid hormone modulates ClC-2 chloride channel gene expression in rat renal proximal tubules. J Endocrinol 178:503–511

Saviane C, Conti F, Pusch M (1999) The muscle chloride channel ClC-1 has a double-barreled appearance that is differentially affected in dominant and recessive myotonia. J Gen Physiol 113:457–468

Schaller S, Henriksen K, Sveigaard C, Heegaard AM, Helix N, Stahlhut M, Ovejero MC, Johansen JV, Solberg H, Andersen TL, Hougaard D, Berryman M, Shiodt CB, Sørensen BH, Lichtenberg J, Christophersen P, Foged NT, Delaisse JM, Engsig MT, Karsdal MA (2004) The chloride channel inhibitor NS3736 [corrected] prevents bone resorption in ovariectomized rats without changing bone formation. J Bone Miner Res 19:1144–1153

Scheel O, Zdebik AA, Lourdel S, Jentsch TJ (2005) Voltage-dependent electrogenic chloride/proton exchange by endosomal CLC proteins. Nature 436:424–427

Schlingmann KP, Konrad M, Jeck N, Waldegger P, Reinalter SC, Holder M, Seyberth HW, Waldegger S (2004) Salt wasting and deafness resulting from mutations in two chloride channels. N Engl J Med 350:1314–1319

Schmidt-Rose T, Jentsch TJ (1997) Reconstitution of functional voltage-gated chloride channels from complementary fragments of CLC-1. J Biol Chem 272:20515–20521

Schwake M, Friedrich T, Jentsch TJ (2001) An internalization signal in ClC-5, an endosomal Cl^{-} channel mutated in Dent's disease. J Biol Chem 276:12049–12054

Schwappach B, Stobrawa S, Hechenberger M, Steinmeyer K, Jentsch TJ (1998) Golgi localization and functionally important domains in the NH_2 and COOH terminus of the yeast CLC putative chloride channel Gef1p. J Biol Chem 273:15110–15118

Scott JW, Hawley SA, Green KA, Anis M, Stewart G, Scullion GA, Norman DG, Hardie DG (2004) CBS domains form energy-sensing modules whose binding of adenosine ligands is disrupted by disease mutations. J Clin Invest 113:274–284

Silva IV, Cebotaru V, Wang H, Wang XT, Wang SS, Guo G, Devuyst O, Thakker RV, Guggino WB, Guggino SE (2003) The ClC-5 knockout mouse model of Dent's disease has renal hypercalciuria and increased bone turnover. J Bone Miner Res 18:615–623

Simon DB, Bindra RS, Mansfield TA, Nelson-Williams C, Mendonca E, Stone R, Schurman S, Nayir A, Alpay H, Bakkaloglu A, Rodriguez-Soriano J, Morales JM, Sanjad SA, Taylor CM, Pilz D, Brem A, Trachtman H, Griswold W, Richard GA, John E, Lifton RP (1997) Mutations in the chloride channel gene, *CLCNKB*, cause Bartter's syndrome type III. Nat Genet 17:171–178

Sintchak MD, Fleming MA, Futer O, Raybuck SA, Chambers SP, Caron PR, Murcko MA, Wilson KP (1996) Structure and mechanism of inosine monophosphate dehydrogenase in complex with the immunosuppressant mycophenolic acid. Cell 85:921–930

Smith RL, Clayton GH, Wilcox CL, Escudero KW, Staley KJ (1995) Differential expression of an inwardly rectifying chloride conductance in rat brain neurons:a potential mechanism for cell-specific modulation of postsynaptic inhibition. J Neurosci 15:4057–4067

Sobacchi C, Frattini A, Orchard P, Porras O, Tezcan I, Andolina M, Babul-Hirji R, Baric I, Canham N, Chitayat D, Dupuis-Girod S, Ellis I, Etzioni A, Fasth A, Fisher A, Gerritsen B, Gulino V, Horwitz E, Klamroth V, Lanino E, Mirolo M, Musio A, Matthijs G, Nonomaya S, Notarangelo LD, Ochs HD, Superti Furga A, Valiaho J, van Hove JL, Vihinen M, Vujic D, Vezzoni P, Villa A (2001) The mutational spectrum of human malignant autosomal recessive osteopetrosis. Hum Mol Genet 10:1767–1773

Speake T, Kajita H, Smith CP, Brown PD (2002) Inward-rectifying anion channels are expressed in the epithelial cells of choroid plexus isolated from ClC-2 'knock-out' mice. J Physiol 539:385–390

Speirs HJ, Wang WY, Benjafield AV, Morris BJ (2005) No association with hypertension of CLCNKB and TNFRSF1B polymorphisms at a hypertension locus on chromosome 1p36. J Hypertens 23:1491–1496

Staley K, Smith R, Schaack J, Wilcox C, Jentsch TJ (1996) Alteration of $GABA_A$ receptor function following gene transfer of the CLC-2 chloride channel. Neuron 17:543–551

Steinmeyer K, Klocke R, Ortland C, Gronemeier M, Jockusch H, Gründer S, Jentsch TJ (1991a) Inactivation of muscle chloride channel by transposon insertion in myotonic mice. Nature 354:304–308

Steinmeyer K, Lorenz C, Pusch M, Koch MC, Jentsch TJ (1994) Multimeric structure of ClC-1 chloride channel revealed by mutations in dominant myotonia congenita (Thomsen). EMBO J 13:737–743

Steinmeyer K, Ortland C, Jentsch TJ (1991b) Primary structure and functional expression of a developmentally regulated skeletal muscle chloride channel. Nature 354:301–304

Steinmeyer K, Schwappach B, Bens M, Vandewalle A, Jentsch TJ (1995) Cloning and functional expression of rat CLC-5, a chloride channel related to kidney disease. J Biol Chem 270:31172–31177

Stobrawa SM, Breiderhoff T, Takamori S, Engel D, Schweizer M, Zdebik AA, Bösl MR, Ruether K, Jahn H, Draguhn A, Jahn R, Jentsch TJ (2001) Disruption of ClC-3, a chloride channel expressed on synaptic vesicles, leads to a loss of the hippocampus. Neuron 29:185–196

Strange K (2003) From genes to integrative physiology:ion channel and transporter biology in *Caenorhabditis elegans*. Physiol Rev 83:377–415

Suzuki T, Rai T, Hayama A, Sohara E, Suda S, Itoh T, Sasaki S, Uchida S (2006) Intracellular localization of ClC chloride channels and their ability to form hetero-oligomers. J Cell Physiol 206:792–798

Tanford C (1983) Mechanism of free energy coupling in active transport. Ann Rev Biochem 52:379–409

Thiemann A, Gründer S, Pusch M, Jentsch TJ (1992) A chloride channel widely expressed in epithelial and non-epithelial cells. Nature 356:57–60

Thompson CH, Fields DM, Olivetti PR, Fuller MD, Zhang ZR, Kubanek J, McCarty NA (2005) Inhibition of ClC-2 chloride channels by a peptide component or components of scorpion venom. J Membr Biol 208:65–76.

Thomsen J (1876) Tonische Krämpfe in willkürlich beweglichen Muskeln in Folge von ererbter psychischer Disposition. Arch Psychatrie Nerv 6:702–718

Toyoshima C, Nakasako M, Nomura H, Ogawa H (2000) Crystal structure of the calcium pump of sarcoplasmic reticulum at 2.6 Å resolution. Nature 405:647–655

Traverso S, Elia L, Pusch M (2003) Gating competence of constitutively open CLC-0 mutants revealed by the interaction with a small organic Inhibitor. J Gen Physiol 122:295–306

Traverso S, Zifarelli G, Aiello R, Pusch M (2006) Proton sensing of CLC-0 mutant E166D. J Gen Physiol 127:51–66

Tsui LC (1991) Probing the basic defect in cystic fibrosis. Curr Opin Genet Dev 1:4–10

Tyteca D, Van Der Smissen P, Mettlen M, Van Bambeke F, Tulkens PM, Mingeot-Leclercq MP, Courtoy PJ (2002) Azithromycin, a lysosomotropic antibiotic, has distinct effects on fluid-phase and receptor-mediated endocytosis, but does not impair phagocytosis in J774 macrophages. Exp Cell Res 281:86–100

Uchida S, Sasaki S (2005) Function of chloride channels in the kidney. Annu Rev Physiol 67:759–778

Uchida S, Sasaki S, Furukawa T, Hiraoka M, Imai T, Hirata Y, Marumo F (1993) Molecular cloning of a chloride channel that is regulated by dehydration and expressed predominantly in kidney medulla. J Biol Chem 268:3821–3824

Uchida S, Sasaki S, Nitta K, Uchida K, Horita S, Nihei H, Marumo F (1995) Localization and functional characterization of rat kidney-specific chloride channel, ClC-K1. J Clin Invest 95:104–113

van den Hove MF, Croizet-Berger K, Jouret F, Guggino SE, Guggino WB, Devuyst O, Courtoy PJ (2006) The loss of the chloride channel, ClC-5, delays apical iodide efflux and induces a euthyroid goiter in the mouse thyroid gland. Endocrinology 147:1287–1296

van Slegtenhorst MA, Bassi MT, Borsani G, Wapenaar MC, Ferrero GB, de Conciliis L, Rugarli EI, Grillo A, Franco B, Zoghbi HY, et al. (1994) A gene from the Xp22.3 region shares homology with voltage-gated chloride channels. Hum Mol Genet 3:547–552

Vandewalle A, Cluzeaud F, Bens M, Kieferle S, Steinmeyer K, Jentsch TJ (1997) Localization and induction by dehydration of ClC-K chloride channels in the rat kidney. Am J Physiol Renal Physiol 272:F678–F688

Vandewalle A, Cluzeaud F, Peng KC, Bens M, Lüchow A, Günther W, Jentsch TJ (2001) Tissue distribution and subcellular localization of the ClC-5 chloride channel in rat intestinal cells. Am J Physiol Cell Physiol 280:C373–C381

Varela D, Niemeyer MI, Cid LP, Sepúlveda FV (2002) Effect of an N-terminus deletion on voltage-dependent gating of the ClC-2 chloride channel. J Physiol 544:363–372

Vij N, Zeitlin PL (2006) Regulation of the ClC-2 lung epithelial chloride channel by glycosylation of SP1. Am J Respir Cell Mol Biol 34:754–759

Vogt K, Mellor J, Tong G, Nicoll R (2000) The actions of synaptically released zinc at hippocampal mossy fiber synapses. Neuron 26:187–196

Waldegger S, Jeck N, Barth P, Peters M, Vitzthum H, Wolf K, Kurtz A, Konrad M, Seyberth HW (2002) Barttin increases surface expression and changes current properties of ClC-K channels. Pflügers Arch 444:411–418

Waldegger S, Jentsch TJ (2000) Functional and structural analysis of ClC-K chloride channels involved in renal disease. J Biol Chem 275:24527–24533

Wang J, Xu H, Morishima S, Tanabe S, Jishage K, Uchida S, Sasaki S, Okada Y, Shimizu T (2006) Single-channel properties of volume-sensitive Cl^- channel in ClC-3-deficient cardiomyocytes. Jpn J Physiol 31:31

Wang SS, Devuyst O, Courtoy PJ, Wang XT, Wang H, Wang Y, Thakker RV, Guggino S, Guggino WB (2000) Mice lacking renal chloride channel, CLC-5, are a model for Dent's disease, a nephrolithiasis disorder associated with defective receptor-mediated endocytosis. Hum Mol Genet 9:2937–2945

Wang T, Weinman SA (2004) Involvement of chloride channels in hepatic copper metabolism:ClC-4 promotes copper incorporation into ceruloplasmin. Gastroenterology 126:1157–1166

Wang Y, Cai H, Cebotaru L, Hryciw DH, Weinman EJ, Donowitz M, Guggino SE, Guggino WB (2005) ClC-5:role in endocytosis in the proximal tubule. Am J Physiol Renal Physiol 289:F850–F862

Weinreich F, Jentsch TJ (2001) Pores formed by single subunits in mixed dimers of different CLC chloride channels. J Biol Chem 276:2347–2353

Wellhauser L, Kuo HH, Stratford FL, Ramjeesingh M, Huan LJ, Luong W, Li C, Deber CM, Bear CE (2006) Nucleotides bind to the carboxy terminus of ClC-5. Biochem J 11:11

Weng TX, Godley BF, Jin GF, Mangini NJ, Kennedy BG, Yu AS, Wills NK (2002) Oxidant and antioxidant modulation of chloride channels expressed in human retinal pigment epithelium. Am J Physiol Cell Physiol 283:C839–C849

White MM, Miller C (1979) A voltage-gated anion channel from the electric organ of *Torpedo californica*. J Biol Chem 254:10161–10166

Wohlke A, Distl O, Drogemuller C (2006) Characterization of the canine CLCN3 gene and evaluation as candidate for late-onset NCL. BMC Genet 7:13

Wollnik B, Kubisch C, Steinmeyer K, Pusch M (1997) Identification of functionally important regions of the muscular chloride channel ClC-1 by analysis of recessive and dominant myotonic mutations. Hum Mol Genet 6:805–811

Wotring VE, Miller TS, Weiss DS (2003) Mutations at the GABA receptor selectivity filter:a possible role for effective charges. J Physiol 548:527–540

Wright EM, Diamond JM (1977) Anion selectivity in biological systems. Physiol Rev 57:109–156

Wrong OM, Norden AG, Feest TG (1994) Dent's disease; a familial proximal renal tubular syndrome with low-molecular-weight proteinuria, hypercalciuria, nephrocalcinosis, metabolic bone disease, progressive renal failure and a marked male predominance. QJM 87:473–493

Yin J, Kuang Z, Mahankali U, Beck TL (2004) Ion transit pathways and gating in ClC chloride channels. Proteins 57:414–421

Yoshikawa M, Uchida S, Ezaki J, Rai T, Hayama A, Kobayashi K, Kida Y, Noda M, Koike M, Uchiyama Y, Marumo F, Kominami E, Sasaki S (2002) CLC-3 deficiency leads to phenotypes similar to human neuronal ceroid lipofuscinosis. Genes Cells 7:597–605

Yusef YR, Zúñiga L, Catalán M, Niemeyer MI, Cid LP, Sepúlveda FV (2006) Removal of gating in voltage-dependent ClC-2 chloride channel by point mutations affecting the pore and C-terminus CBS-2 domain. J Physiol 9:9

Zdebik AA, Cuffe JE, Bertog M, Korbmacher C, Jentsch TJ (2004) Additional disruption of the ClC-2 Cl^- channel does not exacerbate the cystic fibrosis phenotype of cystic fibrosis transmembrane conductance regulator mouse models. J Biol Chem 279:22276–22283

Zhang R, Evans G, Rotella FJ, Westbrook EM, Beno D, Huberman E, Joachimiak A, Collart FR (1999) Characteristics and crystal structure of bacterial inosine-5'-monophosphate dehydrogenase. Biochemistry 38:4691–4700

Zhang XD, Li Y, Yu WP, Chen TY (2006) Roles of K149, G352, and H401 in the channel functions of ClC-0:Testing the predictions from theoretical calculations. J Gen Physiol 127:435–447

Zhou Y, Morais-Cabral JH, Kaufman A, MacKinnon R (2001) Chemistry of ion coordination and hydration revealed by a K^+ channel-Fab complex at 2.0 Å resolution. Nature 414:43–48

Zúñiga L, Niemeyer MI, Varela D, Catalán M, Cid LP, Sepúlveda FV (2004) The voltage-dependent ClC-2 chloride channel has a dual gating mechanism. J Physiol 555:671–682

Rev Physiol Biochem Pharmacol (2007)
DOI 10.1007/112_2006_0607

S. F. J. van de Graaf · R. J. M. Bindels · J. G. J. Hoenderop

Physiology of epithelial Ca^{2+} and Mg^{2+} transport

Published online: 17 February 2007

Abstract Ca^{2+} and Mg^{2+} are essential ions in a wide variety of cellular processes and form a major constituent of bone. It is, therefore, essential that the balance of these ions is strictly maintained. In the last decade, major breakthrough discoveries have vastly expanded our knowledge of the mechanisms underlying epithelial Ca^{2+} and Mg^{2+} transport. The genetic defects underlying various disorders with altered Ca^{2+} and/or Mg^{2+} handling have been determined. Recently, this yielded the molecular identification of TRPM6 as the gatekeeper of epithelial Mg^{2+} transport. Furthermore, expression cloning strategies have elucidated two novel members of the transient receptor potential family, TRPV5 and TRPV6, as pivotal ion channels determining transcellular Ca^{2+} transport. These two channels are regulated by a variety of factors, some historically strongly linked to Ca^{2+} homeostasis, others identified in a more serendipitous manner. Herein we review the processes of epithelial Ca^{2+} and Mg^{2+} transport, the molecular mechanisms involved, and the various forms of regulation.

Introduction

Serum Ca^{2+} and Mg^{2+} levels are maintained within narrow limits by the human body, despite considerable variations in daily intake and excretion. The ion balance of these divalents is mediated by the coordinated action of the intestine, kidney, and bone. When the body is deprived of Ca^{2+} or Mg^{2+}, the (re)absorption activity in these organs increases accordingly, whereas the reverse action occurs when the plasma levels of these ions threaten to surpass acceptable upper limits. The bone acts as a dynamic storage compartment in this process and contributes to maintain the balance by releasing these ions upon Ca^{2+} or Mg^{2+} deprivation. Herein we will describe in detail the regulation of Ca^{2+} and Mg^{2+} transport in epithelia and discuss potential avenues for future breakthroughs.

S. F. J. van de Graaf · R. J. M. Bindels · J. G. J. Hoenderop (✉)
Radboud University Nijmegen Medical Centre, 286 Cell Physiology,
P.O. Box 9101, 6500 HB Nijmegen, The Netherlands
e-mail: j.hoenderop@ncmls.ru.nl · Tel.: +31-24-3610571 · Fax: +31-24-3616413

Epithelial Ca^{2+} transport

The maintenance of the body Ca^{2+} balance is of vital importance for several crucial physiological functions including neuronal excitability, muscle contraction, and skeletal integrity. Skeleton growth demands a positive Ca^{2+} balance, whereas a Ca^{2+} deficit is observed in aging and postmenopausal women, ultimately resulting in a loss of Ca^{2+} from bone. The latter is associated with a higher risk for bone fractures. The average daily Ca^{2+} intake of an adult is approximately 1 g, of which roughly 0.35 g is absorbed in the gastrointestinal tract (Greer and Krebs 2006). Approximately 8 g Ca^{2+} is filtered in the kidney on a daily basis, whereas only a fraction is excreted into the urine to balance Ca^{2+} that is absorbed in the intestine. Paracellular and transcellular transport mediates Ca^{2+} (re)absorption. The paracellular component of epithelial Ca^{2+} transport is passive and directly connects the luminal compartment with the blood compartment, whereas the transcellular component is active and involves the passage of at least two membrane barriers. Importantly, the transcellular pathway is the main target site for specific regulation of Ca^{2+} (re)absorption by various calciotropic hormones.

Localization of transcellular Ca^{2+} transport

Ca^{2+} (re)absorption occurs in several organs including kidney, intestine, and bone. Fish have an additional organ for Ca^{2+} uptake, the gills. Furthermore, in pregnancy and lactation, the placenta and mammary glands, respectively, are important tissues in the balance between Ca^{2+} uptake and output.

Kidney

A large amount of Ca^{2+} is filtered at the glomerulus. The proximal tubules (PT), including the proximal convoluted (PCT) and proximal straight (PST) tubules, are responsible for the absorption of the bulk of the Ca^{2+} from the filtrate. Approximately 65% of the filtered Ca^{2+} is reabsorbed here, as was demonstrated using micropuncture experiments (Edwards et al. 1973; Friedman 1999; Sutton and Dirks 1975; Ullrich et al. 1963). This transport is passive and follows the local Na^+ reabsorption. This site therefore does not provide an independent regulation of Ca^{2+} reabsorption (Suki 1979). In the subsequent segment of the nephron, i.e., the thin descending and ascending loop of Henle (ATL), virtually no Ca^{2+} is reabsorbed (Rocha et al. 1977). However, the thick ascending loop of Henle (TAL) is again permeable for Ca^{2+}, and this segment accounts for approximately 20% of the total Ca^{2+} reabsorption (Bailly et al. 1990; Bourdeau and Burg 1980; Bourdeau et al. 1987; Di Stefano et al. 1989, 1990; Friedman 1988; Friedman and Gesek 1995a; Imai 1978; Ng et al. 1982; Rocha et al. 1977; Suki 1979; Suki et al. 1980; Suki and Rouse 1981). Several studies suggest that Ca^{2+} mainly follows the paracellular pathway in this segment (Bourdeau and Burg 1979; Shareghi and Agus 1982; Wittner et al. 1991, 1993). This was further corroborated when mutations in paracellin-1 (also called claudin-16), localized in the tight junctions of TAL, were associated with renal Ca^{2+} and Mg^{2+} wasting due to impaired paracellular divalent cation reabsorption (Simon et al. 1999). The destination of the filtrated Ca^{2+} remaining at the end of the TAL is determined by the distal part of the nephron, consisting of the distal convoluted tubule (DCT) and connecting tubule (CNT) (Costanzo and Windhager 1978; Costanzo et al. 2000). The CNT is situated distal from the DCT, arising abruptly in rabbits and more gradually in other species (Bulger et al. 1967; Crayen and Thoenes 1978; Kaissling 1982; Loffing et al. 2001). In the CNT and DCT, Ca^{2+} reabsorption takes place against its electrochemical gradient, indicating that the transport is active (Costanzo et al. 2000). Furthermore, the tight junctions in DCT and CNT are relatively impermeable to Ca^{2+}, in line with a predominant role for an active transcellular Ca^{2+} transport pathway. The relative contribution of the initial (DCT1) and later (DCT2) part of DCT and CNT to the active Ca^{2+} reabsorption is not clear. Microperfusion studies by Costanzo and Windhager showed active Ca^{2+} transport in both DCT and CNT (Costanzo and Windhager 1978), whereas studies from Greger and coworkers indicated a predominant role for CNT (Greger et al. 1978). Furthermore, the relative contribution of each segment could be regulated; Imai et al. demonstrated that DCT and CNT do not respond to the same extent to parathyroid hormone (PTH) administration (Imai 1981). The cortical collecting duct (CCD) accounts for maximally 3% of the filtered Ca^{2+} (Ullrich et al. 1963). As net transport occurs against the electrochemical gradient for Ca^{2+}, Ca^{2+} reabsorption must be active here as well.

Intestine

Intestinal Ca^{2+} absorption is crucial in the maintenance of the Ca^{2+} balance, as it primarily mediates the uptake of Ca^{2+} from the diet. Only about 30% of dietary Ca^{2+} is absorbed and the remaining 70% is excreted in the feces. Transcellular Ca^{2+} absorption occurs mainly in duodenum and in the initial part of jejunum and, to a much lesser extent, in ileum and colon, whereas paracellular Ca^{2+} absorption takes place throughout the entire small intestine (Bronner et al. 1986). The relative contribution of paracellular versus transcellular Ca^{2+} absorption depends on the dietary Ca^{2+} content (Bronner 2003; Bronner and Pansu 1999). Chyme moves down the intestinal lumen in approximately 3 h, spending only a short time in the duodenum, but over 2 h in the distal part of the small intestine. When dietary Ca^{2+} intake is low, transcellular Ca^{2+} transport accounts for a substantial fraction of the absorbed Ca^{2+}, and vice versa when Ca^{2+} intake is high. In addition, the contribution of transcellular Ca^{2+} transport is strongly upregulated by calciotropic hormones including vitamin D, as in "Regulation of epithelial Ca^{2+} transport".

Bone

Bone serves as an important storage point for Ca^{2+}, as it contains 99% of the total body Ca^{2+}. There are two types of bone, cortical and trabecular, the former constituting approximately 80% of the total bone mass (Nussey and Whitehead 2001). In the long bones of the skeleton, cortical or compact bone predominates. In the axial skeleton (skull, ribs, vertebrae) there is only a relatively thin layer of circumferential cortical bone with a much greater mass of trabecular or spongy bone. Since the surfaces within bone exposed to the extracellular fluid are higher in trabecular than compact bone it plays a more important role in Ca^{2+} homeostasis. Bone formation and resorption take place at these surfaces. Formation is carried out by active osteoblasts that extrude collagen into the extracellular space and deposit Ca^{2+} (Nussey and Whitehead 2001; Rodan 1992). As the osteoblasts become surrounded by mineralized bone these cells lose their activity and become interior osteocytes. However, they remain in contact with the bone surfaces and are supposed to play a role in regulated Ca^{2+} release in a process termed osteocytic osteolysis (Nussey and Whitehead 2001). Bone resorption also occurs on bone surfaces and is carried out by osteoclasts, which literally tunnel their way into bone, forming resorption pits (Nussey and Whitehead 2001; Teitelbaum 2000). Bone resorption and formation are linked and bone is continuously remodeled (Nussey and Whitehead 2001; Rodan 1992; Rodan and Martin 2000). As the maximum bone density is reached at the age of approximately 30 years, the rate of bone formation exceeds bone resorption until this age (Greer and Krebs 2006; Nussey and Whitehead 2001). Later in life, the reversal of this balance can lead to impaired bone thickness (osteoporosis), where the bone is prone to fracture (Greer and Krebs 2006; Harada and Rodan 2003; Nussey and Whitehead 2001; Rodan and Martin 2000). Bone formation and its remodeling are controlled by several calciotropic hormones that will be discussed later. However, it is still unclear in which way bone resorption and formation contributes to blood Ca^{2+} homeostasis since these processes are relatively slow for the buffering of rapid responses in serum Ca^{2+}. Furthermore, knowledge about the molecular mechanism of bone formation and resorption is limited (Van der Eerden et al. 2005). It is therefore essential to identify and further characterize the Ca^{2+} transport processes in bone.

Placenta

During pregnancy Ca^{2+} absorption in the placenta is solely responsible for the Ca^{2+} supply to the developing fetus. Ca^{2+} is transported across the placenta from the maternal to the fetal circulation via an active transcellular pathway to meet the requirements of the rapidly mineralizing skeleton and to maintain an extracellular level of Ca^{2+} that is physiologically appropriate for the development of fetal tissues (Belkacemi et al. 2002, 2003, 2004; Brunette 1988; Fukuoka and Satoh 1982; Lafond et al. 1991; Moreau et al. 2002a, b, 2003a, b; Pitkin 1985). Ca^{2+} is transported by the syncytiotrophoblasts, cells that form the epithelial layer separating the maternal and fetal circulation (Faulk and McIntyre 1983). It has been postulated that the molecular mechanisms of placental Ca^{2+} transport has considerable similarity with active Ca^{2+} transfer across the intestinal and renal epithelial cells, as the same Ca^{2+} transport proteins are expressed (Belkacemi et al. 2002, 2003, 2004, 2005).

Paracellular Ca^{2+} transport

Movement of Ca^{2+} ions through the tight junctions is a passive process that largely depends on the concentration and electrical gradient across the epithelium. The paracellular transport route must be regulated for the epithelium to remain selectively permeable. Depending on the functional requirements of an epithelium, there may be small or large amounts of solutes flowing passively through this path. Tight epithelia of high resistance can generate and maintain high transepithelial electrical potentials and ionic gradients to form luminal fluids with compositions that deviate significantly from that of interstitial fluid (Schneeberger and Lynch 2004). Furthermore, the paracellular pathway varies in its selectivity for ions and noncharged solutes (Van Itallie et al. 2003).

Tight junctions consist of linear arrays of integral membrane proteins, which include occludin, claudins, and several immunoglobulin superfamily members, such as the junctional adhesion molecule (Ebnet et al. 2003; Goodenough 1999; Martin-Padura et al. 1998). The claudin family consists of at least 20 related integral membrane proteins with four transmembrane domains (TMs) and functions as major structural components of the tight junctional complex, while occludin is an accessory protein involved in tight junction formation of which two isoforms have been described (Furuse et al. 2001; Gonzalez-Mariscal et al. 2003; Hirase et al. 1997; McCarthy et al. 1996; Morita et al. 1999; Saitou et al. 1997, 1998; Schneeberger and Lynch 2004; Tang and Goodenough 2003; Tsukita and Furuse 2000).

The molecular nature of the wide variety in electrical resistance and solute permeability of tight junctions is not completely understood. The existence of only two isoforms of occludin and the limited number of charged amino acid residues in the two extracellular domains of these proteins suggest that occludin is unlikely to contribute directly to the formation of ion-selective pores in the tight junction (Schneeberger and Lynch 2004). Furthermore, disruption of both occludin alleles by homologous recombination resulted in embryonic stem cells that not only differentiated into polarized epithelial cells but also formed an effective barrier to the diffusion of low-molecular-weight molecules (Saitou et al. 1998). Therefore, it is more likely that claudins form the critical component that determines the ion selectivity of tight junctions, particularly considering the large variation of charge in amino acid residues of their extracellular loops.

Although the mechanisms that establish the Ca^{2+} flux via the paracellular pathway are largely unknown, the role of claudins in epithelial Ca^{2+} transport is further supported by the disease mutations documented in these proteins (Cole and Quamme 2000; Wilcox et al. 2001; Wong and Goodenough 1999). Mutations in claudin-16, which are associated with a renal Ca^{2+} and Mg^{2+} wasting syndrome, implicate this particular claudin protein in paracellular reabsorption of Mg^{2+} and Ca^{2+}, but not monovalent ions (Simon et al. 1999). This topic will be more elaborately discussed below (see "Hypomagnesemia with hypercalciuria and nephrocalcinosis"). Mutations in claudin-14 cause nonsyndromic recessive deafness, and this tight junction protein is essential to maintain the electrochemical gradient between the endolymph and its surrounding tissues (Wilcox et al. 2001). However, it is currently not clear whether this phenotype involves changes in paracellular Ca^{2+} permeability. The role of other claudin family members in determining the rate of Ca^{2+} absorption in epithelia also remains elusive. The rate of paracellular Ca^{2+} transport is mainly determined by the electrochemical gradient across the epithelium (Fig. 1). Therefore, hormones and factors affecting this gradient will indirectly influence the passive Ca^{2+} fluxes through the tight junctions. In addition, the tight junction permeability itself is dynamically regulated (Goodenough 1999; Pappenheimer 1987) and subject to modulation by growth factors, cytokines, bacterial toxins, hormones, and other factors (Benais-Pont et al. 2003; Garcia et al. 1998; Gopalakrishnan et al. 2002; Wang et al. 2004b). In addition, paracellular transport can be mediated by protein kinases, as nicely illustrated by the threonine-serine kinase WNK4, which is present in tight junctions (Kahle et al. 2004; Yamauchi et al. 2004). It was recently demonstrated that WNK4 can bind and phosphorylate claudins-1 through -4 and that a human disease-causing mutant of WNK4 hyperphosphorylates claudins and increases paracellular Cl^- permeability (Kahle et al. 2004; Wilson et al. 2001; Yamauchi et al. 2004). Similarly, it was postulated that phosphorylation of specific claudins (i.e., claudin-16) might provide means to regulate the paracellular Ca^{2+} flux. However, because of the low Ca^{2+} specificity of paracellular transport compared to transcellular Ca^{2+} movement, specific regulation of the Ca^{2+} flux through tight junctions is likely to play a minor role in the fine-tuning of the Ca^{2+} balance.

Transcellular Ca^{2+} transport

Transcellular Ca^{2+} transport is the pivotal target for specific regulation of Ca^{2+} (re)absorption by various calciotropic hormones (Bawden 1989; Bouillon et al. 2003; Bronner 2003; Bronner and Pansu 1999; Brunette 1988; Friedman and Gesek 1995a; Hoenderop et al. 2005; Wasserman and Fullmer 1995). Transcellular Ca^{2+} (re)absorption can be divided into three consecutive steps. First, Ca^{2+} enters the cells from the luminal compartment. Second, Ca^{2+} bound to specialized Ca^{2+}-binding proteins diffuses to the basolateral side of the cell. Third, Ca^{2+} is extruded into the interstitial fluid by two extrusion mechanisms (Fig. 1). It is essential that the Ca^{2+} influx and efflux mechanism(s) maintains

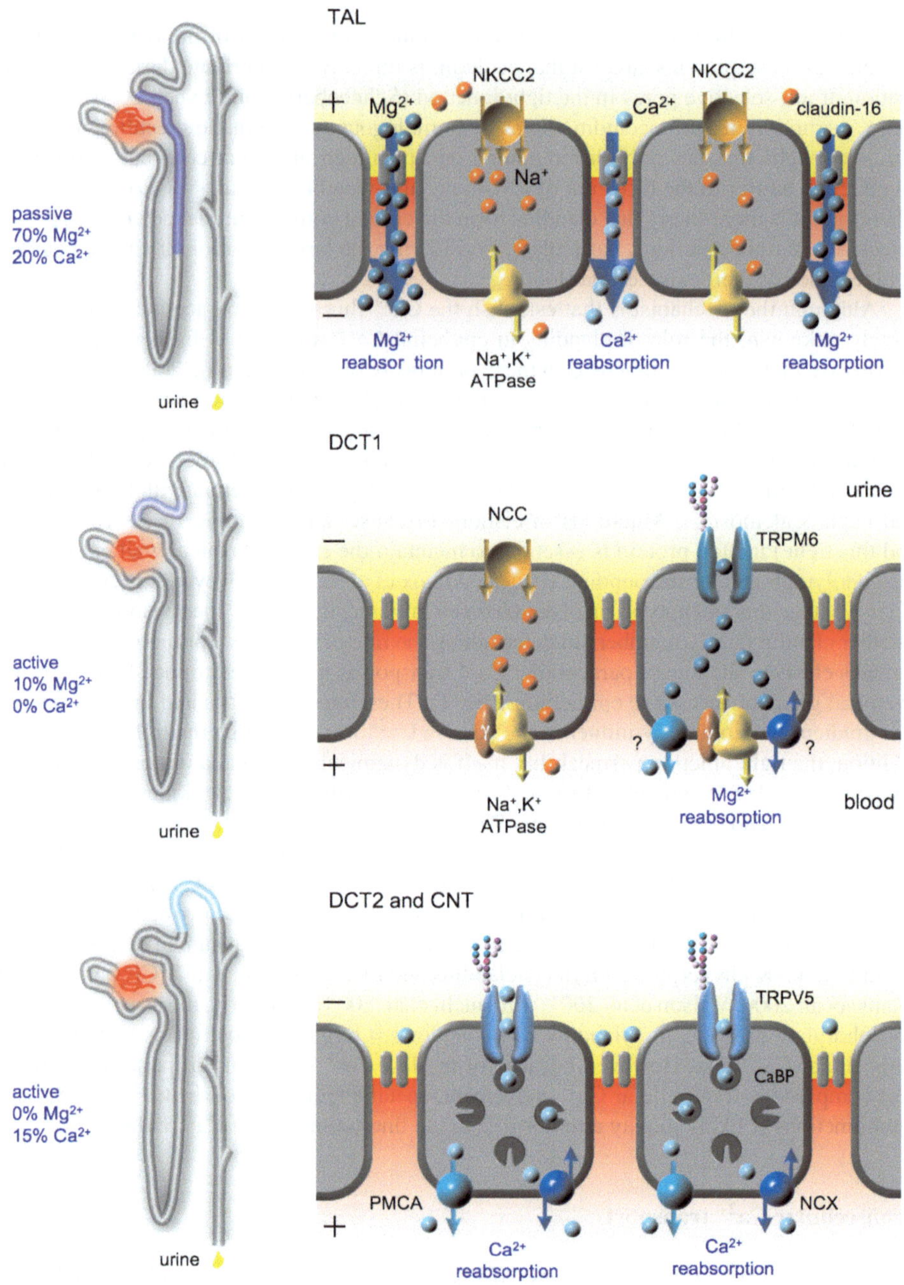

a polar distribution to ensure net Ca^{2+} transport from the apical or luminal side to the basolateral or serosal compartment. In this respect, it is important to study the Ca^{2+} influx pathway in polarized cell models. Until now, two polarized confluent epithelial cell systems representing duodenal and renal active Ca^{2+} (re)absorption have been studied. First, Caco-2 cells spontaneously differentiate under standard culture conditions into a tissue that exhibits functional duodenal transport properties (Giuliano and Wood 1991). These cells form

◀ **Fig. 1** Mechanisms of Ca^{2+} and Mg^{2+} reabsorption in the kidney. In TAL, Ca^{2+} and Mg^{2+} are reabsorbed via passive paracellular transport across the tight junctions. This transport is driven by the electrochemical gradient across the epithelium and requires functional claudin-16. Further downstream, Mg^{2+} and Ca^{2+} reabsorption are localized to distinct nephron segments. TRPM6 colocalizes with NCC in DCT1, where the former mediates apical Mg^{2+} influx. Mechanisms underlying the subsequent steps in transcellular Mg^{2+} reabsorption remain elusive. Mg^{2+} buffering is postulated and energy-consuming basolateral Mg^{2+} extrusion is mediated by a putative Na^+/Mg^{2+} exchanger or ATP-dependent Mg^{2+}-ATPase. The Na^+ gradient driving the putative Na^+/Mg^{2+} exchanger is established by the activity of the Na^+-K^+-ATPase that is modulated by the γ-subunit in DCT. In DCT2, a three-step process facilitating active and transcellular Ca^{2+} transport takes place. The first step is the entry of Ca^{2+} at the luminal side of the cell through the (hetero)tetrameric the epithelial Ca^{2+} channels TRPV5 and TRPV6. Subsequently, calbindin buffers Ca^{2+} and the Ca^{2+} diffuses to the basolateral membrane. At the basolateral membrane, Ca^{2+} is extruded via an ATP-dependent Ca^{2+}-ATPase (PMCA1b) and a Na^+/Ca^{2+} exchanger (NCX1)

a polarized epithelial layer and express several markers that are unique to differentiated small intestinal epithelium (e.g., high sucrase-isomaltase mRNA and protein levels) (Chantret et al. 1988). Furthermore, Caco-2 cells exhibit net apical-to-basolateral Ca^{2+} transport kinetics, and the rate of transport can be enhanced by pretreatment with vitamin D (Fleet et al. 2002; Fleet and Wood 1999; Giuliano et al. 1991). Second, the use of primary cultures and immortalized cell lines originating from the distal part of the nephron greatly facilitated our understanding of regulated renal Ca^{2+} influx in and transport through these cells. Two groups, Bindels et al. (Bindels et al. 1991, 1993; Hoenderop et al. 1998, 1999a, b; Raber et al. 1997; Van Baal et al. 1996a, b; 1999), and Gesek and Friedman (Bacskai and Friedman 1990; Friedman 1988; Friedman et al. 1996; Friedman and Gesek 1993; Friedman and Gesek 1994, 1995b; Gesek and Friedman 1992a, b; Magyar et al. 2002; White et al. 1998), have used immunodissected cell lines from rabbit and mouse kidney, respectively, to investigate hormone-stimulated Ca^{2+} transport. In addition, Bindels et al. demonstrated that primary cultures of rabbit CNT and CCD cells exhibit many characteristics of the original epithelium, including calciotropic hormone-stimulated Ca^{2+} transport from the apical to the basolateral compartments (Bindels et al. 1991).

Luminal Ca^{2+} influx

In order to identify the apical Ca^{2+} influx channel involved in transcellular Ca^{2+} (re)absorption, Hoenderop and coworkers performed functional expression cloning using a complementary DNA (cDNA) library from rabbit primary CNT and CCD. Injection of the total mRNA from this isolation in *Xenopus laevis* oocytes induced a $^{45}Ca^{2+}$ uptake 2–3 times above background. Subsequently, the entire cDNA library was screened for $^{45}Ca^{2+}$ uptake and a single transcript was isolated encoding for a novel epithelial Ca^{2+} channel, named ECaC1 and later renamed as the transient receptor potential channel TRPV5 (Hoenderop et al. 2001b; Montell et al. 2002). Similarly, Hediger and coworkers applied the same approach to screen a cDNA library obtained from rat small intestine. They identified Ca^{2+} transporter 1 (CaT1) that was later renamed into TRPV6 (Montell et al. 2002; Peng et al. 1999).

In the literature TRPV5 is also known as ECaC, ECaC1, and CaT2, whereas TRPV6 has been named previously CaT1, ECaC2, and CaT-Like. TRPV5 and TRPV6 display the defining properties of the long-sought epithelial Ca^{2+} channels, including hormonal regulation, localization, and functional properties as will be discussed below (see "Localization of TRPV5 and TRPV6" through to "Regulation of epithelial Ca^{2+} transport"). TRPV5 and TRPV6 are encoded by two distinct genes, rather than being splice variants (Hoenderop et al. 2001b; Weber et al. 2001a). These genes are juxtaposed on chromosome 7q35, suggesting

Fig. 2 Overview of TRPV5 and TRPV6 regulatory proteins. **a** The epithelial Ca^{2+} channels contain a core domain consisting of six transmembrane segments, with an additional hydrophobic stretch between TM5 and TM6 forming the pore forming region. This core TM segment-containing region is flanked by large amino and carboxyl-termini that face the intracellular compartment. The amino-terminus contains ankyrin repeats, which play a role in channel oligomerization. Calmodulin associates with both the amino and carboxyl-termini. Klotho affects TRPV5 and TRPV6 from the extracellular medium, where it modifies the glycosylated part of the channel. NHERF2 binds to the last three amino acids of the carboxyl-terminus, whereas the remaining channel-associated proteins identified to date, including Rab11a, 80K-H, BSPRY, NHERF4, and the S100A10/annexin 2 complex, bind to more upstream regions. Calbindin-D_{28K} has been shown to translocate to the apical plasma membrane to associate with TRPV5 at a low intracellular Ca^{2+} concentration. This enables the local buffering of Ca^{2+} near the channel, allowing significant Ca^{2+} influx by preventing rapid channel inactivation. **b** The gate and selectivity filter of TRPV5 and TRPV6 are formed by four channel subunits facing the center of the channel with the pore-forming region. This is a hydrophobic region that is flanked by two transmembrane domains (TM5 and TM6) and contains negatively charged amino acids (D542 in rabbit TRPV5) that determine pore size and Ca^{2+} selectivity. *$CaBP_{28K}$* calbindin-D_{28K}, *CaM* calmodulin, NHERF2 Na^{+}/H^{+} exchanger regulatory factor 2, *NHERF4* Na^{+}/H^{+} exchanger regulatory factor 4, *BSPRY* B-box and SPRY domain containing protein, *ANK* ankyrin repeat

a gene duplication event during evolution (Muller et al. 2000; Peng et al. 2000b). An identical situation was observed in mouse, where the two genes are close together on chromosome 6 (Hoenderop et al. 2003a; Weber et al. 2001a). The distinct genes comprise 15 exons encoding proteins of approximately 730 amino acids. TRPV5 and TRPV6 share a predicted topology of six TMs with an additional hydrophobic region between TM5 and TM6, which forms the channel pore (Fig. 2). TRPV5 and TRPV6 have been cloned from a variety of species, including rabbit, rat, mouse, human, and fish (Hoenderop et al. 1999c; Muller et al. 2000; Pan et al. 2005; Peng et al. 1999, 2000a, b). In the latter species, the distinction between TRPV5 and TRPV6 is difficult (Pan et al. 2005; Shahsavarani et al. 2006). The overall similarity of these sequences is approximately 75%. Remarkably, several domains in TRPV5 and TRPV6 are strongly conserved, including the pore region and binding sites involving associated regulatory proteins (Fig. 2). TRPV5 and TRPV6 belong to the superfamily of TRP channels consisting of several groups, including the TRPC, TRPV, and TRPM branches of the family (Clapham 2003). TRPV5 and TRPV6 share the highest homology with members of the TRP vanilloid family (TRPV). This group of cation channels further includes TRPV1–4, which respond to heat, osmolarity, odorants, and mechanical stimuli (for additional information see http://clapham.tch.harvard.edu/trps/TRPVs_2005_web.pdf). The homology of TRPV5 and TRPV6 with these members of the TRP channels is about 30% at the amino acid level.

Structural properties of TRPV5 and TRPV6 Hoenderop and coworkers addressed the oligomerization of TRPV5 and TRPV6 channels. Cross-linking studies, co-immunoprecipitations, and molecular mass determination of TRPV5 and TRPV6 complexes using sucrose gradient sedimentation showed that the epithelial Ca^{2+} channels form homo- and heterotetrameric channel complexes (Hoenderop et al. 2003b). The four subunits surround a single pore. Hellwig et al. recently addressed the homo- and heteromultimerization of TRPV channel by analysis of subcellular colocalization, fluorescence resonance energy transfer, and co-immunoprecipitation. TRPV channel subunits preferentially assemble into homomeric complexes, with the exception of TRPV5 and TRPV6, which also readily form heterotetramers (Hellwig et al. 2005).

The molecular determinants of TRPV6 oligomerization were addressed by Erler and coworkers. They identified the ankyrin repeat at position 116–191 of the amino-terminus as a stringent requirement for physical assembly of TRPV6 subunits (Erler et al. 2004). It was

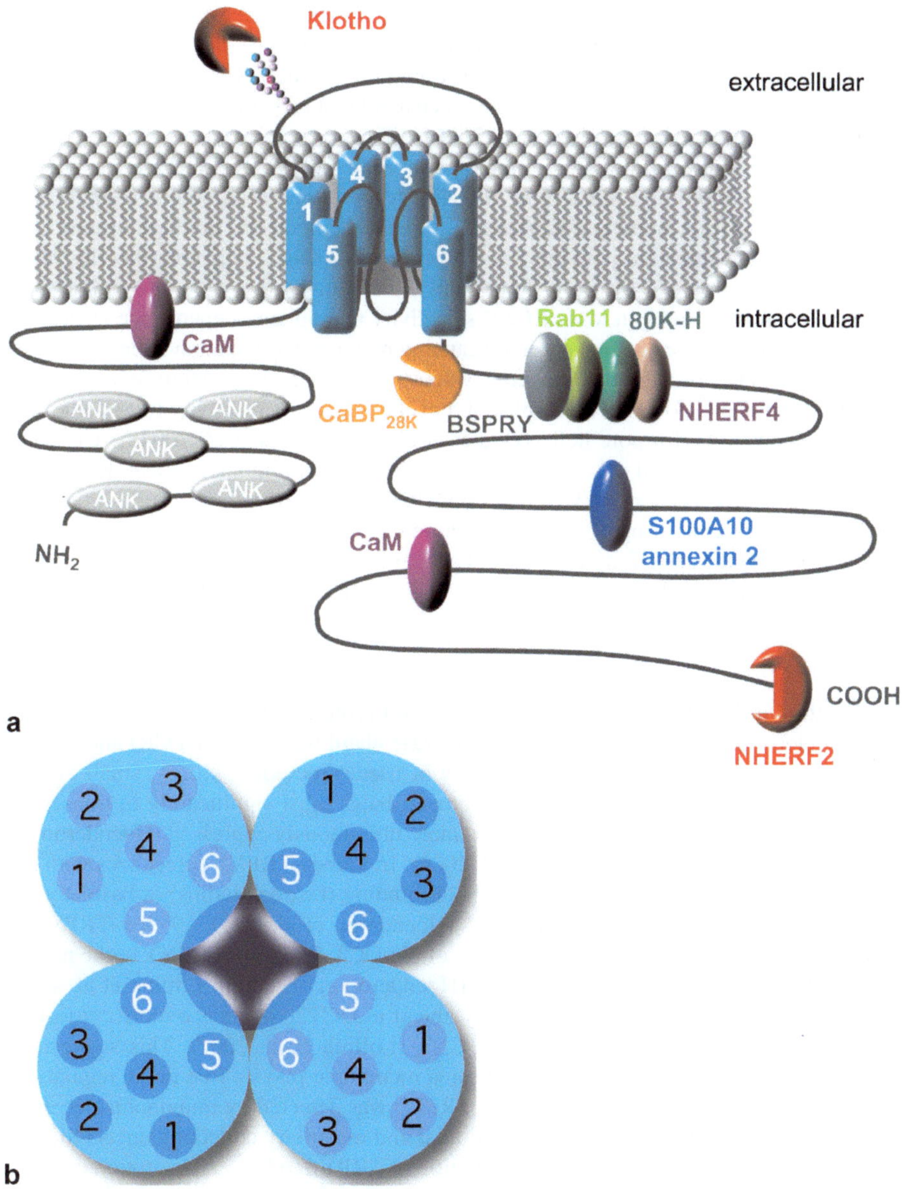

proposed that this repeat initiates a molecular zippering process that proceeds past the last ankyrin repeat and creates an intracellular anchor that is necessary for functional subunit assembly. Structure prediction programs indicated the presence of 3 to 6 ankyrin repeats in the amino-terminus (Erler et al. 2004; Hoenderop et al. 1999c).

Chang and coworkers confirmed the important role of the amino-terminus in channel assembly in general, and in particular demonstrated that a region near the first ankyrin repeat (position 64–77) is critical in the assembly process of TRPV5 (Chang et al. 2004). However, Chang et al. observed slightly increased binding efficiencies using truncated proteins including the first 162 amino acids, indicating that the ankyrin repeat identified for TRPV6 could also be involved in TRPV5 multimerization. Similarly, the pivotal role of ankyrin repeats in the oligomerization was recently also demonstrated for another TRP channel i.e., TRPV4. They showed that a splice variant of this cation channel that lack ankyrin repeats is impaired in its ability to multimerize (Arniges et al. 2006).

Interestingly, a structural model of the outer pore of TRPV5 and TRPV6 was provided by Dodier et al. (2004) and Voets et al. (2004a), respectively. Mutation of a single aspartate residue at position number 542 (D542), a residue crucial for the high-affinity Ca^{2+}-binding, altered the apparent pore diameter (Voets et al. 2004a), indicating that this residue lines the narrowest part of the pore. Importantly, mutation of this aspartate residue in the tetrameric channels results in loss of Ca^{2+} selectivity and voltage-dependence (Hoenderop et al. 2003b). Furthermore, using cysteine-scanning mutagenesis the main structural features of TRPV5 and TRPV6 were determined. Cysteines introduced in a region preceding D542 for TRPV5 and D541 for TRPV6 displayed a cyclic pattern of reactivity to cysteine reacting agents indicative of a pore helix (Dodier et al. 2004; Voets et al. 2004a). The pattern of covalent modification of cysteines supports a 3-D model similar to KcsA K^+ channels (Doyle et al. 1998). The external vestibule in TRPV5 and TRPV6 may build up the three structural domains consisting of a coiled structure that is connected to a 15-amino-acid pore helix followed by the selectivity filter (with D542 and D541 forming the narrowest part) and another coiled structure before the beginning of TM6. These are the first structural models of a TRP channel pore.

Localization of TRPV5 and TRPV6 Expression profiling of TRPV5 and TRPV6 using Northern blotting, reverse transcriptase polymerase chain reaction (RT-PCR) and immunohistochemistry showed expression in a variety of tissues including kidney, small intestine, placenta, and bone (Table 1). TRPV5 likely contributes most significantly to transcellular Ca^{2+} transport in kidney, whereas TRPV6 is more ubiquitously expressed (Hoenderop et al. 2000; Nijenhuis et al. 2003b; Zhuang et al. 2002). In kidney TRPV5 colocalizes with other Ca^{2+} transport proteins involved in intracellular binding (i.e., calbindins) or basolateral extrusion (i.e., NCX1 and PMCA1b) of Ca^{2+} (Hoenderop et al. 2000, 1999c). TRPV5 was localized to the apical domain of the late part of the DCT (DCT2) and CNT in the nephron, matching the sites of active Ca^{2+} transport (Hoenderop et al. 2000). Loffing et al. showed that TRPV5 localization was predominantly apical in DCT2, whereas a more cytosolic localization was observed at the end of the CNT (Loffing et al. 2001). This suggests that channel shuttling to and from the apical plasma membrane plays a role in the regulation of epithelial Ca^{2+} transport. In mouse kidney, TRPV6 was detected by immunohistochemistry at the apical domain of DCT2, CNT, and cortical and medullary collecting ducts (Nijenhuis et al. 2003b). The significance of the contribution of TRPV6 to Ca^{2+} reabsorption in the kidney is currently unknown, although the renal phenotype of TRPV5 knockout mice indicates that TRPV6 cannot compensate for loss of TRPV5 activity in the kidney (Hoenderop et al. 2003a).

Furthermore, several groups demonstrated the expression of TRPV6 in the proximal part of duodenum in line with a prominent role for TRPV6 in intestinal Ca^{2+} absorption (Huybers et al. 2006; Van de Graaf et al. 2003; Walters et al. 2006; Zhuang et al. 2002). However, Wissenbach et al. did not detect the expression of TRPV6 in duodenum and kidney (Wissenbach et al. 2001). TRPV5 and TRPV6 mRNA were both detected in syncytiotrophoblasts (Moreau et al. 2002a). This suggests that TRPV5 and/or TRPV6 mediate basal Ca^{2+} influx in placenta as these cells mediate Ca^{2+} transfer to the fetus. In bone, TRPV5 was detected

Table 1 Localization of TRPV5 and TRPV6 in Ca^{2+}-transporting tissues. TRPV5 forms the predominant Ca^{2+} influx pathway in kidney, whereas TRPV6 is essential for intestinal Ca^{2+} absorption. The relative contribution of TRPV5 and TRPV6 to epithelial Ca^{2+} transport in placenta, bone, and mammary gland remains to be established (Hoenderop et al. 2000)

Tissue	TRPV5	TRPV6	Reference(s)
Kidney	DCT2-CNT	DCT2-CD	Hoenderop et al. 2000; Loffing et al. 2001; Nijenhuis et al. 2003b
Intestine	Duodenum	Duodenum colon	Muller et al. 2000; Weber et al. 2001a; Zhuang et al. 2002
Placenta	Syncytiotrophoblast	Syncytiotrophoblast	Moreau et al. 2002b; Peng et al. 2001a
Bone	Osteoclasts	Osteoclasts and osteoblasts	Van der Eerden et al. 2005; Nijenhuis et al. 2003b; Weber et al. 2001a
Mammary gland	ND	Epithelial cells	Zhuang et al. 2002

DCT2, late distal convoluted tubule;
CNT, connecting tubule;
CD, collecting duct;
ND, not determined

in osteoclasts, in line with a role in Ca^{2+} resorption in these cells (Van der Eerden et al. 2005). Also TRPV6 was detected in bone (Nijenhuis et al. 2003b; Weber et al. 2001a), although little is known about the exact localization and role of this TRP channel here. Finally, the expression of TRPV6 is not restricted to Ca^{2+}-transporting cells and is detected in several exocrine tissues including pancreas, salivary gland, stomach, and prostate (Peng et al. 2000b; Wissenbach et al. 2001; Zhuang et al. 2002). The expression of TRPV6 in prostate is elevated in prostate cancer and correlates with the tumor grade (Fixemer et al. 2003; Peng et al. 2001b; Wissenbach et al. 2001). Although the functional consequence of TRPV6 expression in these tissues is unknown, it was postulated that TRPV6 mediates Ca^{2+} influx under certain conditions during exocytosis or cell proliferation.

Biophysical properties of TRPV5 and TRPV6 TRPV5 and TRPV6 display unique biophysical properties that distinguish them from other TRP channel members (Clapham 2003; Vennekens et al. 2000; Yue et al. 2001). First, TRPV5 and TRPV6 show considerable constitutive activity at low intracellular Ca^{2+} concentrations and physiological membrane potentials. Most other members of the TRP superfamily display smaller constitutive activity, but are activated upon ligand binding, receptor-mediated phospholipase C (PLC) activation, or temperature shifts (Clapham 2003). Furthermore, TRPV5 and TRPV6 are 100 times more selective for Ca^{2+} than for Na^+, making them the most Ca^{2+}-selective TRP channels (Hoenderop et al. 2005; Nilius et al. 2000; Vassilev et al. 2001). At physiological Ca^{2+} concentrations, the currents passing through TRPV5 and TRPV6 are mainly carried by Ca^{2+} (Hoenderop et al. 2005). Single channel conductances of TRPV5 and TRPV6 are 40–70 pS, using Na^+ as a charge carrier (Nilius et al. 2000; Vassilev et al. 2001). In addition, both in the inside-out and in the whole-cell configuration, TRPV5 and TRPV6 show a characteristic inward rectification (Hoenderop et al. 2005). So far, reliable single channel measurements have not been performed in the presence of extracellular Ca^{2+}. Although many of the properties are similar between TRPV5 and TRPV6, there are a number of differences between these channels. First, Ba^{2+} permeates TRPV5 better than TRPV6, e.g., the current ratio for Ba^{2+} over Ca^{2+} (I_{Ba}/I_{Ca}) is approximately 0.9 for TRPV5 and only approximately 0.4 for TRPV6 (Nilius et al. 2002). Furthermore, the rate of channel inactivation is different between TRPV5 and TRPV6, as is detailed below (Nilius et al. 2002). Remarkably, the structural determinants for these two differences are situated in the intracellular linker region between TM2 and TM3 (Nilius et al. 2002). Another distinction between TRPV5 and TRPV6 is the rate of recovery from Ca^{2+}-dependent inactivation (see below), which is roughly three times slower in TRPV5 compared to TRPV6 (Hoenderop et al. 2005). Finally, there are pharmacological differences between TRPV5 and TRPV6, as the latter requires higher concentrations of ruthenium red or Cd^{2+} to be blocked (~100- and ~4-fold increase in IC_{50}, respectively) (Hoenderop et al. 2003b; Nilius et al. 2001b).

TRPV5 and TRPV6 are coexpressed in some tissues, which allows hetero-oligomerization of these channels in vivo (Hoenderop et al. 2003b). As TRPV5 and TRPV6 exhibit different channel kinetics with respect to Ca^{2+}-dependent inactivation, Ba^{2+} selectivity, and sensitivity to inhibition by ruthenium red and Cd^{2+}, heterotetramer composition might influence the functional properties of the formed Ca^{2+} channel (Hoenderop et al. 2003b). This was investigated using concatemeric constructs consisting of four TRPV5 and/or TRPV6 subunits configured in a head-to-tail fashion. A different ratio of TRPV5 and TRPV6 subunits in these concatemers showed that the phenotype resembles the mixed properties of TRPV5 and TRPV6 (Hoenderop et al. 2003b). A high number of TRPV5 subunits in such a concatemer displayed more TRPV5-like properties, indicating that the stoichiometry of TRPV5/6 heterotetramers influences the channel properties. Consequently, regulation of the relative expression levels of TRPV5 and TRPV6 may be a mechanism to fine-tune the Ca^{2+} transport kinetics. For instance, immunohistochemical data in kidney clearly demonstrated coexpression of TRPV5 and TRPV6 in DCT and CNT (Hoenderop et al. 2003b). However, coexpression of TRPV5 and TRPV6 in these tissues has been to date quantified only at the mRNA level, indicating that TRPV6 is 100–10,000 times more expressed than TRPV5. Quantification at the protein level of both channels is certainly important to address the stoichiometry in vivo.

Factors affecting channel gating

Intracellular Ca^{2+} Unlike other members of the TRP superfamily, there are no indications that TRPV5 and TRPV6 require a stimulus or ligand to be activated. However, the constitutive activity of these epithelial channels is regulated by various means. First, TRPV5 and TRPV6 are subject to Ca^{2+}-dependent feedback inhibition (Bodding et al. 2002; Nilius et al. 2001a, 2002; Vennekens et al. 2000, 2001a). Both channels rapidly inactivate during hyperpolarizing voltage steps. This inhibition is dependent on the extracellular Ca^{2+} concentration and occurs also in cells buffered intracellularly with 10 mM BAPTA. Although TRPV5 and TRPV6 are both rapidly inactivated upon an increase in the intracellular Ca^{2+} concentration, the initial inactivation is faster in TRPV6 than in TRPV5 (Nilius et al. 2002). As described above, the intracellular region between TM2 and TM3 was identified as a crucial domain for the fast inactivation of TRPV6 (Nilius et al. 2002). Furthermore, two regions in the carboxyl-terminus of TRPV5 contributed to the Ca^{2+}-dependent inactivation (Nilius et al. 2003). Deletion of the last 30 amino acids of the carboxyl-terminus of TRPV5 (G701X) significantly decreased the Ca^{2+} sensitivity. Detailed mutation analysis revealed that a domain upstream in the carboxyl-terminus (between E649 and C653) forms a second critical stretch for Ca^{2+}-dependent inactivation of TRPV5 (Nilius et al. 2001a). Ca^{2+} influx is a prerequisite for Ca^{2+}-dependent channel inhibition because the Ca^{2+}-impermeable

D542A mutant lacks a monovalent current decay in response to repetitive stimulation (Nilius et al. 2001c). These data suggest that TRPV5 and TRPV6 channels are downregulated by Ca^{2+} influx through the channel, which increases the Ca^{2+} concentration in a microdomain near the pore region. Considering the high affinity of Ca^{2+}-dependent TRPV5 and TRPV6 inhibition, the presence of intracellular Ca^{2+} buffer proteins such as calbindins plays an important role to maintain channel activity (Lambers et al. 2006b). Measurements of the endogenous TRPV5 or TRPV6 activity in native primary cells that express all proteins involved in transepithelial Ca^{2+} transport would therefore provide another stimulus to our knowledge on the Ca^{2+}-dependent regulation of these channels. Due to the low endogenous expression of the channels, such data remain unavailable to date.

Nilius et al. showed that recovery from inhibition occurred both upon washout of extracellular Ca^{2+} (whole-cell configuration) or by removal of Ca^{2+} from the inner side of the channel (inside-out patches) (Nilius et al. 2001a). However, this process does not simply correlate with the removal of intracellular Ca^{2+}, since full recovery occurs much later than restoration of the basal Ca^{2+} level in non Ca^{2+}-buffered cells, or after removing Ca^{2+} from the inner side of excised membrane patches (Nilius et al. 2001a). Therefore, the recovery from the Ca^{2+}-dependent inactivation seems to be mediated by a mechanism distinct from Ca^{2+}-dependent inactivation. It is currently unknown whether the recovery reflects reopening of channels present in the plasma membrane or insertion of new channels into the plasma membrane.

Intracellular Mg^{2+} and PIP_2 An important feature of TRPV5 and TRPV6 is the voltage-dependent blockage by intracellular Mg^{2+} (Hoenderop et al. 2005; Lee et al. 2005; Voets et al. 2003). Both channels show nearly complete inward rectification, as no outward currents are observed in the presence of intracellular Mg^{2+}. However, the rectification is less pronounced in the absence of intracellular Mg^{2+}, although an intrinsic inward rectification remains (Hoenderop et al. 2005; Voets et al. 2003). Furthermore, in the absence of intracellular Mg^{2+}, hyperpolarizing voltage steps activate inward currents without delay. However, when 1 mM Mg^{2+} is present intracellularly, a slowly rising phase ("gating") is observed (Hoenderop et al. 2005; Lee et al. 2005). This is due to "unblock" of the Mg^{2+}-dependent channel inhibition. The amount of Mg^{2+}-dependent blockade relies on the voltage that was present just preceding the hyperpolarization. At less negative potentials, partially blocked channels open in a time-dependent manner due to unblock. However, the "unblock" was less pronounced at highly negative potentials and after strong depolarization. These phenomena can be explained as follows: at mild depolarizing potentials, Mg^{2+} moves toward the pore thereby plugging the permeation pathway for monovalent ions. Unblock occurs at hyperpolarizing voltages. At very large depolarization Mg^{2+} is pushed through the pore, which results in a partial unblock of the channels. These three features (i.e., rectification, gating, and voltage-dependence) only appear in the presence of intracellular Mg^{2+}. In the absence, gating and voltage-dependence disappear, whereas rectification is still present, but diminished (Hoenderop et al. 2005; Lee et al. 2005; Voets et al. 2003). This fast voltage-dependent block of TRPV5 was confirmed by Lee et al. However, they also demonstrated a slower Mg^{2+}-dependent channel blockade (tens of seconds compared to milliseconds; Lee et al. 2005). Interestingly, the aspartate at position 542 in the channel pore is responsible for both the fast and slow component of Mg^{2+}-mediated channel blockade (Lee et al. 2005). Phosphatidylinositol bisphosphate (PIP_2) reduces the sensitivity of TRPV5 for this slow Mg^{2+}-induced inhibition (Lee et al. 2005). A role for PIP_2 in the regulation of TRPV5 was also described by Rohacs et al. Application of PIP_2 to inside out patches of TRPV5 expressing *Xenopus* oocytes resulted in a significant current increase (Rohacs et al. 2005). They illustrated a role for arginine at position 599 of (rat) TRPV5 that seems critical for the direct interaction with PIP_2. Mutation of other positively charged residues in this area including K587Q and R600Q (most likely referring to K600Q and R587Q) resulted in only modest amplitude decreases. Lee et al. indeed demonstrated that the rabbit TRPV5 mutant corresponding to R599Q (of rat TRPV5) was more sensitive to inhibition by intracellular Mg^{2+}, in line with a reduced PIP_2 binding (Lee et al. 2005). The physiological role of the regulation of TRPV5 by intracellular Mg^{2+} in a PIP_2-modulated fashion needs to be further addressed, but could be related to receptor-activated PLC signaling.

Intra- and extracellular protons Early studies in excitable tissues indicated that acidic pH inhibits voltage-gated Na^+ channel activity (Woodhull 1973). Hess and collaborators observed similar findings for L-type Ca^{2+} channels (Prod'hom et al. 1989). Changes in pH have been found to regulate a number of TRP channels. Extracellular acidic pH increases TRPV1 currents (Caterina and Julius 2001) and decreases the activity of TRPP2 (Gonzalez-Perrett et al. 2002), a distant TRP family member of TRPV5 and TRPV6. Acidification of the apical medium inhibits transcellular Ca^{2+} absorption across primary cultures of rabbit CNT and CCD cells, providing a pH-dependent activity of the apical Ca^{2+} influx pathway (Bindels et al. 1994). Importantly, it was previously demonstrated that $^{45}Ca^{2+}$ uptake in TRPV5-expressing *Xenopus laevis* oocytes is inhibited by acidification of the incubation medium (Hoenderop et al. 1999c; Peng et al. 2000a). Indeed, extracellular acidification reduced currents through TRPV5 carried by either monovalent or divalent cations (Peng et al. 2000a; Vennekens et al. 2001b). Recently, the mechanism of proton-dependent modulation of TRPV5 channel properties has been addressed by Huang and coworkers. First, Yeh et al. demonstrated that mutation of the glutamate at position 522, preceding the pore region, to glutamine (E522Q) decreases the inhibition of the channel by extracellular protons. Therefore, this residue may act as the "pH sensor" of TRPV5 (Yeh et al. 2003). In a follow-up paper, the same group demonstrated that also intracellular protons inhibit TRPV5 activity (Yeh et al. 2005). Seventeen amino acids surrounding the putative intracellular entrance of the pore were mutated into a nontitratable amino acid and only mutation of lysine-607 to asparagine (K607N) decreased the sensitivity of the channel to inhibition by intracellular acidification. Measurements of the relative permeability of inorganic monovalent cations to Na^+ indicated that both intra- and extracellular acidification reduces the estimated TRPV5 pore diameter. This is possibly due to a rotation of the pore helix, as shown by measurements of the accessibility of pore residues to a cysteine-reactive agent, which blocks the channel upon covalent binding (Yeh et al. 2005). Extrapolating the pH-dependence of TRPV5 to the in vivo situation suggests that inhibition of TRPV5 by protons may at least in part provide the molecular basis of acidosis-induced calciuresis. At least two additional mechanisms contribute to the calciuresis. The first is pH-dependent modulation of gene expression of Ca^{2+} transport proteins, as will be discussed in "Acidosis and alkalosis" (Nijenhuis et al. 2006). Second, Lambers et al. demonstrated a pH-dependent translocation of TRPV5 to the plasma membrane. This mechanism stimulated TRPV5 activity at higher pH due to an increase in the number of TRPV5 channels at the cell surface (Lambers et al. 2006c; Nilius and Mahieu 2006).

TRPV5 knockout mice Hoenderop and coworkers generated TRPV5 knockout ($TRPV5^{-/-}$) mice that displayed a number of alterations that could be linked to modification of the Ca^{2+} balance (Hoenderop et al. 2003a). First, metabolic studies demonstrated that

TRPV5$^{-/-}$ mice exhibit a robust calciuresis, since significantly more Ca^{2+} was excreted in the urine compared to wild-type littermates. Second, serum analysis showed that TRPV5$^{-/-}$ mice have normal serum Ca^{2+} concentrations, but significantly elevated levels of active vitamin D ($1,25(OH)_2D_3$) compared to wild-type littermates. Third, Ca^{2+} absorption in small intestine was significant increased in TRPV5$^{-/-}$ mice indicating a compensatory role of the small intestine. Fourth, TRPV5$^{-/-}$ mice show a severe bone phenotype.

One of the most appealing aspects of the phenotype of TRPV5$^{-/-}$ mice is their massive renal Ca^{2+} excretion (Hoenderop et al. 2003a). The urinary Ca^{2+} concentration of TRPV5$^{-/-}$ mice reached values 4–10 times higher compared to wild-type mice. This hypercalciuria persists during life as it was observed in mice ranging from 10 to 52 weeks (Van Abel et al. 2006). In vivo micropuncture studies were performed in these transgenic mice to pinpoint the defective site of the Ca^{2+} reabsorption along the nephron. Ca^{2+} reabsorption in TRPV5$^{-/-}$ mice was unaffected up to the last surface loop of the late proximal tubule (LPT). However, mean Ca^{2+} delivery to puncturing sites within DCT and CNT was significantly enhanced in TRPV5$^{-/-}$ mice (Hoenderop et al. 2003a). This defect in Ca^{2+} reabsorption along the DCT and CNT is consistent with the localization of TRPV5 in mice (Loffing et al. 2001). Interestingly, polyuria and polydipsia was consistently observed in TRPV5$^{-/-}$ mice compared to wild-type littermates (Hoenderop et al. 2003a). Polyuria reduces the potential risk of Ca^{2+} precipitations and thereby facilitates the excretion of large quantities of Ca^{2+}. The hypercalciuria-induced polyuria has been observed in humans (Miller and Stapleton 1989) and animal models (Frick and Bushinsky 2003; Puliyanda et al. 2003). Exactly how the increased luminal Ca^{2+} concentration induces polyuria is unknown. It is postulated that the high Ca^{2+} concentration activates the Ca^{2+}-sensing receptor (CaSR) in the apical membrane of the inner medullary collecting duct (IMCD), which could stimulate the retrieval of aquaporin 2 (AQP2) to reduce water reabsorption (Sands et al. 1997). Furthermore, TRPV5$^{-/-}$ mice produced urine that was significantly more acidic compared to wild-type mice. Acidification of the urine could also contribute to the prevention of renal stone formation during hypercalciuria, since Ca^{2+} precipitates will not form at pH 5–6 (Baumann 1998).

A significant increase in the rate of Ca^{2+} absorption in the small intestine was observed in TRPV5$^{-/-}$ mice compared to wild-type littermates, indicating an intestinal compensation for renal Ca^{2+} wasting (Hoenderop et al. 2003a; Renkema et al. 2005). Duodenal Ca^{2+} absorption gradually decreased upon aging to 52 weeks in wild-type and TRPV5$^{-/-}$ mice, but remained elevated in the latter compared to age-matched wild-type mice (Van Abel et al. 2006). Intestinal TRPV6 and calbindin-D_{9K} expression levels were significantly upregulated in TRPV5$^{-/-}$ mice consistent with this increased Ca^{2+} absorption (Hoenderop et al. 2003a). To address the role of $1,25(OH)_2D_3$ in Ca^{2+} hyperabsorption double-knockout mice (TRPV5$^{-/-}$/1α-OHase$^{-/-}$ mice) were generated that lack the TRPV5 and 25-hydroxyvitamin-D_3-1α-hydroxylase genes of which the latter is responsible for the production of $1,25(OH)_2D_3$. Renkema et al. demonstrated that increased serum levels of $1,25(OH)_2D_3$ in TRPV5$^{-/-}$ mice were essential for this compensatory Ca^{2+} hyperabsorption. TRPV5$^{-/-}$/1α-OHase$^{-/-}$ mice have undetectable serum levels of $1,25(OH)_2D_3$ and display a significant hypocalcemia (Renkema et al. 2005). Intestinal TRPV6 and calbindin-D_{9K} expression levels were decreased compared with wild-type mice. The renal Ca^{2+} leak, as demonstrated in TRPV5$^{-/-}$ mice, persisted in TRPV5$^{-/-}$/1α-OHase$^{-/-}$ mice, but a compensatory upregulation of intestinal Ca^{2+} transporters was abolished (Renkema et al. 2005). From these results, it was concluded that hypervitaminosis D is of crucial importance in TRPV5$^{-/-}$ mice to maintain normocalcemia despite impaired renal Ca^{2+} reabsorption.

The molecular mechanisms involved in bone resorption and formation are largely elusive, although both processes likely require transcellular transport of Ca^{2+}. Detailed analyses of femurs demonstrated that trabecular thickness in the femoral head of TRPV5$^{-/-}$ mice was drastically reduced compared with TRPV5$^{+/+}$ mice (Hoenderop et al. 2003a). Furthermore, the cortical bone volume, cortical volume fraction, and cortical bone thickness were decreased in TRPV5$^{-/-}$ versus wild-type mice. Van der Eerden et al. further investigated the functional role of TRPV5 in bone. TRPV5 mRNA was expressed in human and murine bone samples and in osteoclasts along with other genes involved in transcellular Ca^{2+} transport, including calbindin-D_{9K} and calbindin-D_{28K}, Na^+/Ca^{2+} exchanger 1, and plasma membrane Ca^{2+}-ATPase 1b (Van der Eerden et al. 2005). TRPV5 shows predominant localization to the ruffled border membrane of murine osteoclasts. However, TRPV5 was absent in osteoblasts (Van der Eerden et al. 2005). Analyses of femoral bone sections and an in vitro bone marrow culture system revealed increased osteoclast numbers and osteoclast area in TRPV5$^{-/-}$ mice, whereas a urinary bone resorption marker was reduced compared to TRPV5$^{+/+}$ mice. Using a functional resorption pit assay, it was, however, found that bone resorption was nearly absent in osteoclast cultures from TRPV5$^{-/-}$ mice, supporting the impaired resorption observed in vivo (Van der Eerden et al. 2005). Although this study clearly showed that TRPV5 is essential for osteoclastic bone resorption and demonstrates the significance of transcellular Ca^{2+} transport in osteoclasts, it does not explain the bone phenotype of TRPV5$^{-/-}$ mice. As osteoclastic function is impaired, one would expect increased bone thickness. However, TRPV5$^{-/-}$ mice show reduced bone thickness suggesting functional Ca^{2+} resorption from bone. Furthermore, in TRPV5$^{-/-}$/1α-OHase$^{-/-}$ mice, rickets was even more pronounced than observed in single TRPV5 or 1α-OHase$^{-/-}$ mice, suggesting that the high vitamin D levels do not cause the reduced bone thickness in TRPV5$^{-/-}$ mice (Renkema et al. 2005). Perhaps the answer lies in the role of bone formation by osteoblasts. This process remains largely elusive, and the role of transcellular Ca^{2+} transport and TRPV5 and/or TRPV6 therein should be one of the first topics of future investigation.

In summary, targeted ablation of the TRPV5 gene seriously disturbs renal Ca^{2+} handling, causing increased $1,25(OH)_2D_3$ serum levels, Ca^{2+} hyperabsorption, and reduced bone formation. These data from TRPV5$^{-/-}$ mice convincingly demonstrate that TRPV5 is the gatekeeper in active Ca^{2+} reabsorption. Similarly, TRPV6 provides a good candidate as the apical Ca^{2+} influx channel involved in intestinal Ca^{2+} absorption. However, the creation of TRPV6 knockout mice to obtain the most direct and ultimate proof for this hypothesis has not been achieved to date.

Intracellular Ca^{2+} transport

The process of transcellular Ca^{2+} transport places a substantial and continuous challenge on epithelial cells, as substantial amounts of Ca^{2+} traffic through the cytosol, while the cytosolic free Ca^{2+} concentration ($[Ca^{2+}]_i$) needs to be maintained at low levels. Results from mathematical modeling (Bronner and Stein 1988) have indicated that diffusion of free Ca^{2+} at physiological cytosolic concentrations can only account for about 1/70th of the observed rate of transcellular transport (Costanzo and Windhager 1978). To date, two models have been proposed to explain this discrepancy.

Facilitated diffusion In the first model, termed the "facilitated diffusion model," intracellular Ca^{2+}-binding proteins keep the free Ca^{2+} concentration low, while the diffusible cellular Ca^{2+} concentration is high enough to support massive transcellular Ca^{2+} rates (Bronner et al. 1986; Bronner and Stein 1988; Feher 1983; Feher and Wasserman 1979; Feher et al. 1992; Larsson and Nemere 2002). The abundance of these Ca^{2+}-buffering proteins is regulated by calciotropic hormones, i.e., vitamin D and PTH, to ensure sufficient Ca^{2+} (re)absorption capacity in conditions when this is required. Indeed, there are two major subclasses of vitamin D-dependent Ca^{2+}-binding proteins, calbindin-D_{9K} and calbindin-D_{28K}. Calbindin-D_{28K} is highly conserved during evolution and present in kidney, small intestine (only birds), pancreas, placenta, bone, and brain. Calbindin-D_{9K} is present in highest concentrations in small intestine as well as in kidney (only mouse). The expression level of these calbindins in kidney and intestine is closely correlated with the efficiency of Ca^{2+} (re)absorption; calbindins, therefore, play a central role in the facilitated diffusion model. Even though these buffering proteins diffuse more slowly than the Ca^{2+} ion (inversely related to the square root of the molecular weight), the intracellular concentration of calbindins (submillimolar range) is sufficient to raise the total diffusible Ca^{2+} concentration to the level needed to attain the experimental diffusion rates (Bronner and Stein 1988). Importantly, due to the relatively slow binding kinetics of these Ca^{2+}-binding proteins, Ca^{2+} signaling can occur independently of transcellular Ca^{2+} movement mediated by calbindin-D_{9K} and calbindin-D_{28K} (Koster et al. 1995). In addition, the Ca^{2+}-buffering activity of the calbindins plays an important role to maintain TRPV5 and TRPV6 channels in an open conformation. These channels are rapidly inactivated upon a rise in the $[Ca^{2+}]_i$, and therefore it is pivotal to keep $[Ca^{2+}]_i$ low. Recent data show that calbindin-D_{28K} can fulfill this role for TRPV5 by direct interaction with the channel. Therefore, calbindin-D_{28K} provides epithelial cells with both local and general Ca^{2+} buffering to support high rates of transcellular Ca^{2+} transport (Lambers et al. 2006b).

Vesicular transport Previous studies have also proposed a vesicular model in which the Ca^{2+}-transporting cells use lysosomes to sequester Ca^{2+} and facilitate its movement to the basolateral membrane (Larsson and Nemere 2002; Nemere et al. 1986). Formation of Ca^{2+}-enriched vesicles is initiated by influx of Ca^{2+} through Ca^{2+} channels in the apical or luminal membrane. The rapid increase in $[Ca^{2+}]$ in close vicinity to the apical membrane disrupts the actin filaments near the Ca^{2+} channels and initiates the formation of endocytic vesicles. The formed Ca^{2+}-containing vesicles are transported by microtubules and fuse with lysosomes (Larsson and Nemere 2002; Nemere and Norman 1990). While calbindins have been found to associate with lysosomes, the role of these Ca^{2+}-binding proteins in this latter model is at present unclear.

Calbindin-D_{28K} knockout mice Homozygous calbindin-D_{28K} knockout (calbindin-$D_{28K}{}^{-/-}$) mice were previously generated, which developed normally (Airaksinen et al. 1997a, b; Barski et al. 2003; Sooy et al. 1999, 2000). Calbindin-$D_{28K}{}^{-/-}$ mice fed a regular Ca^{2+} diet displayed an approximately twofold increase in the urinary Ca^{2+} excretion compared to wild-type littermates (Sooy et al. 2000; Zheng et al. 2004). However, these mice displayed no significant differences in serum Ca^{2+} or PTH levels. This could indicate that intestinal hyperabsorption compensates for the hypercalciuria induced by calbindin-D_{28K} deficiency. Furthermore, only in mice is calbindin-D_{9K} expressed in renal epithelia. This suggests that calbindin-D_{9K} could mediate a significant amount of Ca^{2+} buffering, not only in intestine (as in all other species), but also in the mouse kidney where it colocalizes with TRPV5 (Hoenderop et al. 2002). In this respect, it is important to note that ablation of the TRPV5 gene results in a greater than sixfold increase in calciuria (Hoenderop et al. 2003a), whereas inactivation of the calbindin-D_{28K} results only in roughly twofold increase

(Zheng et al. 2004), on a high Ca^{2+} diet, indicating that the renal defect in Ca^{2+} handling is submaximal in calbindin-$D_{28K}$$^{-/-}$ mice. This suggests that TRPV5 plays a more pivotal role in Ca^{2+} reabsorption compared to calbindin-D_{28K} that could also be explained by a possible redundancy for this calbindin. The relative contribution of TRPV5 and calbindin-D_{28K} was investigated using mice lacking both the TRPV5 and calbindin-D_{28K} gene (Gkika et al. 2006). This study suggests that TRPV5 is the most crucial factor of the two proteins, as removal of the calbindin-D_{28K} gene did not further deteriorate the Ca^{2+} hyperexcretion in TRPV5-lacking mice. It is, however, also possible that calbindin-D_{9K} now functions as an alternative Ca^{2+} buffer since this protein is also expressed in mice DCT. Mutant mice that lack the calbindin-$D_{9K}$$^{-/-}$ gene have recently been generated (Kutuzova et al. 2006) and they show normal serum Ca^{2+} levels. However, no information about the intestinal Ca^{2+} absorption or urine Ca^{2+} excretion in these animals was provided. Further studies using mice lacking both calbindins would provide further insight into the role of these Ca-buffering proteins in epithelial Ca^{2+} transport. Li et al. had previously demonstrated that genetic inactivation of the vitamin D receptor (VDR) gene leads to a 90% reduction in renal calbindin-D_{9K} expression, but little change in calbindin-D_{28K} (Li et al. 2001). To address whether calbindin-D_{9K} compensates for the role of calbindin-D_{28K} in Ca^{2+} homeostasis, Zheng et al. generated VDR/calbindin-D_{28K} double knockout mice, which expressed no calbindin-D_{28K} and only 10% of calbindin-D_{9K} in kidney. VDR knockout mice suffer from hypocalcemia, secondary hyperparathyroidism, rickets, and osteomalacia (Zheng et al. 2004). However, ablation of the calbindin-D_{28K} gene further deteriorates the phenotype, as the double knockout mice were even more growth-retarded, significantly smaller in body weight than VDR$^{-/-}$ mice, and died prematurely at a few months of age. Compared with VDR$^{-/-}$ mice, the VDR$^{-/-}$/calbindin-$D_{28K}$$^{-/-}$ mice had higher urinary Ca^{2+} excretion and developed more severe secondary hyperparathyroidism and rachitic skeletal phenotype, which were manifested by larger parathyroid glands, higher serum PTH levels, and much lower bone mineral density (Zheng et al. 2004). Using histomorphometry and microcomputer tomography, a recent study from Margolis and coworkers showed that the femora of calbindin-$D_{28K}$$^{-/-}$ mice have significantly increased cortical bone volume compared to wild-type mice (Margolis et al. 2006). These results directly suggest that calbindin-D_{28K} plays a distinct role in maintaining Ca^{2+} homeostasis and skeletal mineralization.

Ca^{2+} extrusion mechanisms

The energy-consuming step of transcellular Ca^{2+} transport lies in the Ca^{2+} efflux pathways. This step transports intracellular Ca^{2+} to the serosal side across the basolateral membrane against a considerable electrochemical gradient. Two Ca^{2+} transporters have been implicated in this extrusion process, a Na^+/Ca^{2+} exchange mechanism (NCX1) and a Ca^{2+}-ATPase (PMCA1b).

Na^+/Ca^{2+} exchanger Counter transport of Na^+ for Ca^{2+} across the plasma membrane accounts for the maintenance of low $[Ca^{2+}]_i$ in a wide variety of cells (Linck et al. 1998). To date, three genes for NCX, designated NCX1, NCX2, and NCX3, have been identified in mammals. Similarities between NCX1–3 include a homology of around 70% sequence identity at the protein level, the presence of an amino-terminal signal sequence, two sets of multiple transmembrane α-helices near the ends of the protein, and a large intracellular loop (Blaustein and Lederer 1999; Schulze et al. 2002). A functional comparison of the three isoforms of NCX stably expressed in baby hamster kidney (BHK) cells failed to detect striking differences (Linck et al. 1998).

The genes encoding NCX1, NCX2, and NCX3 have been mapped to mouse chromosomes 17, 7, and 12, respectively (Nicoll et al. 1996). At the posttranscriptional level, at least 12 NCX1 and 3 NCX3 proteins are generated through alternative splicing (Kofuji et al. 1994). These variants arise from a region of the large intracellular f loop, are encoded by six small exons defined A to F, and are used in different combinations in a tissue-specific manner (Lee et al. 1994; Nakasaki et al. 1993). All splice variants include either exon A or B (Quednau et al. 1997). Excitable tissues, such as those of the brain and heart, are usually characterized by the presence of exon A, whereas kidney, stomach, and skeletal muscle tissues comprise NCX with exon B (Quednau et al. 1997). Reilly and Shugrue identified the sequence of the rabbit kidney NCX1 (Reilly and Shugrue 1992) and in kidney the expression of this transporter is restricted to the distal part of the nephron where it is predominantly localized along the basolateral membrane (Biner et al. 2002; Hoenderop et al. 2000; Loffing et al. 2001). NCX1 is widely distributed in many different mammalian tissues, whereas NCX2 and NCX3 are only expressed in brain and skeletal muscle (Li et al. 1994; Nicoll et al. 1996). Bindels and coworkers suggested that in DCT and CNT NCX1 is the primary extrusion mechanism, whereas only a minor amount of Ca^{2+} is extruded by the plasma membrane Ca^{2+} pump (Bindels et al. 1992; Van Baal et al. 1996b). NCX1 is also expressed in the basolateral membrane of enterocytes (Hildmann et al. 1982; Kikuchi et al. 1988; Van Abel et al. 2003). In fish enterocytes, NCX appears to be the main mechanism by which Ca^{2+} is extruded from the cells at the basolateral surface, whereas in mammals PMCA is the predominant extrusion mechanism (Flik et al. 1990; Hildmann et al. 1982; Van Abel et al. 2003). Together, these studies suggest that in kidney basolateral Ca^{2+} efflux is mainly mediated by NCX1, whereas Na^+/Ca^{2+} exchange seems less important in the small intestine. Recently, it was demonstrated that NCX1 knockout mice do not have a spontaneously beating heart and die in utero (Koushik et al. 2001; Reuter et al. 2002). Unfortunately, this animal model is, therefore, not suitable to verify the importance of NCX1 in renal epithelial Ca^{2+} transport. In addition, K^+-dependent Na^+–Ca^{2+} exchangers (NCKX) could play a role in cellular Ca^{2+} efflux (Blaustein and Lederer 1999; Philipson and Nicoll 2000). Northern blot analysis demonstrated that some isoforms [i.e., NCKX4 (Li et al. 2002) and NCKX6 (Cai and Lytton 2004)] of this family are expressed in epithelia including small intestine and kidney. Therefore, it remains to be established whether the NCKX transporters play a role in epithelial Ca^{2+} transport or NCX1 is indeed the pivotal extrusion mechanism in renal Ca^{2+} reabsorption.

The stoichiometry of NCX is generally accepted to be three Na^+ ions per one Ca^{2+} ion. However, it has recently been demonstrated that ion flux ratio can vary from 1:1 to a maximum of 4:1, depending on the intracellular concentration of Na^+ and Ca^{2+} ions (Fujioka et al. 2000; Kang and Hilgemann 2004). This variation might be most relevant in excitable cells. In resting excitable cells, when $[Ca^{2+}]_i$ rises and the cells require the return of $[Ca^{2+}]_i$ to resting levels (Carafoli 1985), NCX-mediated transport couples the extrusion of Ca^{2+} to the influx of Na^+ ions into the cells down the electrochemical Na^+ gradient. This mode of operation, defined as forward mode (Blaustein and Santiago 1977), maintains the steep Ca^{2+} gradient across the cell membrane. However, when the transmembrane Na^+ electrochemical gradient is reduced, i.e., upon membrane depolarization, NCX can operate in reverse mode and mediates the extrusion of $[Na^+]_i$ and the influx of Ca^{2+} ions (Baker et al. 1969; DiPolo 1979). Although this latter feature is an important topic in the field of neuroscience and cardiac function, the role of NCX in epithelial Ca^{2+} transport is probably limited to Ca^{2+} efflux.

Plasma membrane Ca^{2+}-ATPase PMCAs are high-affinity Ca^{2+} efflux pumps present in virtually all eukaryotic cells, wherein they are responsible for the maintenance and resetting of the resting $[Ca^{2+}]_i$ levels (Blaustein et al. 2002). Four PMCA isoforms (PMCA1–4; human gene symbols ATP2B1–4) have been identified in mammalian tissues (Strehler and Zacharias 2001). In addition, alternative splicing of the transcripts of these genes yields a large variety of splice variants differing mainly in their carboxyl-terminal amino acid sequence (Stauffer et al. 1993; Strehler and Zacharias 2001). PMCA1 and PMCA4 are expressed in all tissues, suggesting that they are housekeeping genes, involved in the maintenance of cellular Ca^{2+} homeostasis (Greeb and Shull 1989; Stauffer et al. 1993). In contrast, the limited tissue distribution of PMCA2 and PMCA3 suggests that they have tissue-specific functions. The relevance for PMCAs as a system for the extrusion of Ca^{2+} in Ca^{2+}-transporting epithelial cells is variable between tissues. In the kidney, PMCA 1b, 2b, and 4b isoforms are present in all nephron segments (Caride et al. 1998; Magosci et al. 1992). Compared to other nephron segments, DCT shows the strongest immunocytochemical reactivity for PMCA protein expression and the highest Ca^{2+}-ATPase activity (Borke et al. 1989; Borke et al. 1987; Doucet and Katz 1982; Magosci et al. 1992). PMCA1b transcripts were demonstrated in rabbit CNT and collecting duct (CD), whereas expression of the PMCA2 isoform was not detected (Hoenderop et al. 2000; Kip and Strehler 2003). In addition, Strehler and coworkers demonstrated in Madin-Darby canine kidney (MDCK) cells that PMCA4b plays a significant role in basolateral Ca^{2+} extrusion (Kip and Strehler 2003, 2004). However, in renal epithelia, NCX mediates the majority (~70%) of the Ca^{2+} efflux in Ca^{2+}-transporting cells in DCT and CNT, whereas PMCA mediates the extrusion of the remaining 30% (Bindels et al. 1992). In addition, PMCA probably serves a general role in the maintenance of cellular Ca^{2+} homeostasis of all other nephron segments. Interestingly, in the small intestine PMCA1b is abundantly expressed, whereas NCX1 is expressed only at low levels. This suggests that PMCA1b is the principal Ca^{2+} extrusion mechanism in intestinal Ca^{2+} absorption (Van Abel et al. 2002; Wasserman and Fullmer 1995).

Recent studies of mice carrying PMCA1, PMCA2, or PMCA4 null mutations reveal the in vivo functions of these isoforms. Mice that are deficient in PMCA3 have not been generated yet. The PMCA1 gene was disrupted by removal of sequences encoding the catalytic phosphorylation site (Okunade et al. 2004). When heterozygous mutant mice were bred, only wild-type and heterozygous mutants were observed at birth, whereas null mutants showed early embryolethality. Heterozygous PMCA1 mutants exhibit no apparent disease phenotype. These observations suggest that PMCA1 serves a critical housekeeping function. Because PMCA1b is ubiquitous (Keeton et al. 1993; Stauffer et al. 1993), it is likely to be the major housekeeping form of the enzyme. Furthermore, the role of PMCA2 was demonstrated using two mouse strains [deafwaddler (Street et al. 1998) and wriggle mouse Sagami (Takahashi and Kitamura 1999)], carrying PMCA2 mutations, and in PMCA2 knockout mice (Kozel et al. 1998), all displaying phenotypes with deafness and having problems keeping their balance related to Ca^{2+} handling in the inner ear. PMCA4 null mutants survive and appear healthy (Okunade et al. 2004; Schuh et al. 2004). However, an in vitro apoptosis phenotype was identified in smooth muscle of isolated portal veins from PMCA4 knockout mice on a mixed 129SvJ and black Swiss background (Okunade et al. 2004), indicating that loss of PMCA4 could lead to Ca^{2+} overload and apoptotic cell death under some conditions, in particularly when the loss of PMCA4 was combined with the loss of a single copy of the PMCA1 gene. This suggests that PMCA1 is the major housekeeping isoform required for maintenance of $[Ca^{2+}]_i$ but that PMCA4 can contribute to this function in certain tissues. Moreover, PMCA4 plays an important role in testis, as male PMCA4 null mutant mice are infertile, because the sperm cells are unable to achieve hyperactivated motility (Okunade et al. 2004; Schuh et al. 2004).

In conclusion, ATP driven Ca^{2+} extrusion and Na^{+}–Ca^{2+} exchange both play a role in epithelial Ca^{2+} extrusion with PMCA1 and NCX1, respectively forming the molecular identity of this process.

Regulation of epithelial Ca^{2+} transport

1,25-dihydroxyvitamin D_3

The vitamin D_3 endocrine system is critical for the proper development and maintenance of the Ca^{2+} balance (Jones et al. 1998). There are two sources of vitamin D_3 in the body. It is either ingested with the diet or synthesized in the skin from its precursor 7-dehydrocholesterol in the presence of sunlight (Neer 1975). Vitamin D_3 itself is physiologically inactive. It will undergo a modification process, starting with 25-hydroxylation in the liver. Subsequently, the active form of vitamin D, 1,25-dihydroxyvitamin D_3 ($1,25(OH)_2D_3$), is synthesized in the PT by the renal cytochrome P450 enzyme 25-hydroxyvitamin D_3-1α-hydroxylase (1α-OHase) (Fraser and Kodicek 1970; Jones et al. 1998).

The biological effects of $1,25(OH)_2D_3$ on target organs are mediated by both genomic and rapid posttranscriptional mechanisms (Jones et al. 1998). $1,25(OH)_2D_3$ transcriptionally controls the expression of a particular set of target genes (Table 2). The genomic mechanism of action is similar to that of other steroid hormones and is mediated by stereospecific interaction of $1,25(OH)_2D_3$ with a nuclear VDR. Upon binding of $1,25(OH)_2D_3$, the VDR undergoes a conformational change and forms a complex with a retinoid X receptor (RXR). This VDR–RXR complex binds to DNA elements in the promoter regions of target genes described as vitamin D response elements (VDREs). Binding to these VDREs controls the rate of gene transcription. Importantly, VDR is expressed in epithelia that play a role in Ca^{2+} (re)absorption. The intestine and kidney are the main target organs for the calciotropic action of this hormone, although vitamin D affects many processes i.e., notably in the skin and immune system, but also directly in bone and parathyroid gland (DeLuca 2004). The rapid response of vitamin D presumably utilizes a VDR-independent signal transduction pathway that is probably linked to putative plasma membrane receptors for $1,25(OH)_2D_3$. The physiological relevance of rapid actions of vitamin D is not well understood.

A number of human diseases are associated with the role of active vitamin D. These disorders are often characterized by defective bone mineralization and clinical features of rickets, poor growth as an infant, and hypocalcemia and can have at least two distinct genetic causes (Jones et al. 1998; Van de Graaf et al. 2004). Vitamin D-dependent rickets type I (VDDR I) is due to an enzymatic defect in synthesis of the active form of vitamin D (Kitanaka et al. 1998; Panda et al. 2001). It has also been referred to as "pseudovitamin D-deficiency rickets" (Prader et al. 1961). VDDR I is an autosomal recessive disorder caused by mutations in the gene encoding 25-hydroxyvitamin D_3-1α-hydroxylase on chromosome 12q14 (Kitanaka et al. 1998). Patients with this disease show markedly decreased serum $1,25(OH)_2D_3$, hyperparathyroidism, and normal serum 25-hydroxyvitamin D_3. A similar disease is caused by end-organ unresponsiveness of active vitamin D due to recessive mutations in the gene encoding the vitamin D receptor (Hughes et al. 1988; Pike et al. 1984). Vitamin D-dependent rickets type II (VDDR II) is caused by a defect in the vitamin D receptor gene. This defect leads to an increase in the circulating ligand, $1,25(OH)_2D_3$. Furthermore, alopecia (hair loss) is often observed in these patients. Often mutations had been

Table 2 Regulation of the epithelial Ca^{2+} channels

Regulatory factor	TRPV5	TRPV6	Affected process	Reference(s)
Vitamin D	+	+	Transcription	Hoenderop et al. 2001a; Hoenderop et al. 2005; Van Cromphaut et al. 2001
PTH	+	=	Transcription	Van Abel et al. 2005
Estrogen	+	+	Transcription	Van Abel et al. 2002, 2003
Dietary Ca^{2+}	+[a]	+[a]	Transcription	Hoenderop et al. 2002; Van Abel et al. 2003
Tacrolimus	–	ND	Transcription	Nijenhuis et al. 2004
Acidosis	–	ND	Transcription/ channel activity/ trafficking	Nijenhuis et al. 2006; Vennekens et al. 2001b; Yeh et al. 2005
Klotho	+	+	Trafficking	Chang et al. 2005
S100A10/annexin 2	+	+	Trafficking	Van de Graaf et al. 2003
Rab11a	+	+	Trafficking	Van de Graaf et al. 2006a
FKBP52	–	ND	Not known	Gkika et al. 2006
Calmodulin	=	+	Channel activity	Lambers et al. 2004; Niemeyer et al. 2001
80K-H	+	ND	Channel activity	Gkika et al. 2004
$[Ca^{2+}]_i$	–	–	Channel activity	Hoenderop et al. 2001b
NHERF2/SGK1	+	ND	Not known	Embark et al. 2004
NHERF4	=	=	Not known	Van de Graaf et al. 2006c
BSPRY	+	ND	Not known	Van de Graaf et al. 2006d

[a] In vitamin D-depleted mice

"+" and "–" indicate stimulation or inhibition, respectively, of the indicated process including transcription, trafficking and channel activity of TRPV5/6. ND means not determined and "=" indicates no effect. TRPV5, transient receptor potential cation channel subfamily V member 5; TRPV6, transient receptor potential cation channel subfamily V member 6; PTH, parathyroid hormone; FKBP52, FK binding protein 52; $[Ca^{2+}]_i$, intracellular Ca^{2+} concentration; NHERF2, Na^+/H^+ exchanger regulatory factor 2; SGK1, serum and glucocorticoid-regulated kinase 1; NHERF4 Na^+/H^+, exchanger regulatory factor 4; BSPRY, B-box and SPRY domain containing protein. See text for explanation

found in the highly conserved amino-terminal DNA-binding domain of the VDR, a location that does not affect the ligand-binding properties of the receptor. This explains that the receptor can sometimes still bind vitamin D, while the calciotropic genomic consequences are absent (Malloy et al. 1997). Furthermore, vitamin D-dependent rickets type II can be associated with a normal vitamin D receptor cDNA sequence. The VDR suppressive effect in these patients was due to overexpression of a heterogeneous nuclear ribonucleoproteins (hnRNPs) that specifically interacted with a DNA response element known to bind retinoid X receptor-VDR heterodimers, interfering with the vitamin D receptor–DNA interaction (Chen et al. 2003).

Targeted deletion of genes encoding 1α-OHase (Dardenne et al. 2001; Panda et al. 2001) and of the nuclear VDR (Li et al. 2001; Takeyama et al. 1997; Van Cromphaut et al. 2001; Yoshizawa et al. 1997) have provided useful mice models of inherited human diseases of VDDR I and VDDR II. St. Arnaud and coworkers generated 1α-OHase knockout (1α-OHase$^{-/-}$) mice that represent a unique animal model for VDDR I since these mice display undetectable $1{,}25(OH)_2D_3$ concentrations, hypocalcemia, and secondary hyperparathyroidism. On a normal diet, 1α-OHase$^{-/-}$ mice have an average lifespan of approximately 12 weeks (Dardenne et al. 2001; Hoenderop et al. 2002). In addition, the 1α-OHase$^{-/-}$ mice developed distinct histological evidence of rickets and osteomalacia (Dardenne et al. 2001; Panda et al. 2001). Previous studies indicated that daily injections of $1,25(OH)_2D_3$ completely rescued these 1α-OHase$^{-/-}$ mice (Dardenne et al. 2003b). Bone histology and histomorphometry confirmed that the rickets and osteomalacia were cured by this $1,25(OH)_2D_3$ supplementation. Blood analysis further revealed that the rescue treatment corrected the hypocalcaemia and secondary hyperparathyroidism.

Interestingly, in this VDDR I mouse model, there was a positive correlative relationship between the expression level of TRPV5, calbindin-D_{28K}, and NCX1 proteins in kidney, TRPV6, calbindin-D_{9K}, and PMCA1b in duodenum, and the serum Ca^{2+} concentration (Hoenderop et al. 2002; Van Abel et al. 2002, 2003). Normalization of the serum Ca^{2+} concentration by $1,25(OH)_2D_3$ supplementation was associated with a restoration of the expression level of the Ca^{2+} transporters, confirming the essential role of these proteins in active $1,25(OH)_2D_3$-mediated Ca^{2+} (re)absorption. Analogous observations were made from experiments performed with VDR knockout mice (Van Cromphaut et al. 2001; Weber et al. 2001a). In these hypocalcemic mice, urinary Ca^{2+} excretion is inappropriately high, suggesting renal Ca^{2+} wasting due to disturbed Ca^{2+} reabsorption. Furthermore, it has been demonstrated in this mouse model that duodenal TRPV5 and TRPV6 levels are dramatically downregulated (Van Cromphaut et al. 2001; Weber et al. 2001a). Calbindin-D_{9K} expression was also downregulated, although to a lesser extent.

Finally, more recent evidence for a role of vitamin D in the positive regulation of the epithelial Ca^{2+} channel comes from two distinct double knockout models. First, it was shown that the increased vitamin D levels observed in TRPV5$^{-/-}$ mice are pivotal for the compensatory intestinal hyperabsorption seen in these mice. This was demonstrated using mice TRPV5$^{-/-}$/1α-OHase$^{-/-}$ double knockout mice (Renkema et al. 2005). Second, 1α-OHase$^{-/-}$ and PTH double knockout mice were created to eliminate a possible role of PTH during vitamin D administration. Administration of $1,25(OH)_2D_3$ upregulated mRNA and protein levels of the renal TRPV5, calbindin-D_{28K}, calbindin-D_{9K} and NCX1, increased serum Ca^{2+} concentration and stimulated bone formation (Xue et al. 2006).

The correlation between vitamin D and the expression level of the Ca^{2+} transport proteins has also been addressed in several cell models. Wood and coworkers observed the correlation between the $1,25(OH)_2D_3$-induced expression of TRPV6, calbindin-D_{9K}, and PMCA1b and transcellular Ca^{2+} transport in Caco2 cells, a model duodenal cell line (Fleet et al. 2002; Wood et al. 2001). Furthermore, in controlled tissue culture conditions using primary cultures from the distal part of the nephron including DCT and CNT, a direct relationship between $1,25(OH)_2D_3$-induced expression of Ca^{2+} transport proteins and transcellular Ca^{2+} transport was also shown (Bindels et al. 1991; Van Baal et al. 1996b). In contrast, Barley et al. could not confirm the generally observed vitamin D-dependent sensitivity of TRPV6 in duodenal biopsies from 20 normal volunteers. However, samples were taken from individuals that formed a very variable population of men and women of age 25–71 years (Barley et al. 2001). It is hypothesized that both vitamin D levels and the expression of Ca^{2+} transport proteins is age-dependent and possibly gender-sensitive, which could explain the tenfold variation between the lowest and the highest level of TRPV6 expression. This variation made it hard to disclose a relationship between TRPV6 expression and vitamin D metabolites in this study. TRPV5 and TRPV6 promoter analysis indicated that there are functional VDREs located upstream of the start codon (Hoenderop et al. 2001a; Meyer et al. 2006; Peng et al. 2000b; Weber et al. 2001a). Mutagenesis of the VDREs within the –2.1-kb and –4.3-kb region and the VDRE at –1.2 kb abrogated all response to $1,25(OH)_2D_3$ when examined within the TRPV6 promoter (Meyer et al. 2006). Taken together, vitamin D-deficient animal models and epithelial cell lines demonstrated a consistent

$1{,}25(OH)_2D_3$-sensitivity of TRPV5, TRPV6, and the calbindins and to a lesser extent the basolateral extrusion systems NCX1 and PMCA1b.

Parathyroid hormone

The parathyroid glands play a key role in maintaining the extracellular Ca^{2+} concentration ($[Ca^{2+}]_e$) through their secretion of PTH (Potts 2005). PTH controls the extracellular Ca^{2+} balance by activation of the PTH receptor, regulating concerted Ca^{2+} transport in bone, intestine, and kidney. Parathyroid cells sense decreases in $[Ca^{2+}]_e$ by means of the CaSR, to increase PTH secretion. Brown et al. identified CaSR by expression cloning in *Xenopus laevis* oocytes using parathyroid cell cDNA (Brown et al. 1993). CaSR comprises a large extracellular region, forming the Ca^{2+} binding site, followed by a seven-membrane-spanning domain typical for members of the G protein-coupled receptor superfamily. In addition to parathyroid tissue, the receptor was also expressed in regions of the kidney involved in regulated Ca^{2+} and Mg^{2+} reabsorption (Lee et al. 1996; Yang et al. 1997). PTH itself acts primarily on kidney and bone, where it activates the PTH receptor (Juppner et al. 1991; Mannstadt et al. 1999). This receptor can also bind the parathyroid-related protein (PTHrP). PTHrP was isolated from a human lung cancer cell line (Moseley et al. 1987). Although a separate gene encodes PTHrP, eight of the first 13 amino acids in the mature peptide are identical to those of PTH. Distinct pathophysiological manifestations are associated with changes in PTH levels in the blood. Garfield and Karaplis reviewed the various causes and clinical forms of hypoparathyroidism (Garfield and Karaplis 2001). Hypoparathyroidism is characterized by hypocalcemia and hyperphosphatemia, the first causing the most severe symptoms, including tetany (Potts 2005). In contrast, hyperparathyroidism is characterized by hypercalcemia and often severe bone loss (Potts 2005). Ca^{2+} handling by the kidney is also abnormal in individuals with hyperparathyroidism, who fail to show a normal hypercalciuric response to hypercalcemia. There are multiple factors that can lead to hyper- or hypoparathyroidism. Hypoparathyroidism manifests when insufficient PTH is secreted from the parathyroid glands to maintain normal $[Ca^{2+}]_e$ or less commonly when PTH is unable to function optimally in target tissues, despite adequate circulating levels. The latter is due to mutations in the PTH-receptor (Pearce et al. 1995, 1996; Pollak et al. 1993). Low plasma PTH levels are caused by trauma to the parathyroids during neck surgery or mutations of the PTH or CaSR (*CASR*) genes (Lovlie et al. 1996; Parkinson and Thakker 1992). Furthermore, the gene causing the X-linked recessive form of hypoparathyroidism remains to be identified, although Bowl et al. recently suggested a role for the transcription factor SOX3 in the development of this disease (Bowl et al. 2005). Hyperparathyroidism is observed in patients with chronic renal insufficiency. Furthermore, excessive PTH secretion can be due to parathyroid tumor. In addition, PTHrP is responsible for most cases of hypercalcemia of malignancy of other cells (Strewler 2000). Individuals with mutations in CaSR have an altered relation between PTH secretion and serum Ca^{2+} concentration. Mutations in CaSR can result in reduced PTH secretion or in hypersecretion of PTH, depending whether the mutations activate or inactivate the receptor, respectively.

Recent investigations on PTH focus also on its role as a therapy in osteoporosis (Hodsman et al. 2003; Potts 2005). Paradoxically, although hyperparathyroidism is associated with severe bone loss, administration of PTH restores bone mass and strength and reduces fracture incidence in the treatment of postmenopausal osteoporosis (Potts 2005). This discrepancy might lie in delicate timing differences. Although PTH ultimately activates osteoclasts, these cells express no PTH receptors. Instead, PTH is sensed by osteoblasts, which in turn increase osteoclast activity. It is hypothesized that transient elevations of PTH (due to injection) favor the osteoblasts anabolic action on bone, whereas chronically high PTH levels favor the catabolic osteoclast activity (Dobnig and Turner 1997).

In addition to the effects on bone, PTH stimulates the activity of 1α-OHase in proximal tubules (Fraser and Kodicek 1973). Thereby, PTH increases the $1,25(OH)_2D_3$-dependent (re)absorption of Ca^{2+}. In addition, activation of the PTH/PTHrP receptor directly enhances the Ca^{2+} (re)absorption in kidney and intestine. Immunohistochemical analysis of rat duodenal sections showed localization of the PTH/PTHrP receptor in epithelial cells along the villus (Gentili et al. 2003). Interestingly, the receptor is absent in goblets cells. The first indication of a direct effect of PTH on the intestine was accomplished by a perfusion experiment of isolated duodenal loops, showing increased Ca^{2+} transport with addition of PTH (Nemere and Norman 1986; Nemere and Szego 1981). These findings were confirmed by Picotto and coworkers who demonstrated that PTH directly stimulates enterocyte Ca^{2+} influx (Picotto et al. 1997). Several groups localized PTH/PTHrP receptor mRNA in rat kidney to glomerular podocytes, PCT, PST, cortical segment of the TAL (cTAL), and DCT, but the receptor was not detected in the thin limb of Henle's loop or in CD (Lee et al. 1996; Yang et al. 1997). PTH directly stimulates active Ca^{2+} reabsorption in the distal part of the nephron (Greger et al. 1978). In TAL, it was shown that PTH increases the transepithelial driving force for Ca^{2+} reabsorption, enhancing paracellular Ca^{2+} transport (Wittner et al. 1993). Various mechanisms of PTH action have been proposed for the effect in DCT, including membrane insertion of apical Ca^{2+} channels (Bacskai and Friedman 1990), opening of basolateral chloride channels resulting in cellular hyperpolarization (Friedman and Gesek 1994), and modulation of PMCA activity (Tsukamoto et al. 1992). Van Abel et al. recently reported that PTH stimulates renal Ca^{2+} reabsorption through the coordinated expression of renal transcellular Ca^{2+} transport proteins. They showed that parathyroidectomy in rats resulted in decreased serum PTH levels and hypocalcemia, which was accompanied by reduced levels of TRPV5, calbindin-D_{28K}, and NCX1 (Van Abel et al. 2005). Supplementation with PTH restored serum Ca^{2+} concentrations and abundance of the Ca^{2+} transport proteins (Table 2). Similarly, infusion of a calcimimetic compound (chemical that activates CaSR at low serum Ca^{2+} concentrations) decreased PTH levels, resulted in reduced expression of TRPV5, calbindin, and NCX1, which is consistent with diminished Ca^{2+} reabsorption, and in line with the observed hypocalcemia in these mice (Van Abel et al. 2005). Importantly, serum $1,25(OH)_2D_3$ levels and renal VDR or CaSR mRNA abundance did not significantly change during these treatments (Van Abel et al. 2005). Furthermore, PTH injection in mice increased both TRPV5 and TRPV6 mRNA expression in kidney (Okano et al. 2004). This demonstrates the important role for PTH in epithelial Ca^{2+} transport.

Estrogen

Previous studies indicated that estrogen affects Ca^{2+} handling by kidney, intestine, and bone. Estrogen deficiency results in a negative Ca^{2+} balance and has been strongly associated with bone loss in postmenopausal women (Nordin et al. 1979, 1991; Prince et al. 1995; Young et al. 1968; Young and Nordin 1967). In addition, estrogen plays an essential role in bone handling in men (Carani et al. 1997; Lorentzon et al. 2006; Smith et al. 1994). Aromatase is the key enzyme in the conversion of testosterone to estradiol. A malfunctioning aromatase enzyme has been shown to impair the normal development of the (male) skeleton (Carani et al. 1997). Furthermore, mutations in the estrogen receptor gave rise to a similar phenotype, including low bone mineral density and long stature. The latter phenotype points to a role of estrogen in the determination of bone size (Lorentzon et al. 2006).

It has been generally described that the rise in serum and urine Ca^{2+} upon estrogen deficiency are secondary to an increase in bone resorption. However, there is increasing evidence that, besides bone, the intestine and kidney are also sites for estrogen action on Ca^{2+} handling and regulation. Menopausal estrogen deficiency in humans is associated with reduced duodenal Ca^{2+} absorption (Heaney et al. 1989), whereas estrogen replacement therapy helps prevent bone loss in postmenopausal women and corrects a decline in Ca^{2+} absorption efficiency at the onset of menopause (Gennari et al. 1990). However, the mechanism by which 17β-estradiol (17βE_2) stimulates Ca^{2+} absorption could be direct via estrogen receptors (ERα and ERβ) or indirect via increasing 1,25$(OH)_2D_3$ or the VDR. Several studies were carried out to examine the mechanism of action of estrogen on intestinal Ca^{2+} absorption. Ten Bolscher et al. treated ovariectomized rats with estradiol or 1,25$(OH)_2D_3$ and measured intestinal Ca^{2+} absorption in vivo using single pass perfusion of the duodenum. A pharmacological dose of estradiol caused a significant increase in intestinal absorption of Ca^{2+} (Ten Bolscher et al. 1999). This estrogen-induced rise in intestinal Ca^{2+} absorption was completely blocked by an ER antagonist, whereas this antagonist did not block vitamin D-enhanced intestinal Ca^{2+} absorption (Ten Bolscher et al. 1999). This suggests a direct effect of estrogen on duodenal Ca^{2+} absorption. In contrast, Cotter et al. did not observe an increase in Ca^{2+} uptake in Caco2 cells upon estrogen treatment (Cotter and Cashman 2006). Further evidence for a direct role of estrogen in kidney and intestine was recently provided by Van Abel et al. and Van Cromphaut et al. Van Abel et al. demonstrated that estrogen regulates the expression of TRPV5 in kidney in a 1,25$(OH)_2D_3$-independent manner (Table 2). Estrogen replacement in ovariectomized rats resulted in significant increased renal mRNA levels of TRPV5, calbindin-D_{28K}, NCX1, and PMCA1b and increased the protein abundance of TRPV5 (Van Abel et al. 2002). Furthermore, 17βE_2 upregulated TRPV5 mRNA and protein expression in 1α-OHase$^{-/-}$ mice, demonstrating a vitamin D-independent regulation by estrogen (Van Abel et al. 2002). Moreover, 17βE_2 treatment partially restored serum Ca^{2+} levels in these hypocalcemic mice, suggesting that 17βE_2 is directly involved in renal Ca^{2+} reabsorption via the upregulation of TRPV5 and possibly other Ca^{2+} transport proteins. Van Cromphaut et al. corroborated the direct role of estrogen on transepithelial Ca^{2+} transport. They showed that duodenal TRPV6 expression was reduced in ERα knockout mice and induced by estrogen treatment, pregnancy, or lactation (Van Cromphaut et al. 2003). These latter effects occurred both in VDR knockout and wild-type mice. Therefore, estrogens or hormonal changes during pregnancy or lactation have distinct, vitamin D-independent effects at the genomic level on active duodenal Ca^{2+} absorption mechanisms, mainly through a major upregulation of the Ca^{2+} influx channel TRPV6. The expression of TRPV6 was not altered in ERβ knockout mice, suggesting that the estrogen effects on duodenum are mediated by ERα (Van Cromphaut et al. 2003).

Dietary Ca^{2+} intake

Diets containing high amounts of Ca^{2+} have been implicated in the reduction of risk in osteoporosis (Greer and Krebs 2006). Obesity (Dixon et al. 2005) and hypertension (Karppanen et al. 2005) are some less well-known areas in which increasing dietary Ca^{2+} has a positive outcome. Even in cases of kidney stone formation, restricted Ca^{2+} intake is generally not advised (Borghi et al. 2002; Martini and Wood 2000; Moe 2006; Straub and Hautmann 2005), illustrating the importance of adequate Ca^{2+} intake. Recommended daily intake of Ca^{2+} is 1,000 mg/day for adults and 1,300 mg/day for adolescents, although both groups often do not reach these values, which could have negative consequences on bone density (Greer and Krebs 2006). The power of Ca^{2+} supplementation is best illustrated by the use of VDR and 1α-OHase knockout models. The bone phenotype of VDR-ablated mice can be completely rescued by feeding the animals a high Ca^{2+}, high phosphorus, high lactose diet (Van Cromphaut et al. 2001). Similarly, healing of rickets was demonstrated in a patient with vitamin D resistance by long-term nocturnal Ca^{2+} infusions (Balsan et al. 1986). In addition, the VDDR I phenotype of mice deficient for the 1α-OHase gene has been rescued by feeding them with a high Ca^{2+} diet. Dietary Ca^{2+} normalized the hypocalcemia, secondary hyperparathyroidism, and biomechanical properties of the bone tissue (Dardenne et al. 2003a; Hoenderop et al. 2002, 2004). Other studies indicated, however, that exogenous Ca^{2+} may not entirely compensate for $1,25(OH)_2D_3$ deficiency in mice and piglets (Goltzman et al. 2004; Schlumbohm and Harmeyer 2004).

To investigate the mechanism(s) underlying the effects of dietary Ca^{2+}, the expression level of several Ca^{2+} transport proteins was studied in various mice models. Importantly, the reduced expression level of renal TRPV5, calbindin-D_{28K}, and NCX1 in 1α-$OHase^{-/-}$ mice was restored by high dietary Ca^{2+} intake and accompanied by normalization of the serum Ca^{2+} concentration (Hoenderop et al. 2002). Likewise, the expression of the intestinal Ca^{2+} transport proteins, TRPV6, calbindin-D_{9K}, and PMCA1b, was normalized by this rescuing Ca^{2+} diet (Van Abel et al. 2003). Comparable observations were made in VDR knockout mice, where dietary Ca^{2+} upregulated duodenal TRPV5 and TRPV6 mRNA levels (Van Cromphaut et al. 2001). These findings suggest that dietary Ca^{2+} can affect active Ca^{2+} (re)absorption via vitamin D-independent modulation of the expression of Ca^{2+} transport proteins. However, the molecular mechanism of this vitamin D-independent Ca^{2+}-sensitive pathway remains to be further elucidated.

Diuretics

Thiazide diuretics are commonly used in the treatment of patients with hypertension. These diuretics enhance renal Na^+ excretion through inhibition of the Na^+/Cl^- cotransporter (NCC) present in the apical membrane of DCT cells (Gamba et al. 1993). Importantly, these diuretics have, in contrast to loop diuretics, the unique characteristic of decreasing Na^+ reabsorption, while increasing Ca^{2+} reabsorption resulting in hypocalciuria (Costanzo et al. 2000; Lamberg and Kuhlback 1959; Seitz and Jaworski 1964). This hypocalciuric effect provides therapeutic opportunities, for instance, in idiopathic hypercalciuria and nephrolithiasis. Furthermore, long-term effects of thiazides include increased bone mineral density and decreased fracture risk (Ray et al. 1989). Mutations in the gene encoding NCC were shown to cause Gitelman's syndrome, a recessive disorder with a phenotype resembling chronic thiazide administration including hypocalciuria (Ellison 2000; Gitelman et al. 1966; Ray et al. 1989; Reilly and Ellison 2000; Simon et al. 1996). Gitelman's syndrome is characterized by hypernatriuria and hypocalciuria, paralleled by an increase in bone mineral density. These symptoms are also present in NCC knockout mice, which therefore form a suitable mouse model for Gitelman's syndrome (Loffing et al. 2004; Schultheis et al. 1998).

However, the exact molecular mechanism responsible for thiazide-induced hypocalciuria has been debated strongly. The hypocalciuric effect was suggested to result from direct stimulation TRPV5-mediated Ca^{2+} entry in transcellular Ca^{2+} transport in the DCT (Friedman 1998; Reilly and Ellison 2000). Alternatively, hypocalciuria was proposed to result from enhancement of passive paracellular Ca^{2+} reabsorption in the PT secondary to extracellular volume (ECV) contraction (Biner et al. 2002; Friedman 1998; Friedman and Bushinsky 1999; Loffing et al. 2001), distinct from any effect on transcellular Ca^{2+} transport (Wilson and Freis 1959). Recent evidence strongly supports the latter model. First, it was reported that thiazide-induced hypocalciuria occurs in spite of reduced renal expression of

Ca^{2+} transport proteins in rat (Nijenhuis et al. 2003a). In addition, Nijenhuis et al. showed that the thiazide-induced hypocalciuria was accompanied by a significant decrease in body weight compared to controls, illustrating that extra cellular volume contraction occurred. Furthermore, it was shown that ECV contraction mimics the hypocalciuria, and volume repletion completely reverses thiazide-induced hypocalciuria in these rats (Nijenhuis et al. 2005). Recent micropuncture experiments demonstrated that reabsorption of Na^+ and, importantly, Ca^{2+} in the PT is increased during chronic hydrochlorothiazide (HCTZ) treatment, whereas Ca^{2+} reabsorption in DCT and CNT appeared unaffected (Nijenhuis et al. 2005). Importantly, chronic HCTZ administration still induces hypocalciuria in TRPV5$^{-/-}$ mice, in which active Ca^{2+} reabsorption is abolished. HCTZ did not affect renal expression of the proteins involved in active Ca^{2+} transport, including TRPV5 mRNA and protein expression in wild-type mice (Nijenhuis et al. 2005). Lee et al. confirmed that thiazide treatment in mice does not affect renal TRPV5 expression, except when thiazide treatment is combined with salt repletion (Lee et al. 2004). However, salt repletion alone induced TRPV5 mRNA expression to a similar extent, suggesting that this effect is not thiazide-specific. Loffing and coworkers recently demonstrated that renal TRPV5 expression is unaffected in NCC knockout mice (Loffing et al. 2004). In accordance, micropuncture experiments in these mice showed that active Ca^{2+} reabsorption is unaltered in DCT and CNT, and indicated increased fractional absorption of both Na^+ and Ca^{2+} upstream of DCT (Loffing et al. 2004). In conclusion, these studies demonstrated that chronic thiazide treatment induces hypovolemia that subsequently stimulates proximal Na^+ and Ca^{2+} reabsorption. The latter explains the Ca^{2+}-sparing during thiazide treatment and Gitelman's syndrome.

Acidosis and alkalosis

Acid-base homeostasis is known to affect renal Ca^{2+} handling (Canzanello et al. 1990; Sutton et al. 1979). For instance chronic metabolic acidosis is associated with increased renal Ca^{2+} excretion. Long-standing metabolic acidosis can lead to Ca^{2+} loss from bone and ultimately results in metabolic bone disorders, including osteomalacia and osteoporosis (Lemann et al. 2003). In contrast, chronic metabolic alkalosis is known to decrease urine Ca^{2+} excretion (Bushinsky et al. 1989; Sutton et al. 1979). However, the molecular mechanisms that explain the altered renal divalent excretion during these disturbances of acid–base balance remain unknown.

It has been shown by several groups that extracellular protons inhibit TRPV5 channel activity (Peng et al. 2000a; Vennekens et al. 2001b; Yeh et al. 2005; Yeh et al. 2003). Furthermore, earlier studies, including micropuncture experiments, suggested that systemic acid–base disturbances specifically affect Ca^{2+} reabsorption in DCT and CNT (Sutton et al. 1979; Wong et al. 1986). It has, therefore, been suggested that acidification of the DCT and CNT luminal fluid during chronic metabolic acidosis and subsequent inhibition of TRPV5 explains the decreased Ca^{2+} reabsorption in vivo (Vennekens et al. 2001b; Yeh et al. 2003). Nijenhuis et al. recently addressed the mechanism(s) underlying acid–base balance on renal Ca^{2+} handling in more detail. Metabolic alkalosis was induced by oral $NaHCO_3$ loading and metabolic acidosis by NH_4Cl loading, as well as by acetazolamide administration in wild-type and TRPV5$^{-/-}$ mice (Nijenhuis et al. 2006). Acetazolamide specifically inhibits proximal tubular bicarbonate reabsorption, resulting in a self-limiting metabolic acidosis with, in contrast to NH_4Cl loading, an alkaline urine pH (Dirks et al. 1966; Soleimani 2002; Soleimani and Aronson 1989). This further enabled evaluation of the role of luminal pH during acidosis.

Chronic metabolic acidosis that was induced by NH_4Cl loading enhanced Ca^{2+} excretion and decreased the expression of the epithelial Ca^{2+} channel TRPV5 and calbindin-D_{28K} in wild-type mice. Importantly, although 0.14 M NH_4Cl administration induced a similar metabolic acidosis in $TRPV5^{-/-}$ mice compared with wild-type mice, it did not further increase Ca^{2+} excretion in these knockout mice. These results point to a primary renal Ca^{2+} leak, in contrast to increased Ca^{2+} mobilization from bone as a previously suggested explanation for the Ca^{2+} wasting (Krieger et al. 2004; Lemann et al. 2003). Furthermore, this indicates that, besides a direct effect on TRPV5 activity, downregulation of Ca^{2+} transport proteins that are present in DCT and CNT could be an explanation for the observed acidosis-induced hypercalciuria. Importantly, acetazolamide-induced acidosis also downregulated the expression of the Ca^{2+} transport proteins.

Acetazolamide treatment induces metabolic acidosis by diminishing the proximal tubular bicarbonate-reabsorptive capacity (Dirks et al. 1966; Soleimani and Aronson 1989). This results in an increased luminal pH at more distal nephron segments, including DCT and CNT, which is reflected by urinary alkalinization in contrast to NH_4Cl loading-induced urine acidification. Despite the alkaline pH at the site of TRPV5 expression, this treatment resulted in a significant hypercalciuria. Therefore, luminal pH in DCT and CNT does not seem to be a dominant factor in the long-term in vivo hypercalciuric effect of chronic metabolic acidosis. Instead, acidosis-induced hypercalciuria is in accordance with downregulation of Ca^{2+} transport proteins, including TRPV5. The mechanism underlying the Ca^{2+}-sparing action of chronic metabolic alkalosis is not simply the reverse of this effect. Chronic metabolic alkalosis increased renal expression of the Ca^{2+} transport proteins in wild-type mice. However, chronic $NaHCO_3$ administration induced metabolic alkalosis in wild-type as well as $TRPV5^{-/-}$ mice (Nijenhuis et al. 2006). These findings suggest that upregulation of Ca^{2+} transport proteins in DCT and CNT is not the sole explanation for the alkalosis-induced hypocalciuria. Li^+ clearance studies suggested that $NaHCO_3$-treated mice show increased proximal tubular Na^+ reabsorption and therefore possibly display enhanced passive Ca^{2+} reabsorption compared with NaCl-treated controls. Therefore, the Ca^{2+} sparing effect during alkalosis could be explained by increased passive Ca^{2+} reabsorption independent of DCT and CNT.

Klotho: novel insight in hormonal regulation of Ca^{2+} reabsorption

In 1997 Kuro-o et al. described a transgenic mouse with several age-related disorders caused by the single insertion of a transgene. The affected gene was named klotho, for one of the Fates, the Greek goddess who spins the thread of life (Kuro-o et al. 1997). Mice homozygous for the affected klotho gene (effectively klotho knockout mice) show a phenotype resembling those in patients with premature-aging syndromes: arteriosclerosis, osteoporosis, age-related skin changes, and ectopic calcifications, together with short lifespan and infertility (Kuro-o et al. 1997). Re-introduction of the klotho gene normalized the phenotype. Furthermore, overexpression of klotho in mice resulted in a significant extension of lifespan and a suppression of phenotypes associated with aging (Kurosu et al. 2005). In humans, allelic variations are related to life expectancy and coronary artery disease (Arking et al. 2005; Arking et al. 2002, 2003). The klotho gene encodes a single-pass transmembrane protein of 1,014 amino acids with a putative signal sequence at the amino-terminus and a single transmembrane helix near the carboxyl-terminus. Klotho is secreted and activated by cleavage of the amino-terminal extracellular domain, and this secreted form of klotho exhibits β-glucuronidase activity (Imura et al. 2004; Tohyama et al. 2004).

Several observations connect klotho to a role in Ca^{2+} metabolism. First, klotho-deficient mice have a slight hypercalcemia that was associated with high levels of $1,25(OH)_2D_3$, caused by increased expression of renal 1α-hydroxylase (Yoshida et al. 2002). Furthermore, administration of $1,25(OH)_2D_3$ induces klotho expression in the kidney (Tsujikawa et al. 2003). Second, klotho$^{-/-}$ mice show bone abnormalities including an approximately 20% lower bone mineral density than control mice (Kuro-o et al. 1997). In humans, allelic variants of klotho are associated with osteoporosis, confirming this phenotype (Kawano et al. 2002; Ogata et al. 2002). Third, klotho is strongly expressed in DCT of the kidney and the parathyroid gland, further supporting a role in epithelial Ca^{2+} handling (Chang et al. 2005; Kuro-o et al. 1997). Fourth, Chang et al. recently demonstrated a novel mechanism employed by klotho to directly stimulate active Ca^{2+} reabsorption. They demonstrated that incubation of TRPV5-expressing HEK293 cells with preconditioned culture medium from klotho-expressing cells resulted in strongly increased TRPV5-mediated Ca^{2+} uptake (Chang et al. 2005). This effect was mimicked by β-glucuronidase indicating that the enzymatic activity of klotho is responsible for the increased TRPV5 activity. Mutation of the conserved N-glycosylation site of TRPV5 (N358Q) abolished both klotho- and β-glucuronidase-mediated activation of TRPV5, indicating that klotho may work by affecting the extracellular glycosylation status of the channel (Chang et al. 2005). Membrane protein biotinylation indicated a significant increase in plasma membrane localization of TRPV5 after klotho or β-glucuronidase stimulation. Together, these indicate that klotho traps the channels in the plasma membrane, thereby increasing TRPV5-mediated Ca^{2+} influx activity. Interestingly, disruption of the klotho gene in mice is also associated with hyperphosphatemia (Kuro-o et al. 1997). It has been reported that klotho increases the affinity of fibroblast growth factor 23 (FGF-23) binding to its receptor in the proximal tubule (Razzaque et al. 2006). Knockout of the FGF-23 gene or overexpression of FGF-23 in mice resulted in significant alteration of phosphate (P_i) transport (Razzaque et al. 2006). In humans, fibroblast growth factor 23 mutations are responsible for hypophosphatemic rickets (Jonsson et al. 2003). This suggests that klotho is also a novel phosphaturic factor, possibly forming a hormonal link between Ca^{2+} and P_i homeostasis.

Concerted regulation of epithelial Ca^{2+} transport proteins

An increasing number of studies combine the investigation of multiple Ca^{2+} transport proteins under various (patho)physiological and pharmacological circumstances. Remarkably, virtually all of the results point to a concerted regulation of these Ca^{2+} transport proteins including TRPV5, TRPV6, NCX1, PMCA, and the calbindins. The concomitant regulation of the renal Ca^{2+} transport proteins was observed in studies exploring the regulatory role of $1,25(OH)_2D_3$, estrogens, PTH, and dietary Ca^{2+} (Hoenderop et al. 2002; Van Abel et al. 2002, 2003, 2005). Furthermore, calbindin-D_{28K} and NCX1 showed downregulation in kidneys of TRPV5$^{-/-}$ mice despite elevated levels of $1,25(OH)_2D_3$ (Hoenderop et al. 2003a). This suggests that TRPV5 is primarily involved in the regulation of the Ca^{2+} transport proteins expression in kidney independent of $1,25(OH)_2D_3$ (Van Abel et al. 2005). Furthermore, chronic metabolic alkalosis enhances calbindin-D_{28K} abundance in wild-type mice, but not in TRPV5$^{-/-}$ mice, in line with TRPV5-dependent regulation of this Ca^{2+}-transporting protein (Nijenhuis et al. 2006). An interesting question is how the expression of TRPV5 specifically coordinates the Ca^{2+} transport machinery. Because TRPV5 is the gatekeeper controlling apical Ca^{2+} influx in the kidney, it was postulated that the magnitude of Ca^{2+} influx through TRPV5 determines the expression of the other Ca^{2+} transport proteins (Lambers et al. 2006a; Van Abel et al. 2005). This hypothesis was recently investigated using primary cultures of rabbit CNT and CCD. Long-term exposure to PTH stimulated transepithelial Ca^{2+} transport in these epithelial cells and concomitantly elevated the expression of TRPV5, calbindin-D_{28K}, and NCX1 (Van Abel et al. 2005). Inhibition of TRPV5 channel activity by ruthenium red eliminated this PTH-stimulated transepithelial Ca^{2+} transport, which was accompanied by a reduction in NCX1 and calbindin-D_{28K} expression (Van Abel et al. 2005). These findings support the hypothesis that the magnitude of the Ca^{2+} influx through TRPV5 modulates the expression of the other proteins that are required for transepithelial Ca^{2+} transport. An important question that remains to be answered is how the flux of Ca^{2+} through TRPV5 adjusts the expression of the other Ca^{2+} transport proteins. It is possible that Ca^{2+}-sensitive transcription factors/promoters play a role in this process (Ashby and Tepikin 2002). A 40-bp Ca^{2+}-responsive element has been identified in the promoter sequence of calbindin-D_{28K} that partly underlies the Purkinje cell-specific expression of calbindin-D_{28K} (Arnold and Heintz 1997). This element is also present in the calmodulin II promoter. However, future studies are needed to test whether this element is active in the kidney and/or additional intracellular signaling molecules are involved.

TRPV5/6 regulatory proteins

Recently, several proteins have been identified that associate with TRPV5, TRPV6, or both. These include calmodulin (CaM), S100A10-annexin 2, Rab11a, 80K-H, NHERF2, NHERF4, FKBP52, and BSPRY (Van de Graaf et al. 2006b). The identification of these TRPV5/TRPV6-binding proteins has significantly improved our knowledge of the molecular pathways modulating epithelial Ca^{2+} transport, as will be discussed below.

Calmodulin CaM is a ubiquitous protein encoded by three separate genes, all resulting in the same protein (Means and Dedman 1980). CaM consists of four Ca^{2+}-binding structures, localized in the amino- and carboxyl-terminus. Ca^{2+} binding to CaM is highly cooperative with Ca^{2+} binding first to the carboxyl-terminal EF-hands, which have the highest affinity for Ca^{2+}, followed by Ca^{2+} binding to lower affinity sites located in the amino-terminus (Wang 1985).

CaM is involved in a plethora of processes, many related to Ca^{2+}-related signaling. The protein is well-known to be involved in Ca^{2+}-dependent feedback regulation of several ion channels (Levitan 1999), including multiple TRP channels (Harteneck 2003). The first connection between CaM and the canonical TRPs was established upon the expression cloning of the second TRPC (TRP-Like) from *Drosophila* photoreceptors, for which CaM was used as the probe (Phillips et al. 1992). Upon Ca^{2+} influx, CaM inactivates the TRPL-mediated currents (Scott et al. 1997). Subsequently, it has been shown that many members of the TRPC family bind to and/or are regulated by CaM (Abeele et al. 2003; Boulay 2002; Ordaz et al. 2005; Scott et al. 1997; Shi et al. 2004; Singh et al. 2002; Tang et al. 2001; Trost et al. 2001; Yildirim et al. 2003; Zhang et al. 2001).

Importantly, CaM also binds to several TRPV channels, including TRPV6. Niemeyer and coworkers initially demonstrated that TRPV6 interacts in a Ca^{2+}-dependent manner with CaM (Niemeyer et al. 2001). They showed that CaM binding to a 21-amino-acid sequence in the carboxyl-terminus of TRPV6 is competitively regulated by protein kinase C (PKC) phosphorylation of a threonine residue. This phosphorylated residue is not conserved and absent from mouse TRPV6. However, Hirnet et al. demonstrated that mouse TRPV6 protein is also capable of Ca^{2+}-dependent CaM binding, using a synthetic peptide encompassing the stretch of amino acid residues of the mouse protein corresponding to the CaM binding site within the human TRPV6 protein (Hirnet et al. 2003). The apparent dissociation constant of CaM with the carboxyl-terminal peptide of mouse TRPV6 (43 nM) corresponds to the value obtained for human TRPV6 (65 nM) (Hirnet et al. 2003). Furthermore, Lambers et al. showed that CaM associates in a Ca^{2+}-dependent manner with specific regions in TRPV5 and TRPV6.

Two separate approaches demonstrate a functional role for CaM in the regulation of TRPV6 (Lambers et al. 2004; Niemeyer et al. 2001). First, removal of the CaM binding-site in the carboxyl-terminus of TRPV6 resulted in a significant reduction of the slow component of channel inactivation, revealing a role of CaM in TRPV6 regulation (Niemeyer et al. 2001). Second, HEK293 cells heterologously coexpressing Ca^{2+}-insensitive CaM mutants along with TRPV5 or TRPV6 showed a significantly reduced Ca^{2+} current through TRPV6. Remarkably, no functional effect was demonstrated on TRPV5 channel activity (Lambers et al. 2004), even though TRPV5 contains CaM-binding sites at similar locations. This functional effect on TRPV6 was localized to the high Ca^{2+}-affinity EF-hand structures of CaM.

These data demonstrated a regulatory role of CaM in TRPV6-mediated Ca^{2+} influx. It remains to be established whether CaM functions as a general Ca^{2+} sensor in TRPV5 and TRPV6 channels or, alternatively, can explain the differences in Ca^{2+}-dependent inactivation between the epithelial Ca^{2+} channels.

80K-H Gkika et al. identified 80K-H in a microarray screen designed to identify proteins that respond similarly to vitamin D and/or altered dietary Ca^{2+} intake as TRPV5 (Gkika et al. 2004). 80K-H was originally cloned as a PKC substrate of 80 kDa (Sakai et al. 1989) and was further shown to form a component of a cytosolic signal transduction complex (Goh et al. 1996; Kanai et al. 1997), a receptor for advanced glycation end products (Li et al. 1996) and the β-subunit of endoplasmic reticulum glucosidase II (Trombetta et al. 1996, 2001). In addition, 80K-H has been implicated in the insulin-stimulated translocation of the glucose transporter 4 (GLUT4)-containing vesicles to the plasma membrane cells. This effect is mediated by the interaction of 80K-H with a complex of PKCζ and Munc18c in an insulin-dependent manner (Hodgkinson et al. 2005). Importantly, mutations in 80K-H were suggested as the probable cause of polycystic liver disease (Drenth et al. 2003, 2004; Peces et al. 2005), a dominantly inherited condition characterized by the presence of multiple liver cysts of biliary epithelial origin. This is the reason that 80K-H is also referred to as hepatocystin (Drenth et al. 2004).

80K-H contains two putative EF-hand structures, a highly negatively charged glutamate stretch, and a putative ER-targeting signal (His–Asp–Glu–Leu). Using glutathione S-transferase (GST) pull-down assays and coimmunoprecipitations, a physical interaction between 80K-H and TRPV5 was demonstrated (Gkika et al. 2004). Furthermore, both proteins colocalized in a subset of tubular segments in the kidney indicating that regulation of TRPV5 by 80K-H could occur in vivo. Furthermore, similar transcriptional regulation of both proteins by $1,25(OH)_2D_3$ and dietary Ca^{2+} was shown (Gkika et al. 2004). Electrophysiological studies using 80K-H mutants demonstrated that three domains of 80K-H (the two EF-hand structures, the glutamate stretch, and the His–Asp–Glu–Leu sequence) are critical determinants of TRPV5 activity (Gkika et al. 2004). The Ca^{2+} binding properties of 80K-H are abolished upon inactivation of its two EF-hand structures. Importantly, this modification of the EF-hand pair in 80K-H also reduces the TRPV5-mediated Ca^{2+} current and increased the TRPV5 sensitivity to intracellular Ca^{2+}, accelerating the feedback inhibition of the channel (Gkika et al. 2004). Therefore, it is hypothesized that 80K-H acts a Ca^{2+} sensor to regulate TRPV5 activity at the plasma membrane.

B-box and SPRY-domain containing protein Recently, we identified BSPRY (B-box and SPRY-domain containing protein) as a novel factor involved in the control of TRPV5 activity (Van de Graaf et al. 2006d). BSPRY contains a B-box and SPRY domain, whose tentative functions are protein–protein interaction modules (Borden 1998; Ponting et al. 1997). RT-PCR and Northern blot analysis showed expression of this novel protein in several tissues including kidney, small intestine, prostate, lung, and uterus in mice. BSPRY was less abundantly expressed in heart, whereas skeletal muscle and liver were negative (Van de Graaf et al. 2006d). Rat BSPRY (also called zetin 1) has a shorter amino-terminus compared to mouse and human BSPRY and is ubiquitously expressed in a variety of tissues, with highest expression being found in testis. In adult brain, high levels of BSPRY mRNA were observed in the hippocampus, cerebral cortex, and piriform cortex (Birkenfeld et al. 2003). Using an antibody directed against a conserved peptide in the carboxyl-terminus of BSPRY, the presence of BSPRY was demonstrated along the apical domain of all TRPV5-immunopositive tubules in mouse kidney (Van de Graaf et al. 2006d). These tubules were previously identified as the second part of DCT2 and CNT (Hoenderop et al. 2000). Furthermore, expression of BSPRY in Madin-Darby canine kidney cells stably expressing TRPV5 resulted in a significant reduction of the Ca^{2+} influx without affecting channel cell surface abundance (Van de Graaf et al. 2006d).

As described in "1,25-dihydroxyvitamin D_3" above, TRPV5 expression is strongly regulated by $1,25(OH)_2D_3$. Therefore, the role of vitamin D on the abundance of BSPRY was assessed in wild-type and 1α-OHase$^{-/-}$ mice, which are unable to synthesize $1,25(OH)_2D_3$. Quantitative real-time PCR and computerized analysis of the immunohistochemical BSPRY staining in kidney showed significantly enhanced BSPRY mRNA expression in the 1α-OHase$^{-/-}$ mice compared to wild-type mice, demonstrating the inverse regulation of BSPRY expression by circulating vitamin D (Van de Graaf et al. 2006d). Together with the inhibitory function of BSPRY on TRPV5 activity and the striking colocalization of both proteins, this suggests that BSPRY operates as a negative modulator for TRPV5, and that this mechanism will be downregulated when vitamin D levels increase to stimulate active Ca^{2+} transport (Van de Graaf et al. 2006d). These data provide the first evidence of a functional role of BSPRY. However, the mechanism of these functional effects remains unclear. So far, only two other studies provided information about BSPRY. BSPRY was initially identified in a yeast two-hybrid screen using zyxin as bait (Schenker and Trueb 2000). In epithelial cells zyxin is involved in the formation of cell–cell contacts, which require actin cytoskeleton rearrangements (Vasioukhin et al. 2000). This might hint at a role of the cytoskeleton in the BSPRY-mediated regulation of TRPV5. Second, it was shown that BSPRY interacts with 14-3-3 proteins (Birkenfeld et al. 2003). It has been demonstrated that 14-3-3 proteins bind to specific motifs containing a phosphorylated serine residue and have been implicated in the binding to and activation of signaling proteins (Muslin et al. 1996; Yaffe et al. 1997). Furthermore, a role of 14-3-3 proteins in K^+ channel trafficking was postulated (O'Kelly et al. 2002). However, cell surface biotinylation did not provide evidence for TRPV5 trafficking as an explanation for the observed inhibitory function of BSPRY (Van de Graaf et al. 2006d). Therefore, it is currently hypothesized that BSPRY is involved in inhibitory signaling cascades controlling the activity of the epithelial Ca^{2+} channels at the cell surface.

S100A10 Van de Graaf et al. identified S100A10 (also known as p11 or annexin 2 light chain) as an auxiliary protein for TRPV5 and TRPV6 using a yeast two-hybrid system (Van de Graaf et al. 2003). S100A10 is a member of the S100 superfamily that is present in a large number of organisms including vertebrates, insects, nematodes, and plants. The two EF-hands of S100A10 carry deletions and substitutions that render it Ca^{2+} insensitive. S100A10 is predominantly present as a heterotetrameric complex with annexin 2, which has been implicated in several biological processes including endocytosis, exocytosis, and membrane-cytoskeleton interactions (Gerke et al. 2005).

It was shown that S100A10, annexin 2, and TRPV5 or TRPV6 are coexpressed in Ca^{2+}-transporting cells of the kidney and small intestine (Van de Graaf et al. 2003). The association of S100A10 with TRPV5 and TRPV6 was restricted to a short peptide sequence, VATTV, located in the carboxyl-termini of these channels (Van de Graaf et al. 2003). This stretch is conserved among all identified species of TRPV5 and TRPV6. Interestingly, the TTV sequence in the S100A10-binding site resembles an internal type I PDZ (postsynaptic density 95/disk-large/zonula occludens-1) consensus binding sequence, which is S/TXV (Songyang et al. 1997). However, S100A10 does not contain PDZ domains, indicating that the TRPV5-S100A10 interaction is structurally distinct. The first threonine of the S100A10 interaction motif is a crucial residue. Both the S100A10 binding capacity and the activity of TRPV5 and TRPV6 are largely abolished when this particular threonine is mutated, demonstrating that this motif is essential for channel function (Van de Graaf et al. 2003). Malfunctioning of these mutant channels is accompanied by a major disturbance in their subcellular localization, indicating that the S100A10-annexin 2 heterotetramer facilitates the translocation of TRPV5 and TRPV6 channels to the plasma membrane.

The importance of annexin 2 in this process was demonstrated by small interference RNA. Downregulation of annexin 2 significantly inhibited the currents through TRPV5 and TRPV6 (Van de Graaf et al. 2003). The expression of S100A10 was also downregulated by this approach, indicating that annexin 2, in conjunction with S100A10, is crucial for TRPV5 activity. The association of annexin 2 with TRPV5 was only detectable in the presence of S100A10, demonstrating that annexin 2 binds indirectly to the channel, with S100A10 most likely operating as a molecular bridge between TRPV5 and annexin 2 (Van de Graaf et al. 2003). These findings provide the first functional evidence for a regulatory role of S100A10/annexin 2 controlling Ca^{2+} channel trafficking and therefore the Ca^{2+} balance. Interestingly, previous and later studies indicated that several ion channels and receptors associate with S100A10. It was reported that the background K^+ channel (TASK1) is associated with S100A10 via its carboxyl-terminal sequence SSV (Girard et al. 2002). The S100A10 interaction blocks an ER-retention signal that promotes the translocation of TASK1 to the plasma membrane producing functional K^+ channels (Girard et al. 2002). This sequence resembles the binding motif in TRPV5 and TRPV6 identified in the present study, suggesting a shared structural S100A10 binding pocket. However, another study suggests that S100A10 binding is located at a different binding site in TASK1 formed by a 40-amino-acid region in the proximal carboxyl-terminus, and the authors proposed that S100A10 binding inhibits TASK-1 targeting to the plasma membrane (Renigunta et al. 2006).

In addition to TRPV5 and TASK1, a number of other ion channels have recently been shown to be regulated by S100A10 binding. All these proteins show a requirement for S100A10 binding for their trafficking toward the plasma membrane (Donier et al. 2005; Okuse et al. 2002; Svenningsson et al. 2006). Initially, Okuse et al. identified the tetrodotoxin-insensitive voltage-gated Na^+ channel (Nav1.8), as the first ion channel to associate with S100A10 (Okuse et al. 2002). Nav1.8 was shown to bind S100A10 via its amino-terminus (Poon et al. 2004). The binding of S100A10 is essential for plasma membrane trafficking of this Na^+ channel (Okuse et al. 2002). Furthermore, Donier et al. have recently demonstrated the association of S100A10 with an acid-sensing ion channel (ASIC1) and confirmed this association in rat dorsal root ganglion neurons by coimmunoprecipitation (Donier et al. 2005). Finally, Svenningsson recently reported the interaction of S100A10 with the serotonin 1B (5-HT1B) receptor. S100A10 increases cell surface abundance of 5-HT1B receptors (Svenningsson et al. 2006). The interaction between S100A10 and the 5-HT1B receptor is associated with the pathophysiology of depression, as was shown using S100A10 knockout mice or S100A10 overexpressing mouse models (Svenningsson et al. 2006). On the whole, the S100A10-annexin 2 complex seems a significant component for the regulation of cell surface abundance of several ion channels and receptors, including TRPV5 and TRPV6.

Rab11a Rab11a was recently identified as a novel TRPV5- and TRPV6-associated protein (Van de Graaf et al. 2006a). The Rab family of small guanosine triphosphatases (GTPases) has a well-recognized role in membrane trafficking (Zerial and McBride 2001). They are localized to specific organelles within the cell and have been implicated in distinct transport steps including vesicle budding, targeting, and tethering. Although the role of Rab GTPases in protein trafficking has long been recognized, the underlying mechanism is far from understood. Rab11a is a small GTPase involved in cargo trafficking via recycling endosomes (Brown et al. 2000; Casanova et al. 1999; Wang et al. 2000). Van de Graaf et al. demonstrated that Rab11a colocalizes with TRPV5 and TRPV6 in Ca^{2+}-transporting epithelial cells of the kidney. Here, both TRPV5 and Rab11a are present in vesicular structures below the apical plasma membrane (Van de Graaf et al. 2006a). Using a combination of GST pull-down and coimmunoprecipitation assays, the direct and specific interaction between Rab11a and the epithelial Ca^{2+} channels was shown (Van de Graaf et al. 2006a). Association of cargo with Rab GTPases has recently received much attention with the identification of an association between Rab3 and the polymeric IgA receptor (Smythe 2002; van IJzendoorn et al. 2002) and between Rab11a and the thromboxane A2 receptor (TPβ) (Hamelin et al. 2005). The binding of TRPV5 and TRPV6 to Rab11a provides the first evidence for an ion channel directly associating with a Rab GTPase. Furthermore, it was demonstrated that TRPV5 and TRPV6 preferentially interact with Rab11a in its GDP-bound conformation (Van de Graaf et al. 2006a). Interestingly, although the identified Rab11a-binding regions in TRPV5 and TPβ are not homologous, both proteins interact with the GDP-bound form of Rab11, suggesting a common binding mechanism. Expression of a mutant Rab11a protein, locked in the GDP-bound state, results in a marked decrease of TRPV5 and TRPV6 channels at the cell surface, indicating a direct role of Rab11a in the trafficking of TRPV5 and TRPV6 toward the plasma membrane (Van de Graaf et al. 2006a). Similarly, it was demonstrated that direct binding of Rab11a is a determinant factor in controlling the recycling to the cell surface of TPβ (Hamelin et al. 2005). The association with Rab11a is essential in directing the intracellular trafficking of the receptor from the Rab5-positive intracellular compartment to the perinuclear recycling endosome. It is possible that also TRPV5/6 channels, present on the (apical) plasma membrane, are continuously exchanged with TRPV5 and TRPV6 channels from the intracellular (recycling) endosomes in a Rab11-dependent manner. The molecular mechanisms determining the distribution of TRPV5 between the plasma membrane and the intracellular pool are currently unknown.

NHERF family members The Na^+-H^+ exchanger regulatory factors 1 and 2 (NHERF1 and NHERF2) form a family of adaptor proteins characterized by the presence of two tandem PDZ protein interaction domains and a carboxyl-terminal domain that binds the cytoskeleton proteins ezrin, radixin, moesin, and merlin (Weinman et al. 2006a). These proteins were initially characterized as facilitating the formation of a multiprotein complex that mediates protein kinase A (PKA) phosphorylation of the renal Na^+-H^+ exchanger 3 (NHE3) and downregulation of its activity (Lamprecht et al. 1998; Weinman et al. 1995; Yun et al. 1997; Zizak et al. 1999). In general, these proteins are known to operate as adapter proteins responsible for organizing a multiprotein complex involved in the regulation of receptors, including the PTH receptor (Mahon et al. 2002) and ion channels, for instance the cystic fibrosis transmembrane conductance regulator CFTR (Li et al. 2005; Liedtke et al. 2002; Sun et al. 2000) and several TRP channels (Obukhov and Nowycky 2004; Odell et al. 2005; Tang et al. 2000).

Interestingly, the first hint for a physiological role of role of NHERF in the regulation of TRPV5 and/or TRPV6 came from studies using NHERF1 knockout (NHERF1$^{-/-}$) mice. NHERF1$^{-/-}$mice display a threefold increase in urinary phosphate excretion compared with wild-type animals. In addition, NHERF1$^{-/-}$ mice display increased urine Ca^{2+} excretion compared with wild-type controls, which persists during life (Shenolikar et al. 2002; Weinman et al. 2006b). The molecular mechanism underlying the effect on Ca^{2+} handling is currently unknown, although a secondary effect resulting from hyperphosphaturia seems the most straightforward explanation (Beck et al. 1998; Shenolikar et al. 2002). Subsequently, Embark and coworkers demonstrated that TRPV5 activity increases upon coexpression with NHERF2 and SGK1 or SGK3 (serum and glucocorticoid inducible kinase 1 or 3) in *Xenopus laevis* oocytes. Coexpression of TRPV5 with NHERF2 or SGK1/3 alone did not stimulate TRPV5-mediated currents, indicating that both NHERF2 and SGK1/3 are required (Embark et al. 2004).

Deletion of the second, but not the first, PDZ domain in NHERF2 abrogates the stimulating effect of SGK1/3/NHERF2 on TRPV5 activity (Palmada et al. 2005). Furthermore, TRPV5 activity was not stimulated with a kinase-dead point mutant of SGK1 (K127N), suggesting a phosphorylation-mediated effect (Embark et al. 2004). Using GST pull-down and overlay assays, the specific interaction of NHERF2 with the last three amino acids (YHF) of the carboxyl-terminus of TRPV5 was demonstrated (Van de Graaf et al. 2006c). Furthermore, TRPV6 did not bind NHERF2 (Palmada et al. 2005). These findings suggest that regulation of the epithelial Ca^{2+} channels by NHERF2-SGK is limited to TRPV5 and operates via direct interaction with the channel.

The coexpression of SGK1 with NHERF2 also stimulated the activity of the renal outer medullary K^+ channel (ROMK1), a K^+ channel involved in renal K^+ handling. This results from a stabilization of ROMK1 in the plasma membrane (Yun et al. 2002). Therefore, it is postulated that NHERF2/SGK stimulates the activity of TRPV5 via a comparable mechanism. Furthermore, it was shown that TRPV5 and TRPV6 associate with another PDZ domain-containing protein, NHERF4 (Van de Graaf et al. 2006c). Other names for this protein are intestine and kidney enriched PDZ protein (IKEPP), PDZK1, and Napi-Cap2 (Gisler et al. 2001; Scott et al. 2002).

In contrast to NHERF1 and 2, this PDZ protein possesses four PDZ domains. Yeast two-hybrid, GST pull-down and coimmunoprecipitation assays identified NHERF4 is a novel auxiliary protein for both TRPV5 and TRPV6 (Van de Graaf et al. 2006c). NHERF4 utilizes PDZ domain 1 and 4 to bind the carboxyl-terminus of TRPV5 at a site distinct from NHERF2. Furthermore, NHERF4 is coexpressed with TRPV6 in Caco-2 cells, an intestinal epithelial cell line, whereas limited colocalization with TRPV5 was observed in the kidney (Van de Graaf et al. 2006a). This suggests that NHERF2 predominantly regulates TRPV5, whereas NHERF4 operates on TRPV6.

Calbindin-D_{28K} As discussed above, in Ca^{2+}-transporting, epithelial calbindins act as cytosolic Ca^{2+} buffers, facilitating the intracellular diffusion of Ca^{2+}, while keeping the free

Ca^{2+} concentration at physiological levels. Calbindin-D_{28K} is highly expressed in Ca^{2+}-transporting epithelia where it colocalizes with TRPV5 (Hoenderop et al. 2000; Hoenderop et al. 1999c). Interestingly, several animal studies exploring the effect of various treatments (including vitamin D, dietary Ca^{2+} or PTH depletion or administration, chronic acidosis) showed that the expression of TRPV5 and calbindin-D_{28K} is concomitantly regulated (Hoenderop et al. 2002; Nijenhuis et al. 2006; Van Abel et al. 2005). In a recent study, Lambers et al. demonstrated that calbindin-D_{28K} is a dynamic Ca^{2+} buffer that is functionally and physically tangled together with TRPV5 (Lambers et al. 2006b). The dynamic nature of the buffer was illustrated by evanescent wave microscopy, used to excite fluorophore-coupled calbindin only in the vicinity of the plasma membrane. They showed that calbindin-D_{28K} translocates toward the plasma membrane upon decreases in $[Ca^{2+}]_i$. This effect was only observed when TRPV5 was present in these cells, and confirmed using endogenous expression of TRPV5 and calbindin-D_{28K}. Importantly, calbindin-D_{28K} directly associated with TRPV5 under conditions of low $[Ca^{2+}]_i$ (Lambers et al. 2006b). The functional relevance of this dynamic association between TRPV5 and calbindin-D_{28K} was elucidated using three approaches (Lambers et al. 2006b). First, coexpression of calbindin-D_{28K} with TRPV5 increased the TRPV5-mediated $^{45}Ca^{2+}$ influx in HEK293 cells. This effect was due to the Ca^{2+} buffering capacity of calbindin-D_{28K}, as a mutant protein with affected EF-hands was not able to increase TRPV5 activity. Second, using a photolysable Ca^{2+} chelator to experimentally control $[Ca^{2+}]_i$, it was demonstrated that coexpression of calbindin-D_{28K} with TRPV5 does not alter the Ca^{2+}-dependent channel inactivation characteristics, but likely operates as a local Ca^{2+} buffer to keep the $[Ca^{2+}]_i$ at the vicinity of the channel pore low. Third, overexpression of Ca^{2+}-binding-deficient mutant calbindin resulted in a reduced Ca^{2+} transport in primary cultures of rabbit CNT and CCD, a model for transepithelial Ca^{2+} transport. This Ca^{2+}-binding-deficient calbindin mutant associated with TRPV5, even at high $[Ca^{2+}]_i$. Together these findings constitute the first direct evidence that calbindin-D_{28K} operates as a dynamic channel-associated Ca^{2+} buffer, essential for transepithelial Ca^{2+} transport (Lambers et al. 2006b). At a low $[Ca^{2+}]_i$, calbindin-D_{28K} translocates toward the plasma membrane and associates with TRPV5. At the apical plasma membrane, it buffers Ca^{2+} that enters the cell via TRPV5, thereby avoiding local accumulation of free Ca^{2+} near the pore and subsequent inactivation of the channel. Upon Ca^{2+}-binding, calbindin-D_{28K} releases from TRPV5 and subsequently facilitates diffusion of Ca^{2+} to the basolateral membrane. This illustrates a mechanism of localized dynamic Ca^{2+} buffering mediated by protein–protein interaction, which could operate in various tissues where Ca^{2+} transport or signaling is abundant (Lambers et al. 2006b).

The binding site(s) of TRPV5/6 auxiliary proteins A remarkable feature of the identified auxiliary proteins is the promiscuous binding site in TRPV5/6. The binding of Rab11 was localized to a conserved stretch in the carboxyl- terminus of TRPV5/6 in close proximity of the last TM. Five amino acid residues (MLERK) within this area were identified as a critical region for binding of Rab11 (Van de Graaf et al. 2006a). However, the interaction with 80K-H is localized to the same region, suggesting a possible common binding site in TRPV5/6 (Gkika et al. 2004). Remarkably, Chang and coworkers have also identified this 5-amino-acid motif as one of the components that plays a role in assembly of the tetrameric channel (Chang et al. 2004). Furthermore, this region was shown to be involved in the binding of PIP_2 to TRPV5, as was demonstrated using point-mutants (Rohacs et al. 2005). Finally, the binding of NHERF4 was also impaired upon deletion of this region (Van de Graaf et al. 2006c), whereas S100A10, NHERF2, and CaM associated with distinct regions of the carboxyl-terminus (Lambers et al. 2004; Niemeyer et al. 2001; Van de Graaf et al. 2003, 2006c). This suggests that the MLERK region is either a critical element for the general folding of the carboxyl-terminus of TRPV5/6 or that several of these associated proteins indeed recognize motifs within this region. The latter hypothesis could imply that these proteins compete for binding at this region. The occupancy of this site would depend on the relative concentrations and affinities of the proteins. Future examination of the channel structure or studies that investigate binding competition among multiple channel associated proteins are required to clarify this issue.

Epithelial Mg^{2+} transport

Magnesium (Mg^{2+}) is the second most abundant intracellular cation, the fourth most abundant extracellular cation, and a cofactor in more than 300 enzymatic reactions varying from energy metabolism to protein and nucleic acid synthesis (Elin 1994; Flatman 1984). Approximately 50% of the total body Mg^{2+} is present intracellularly in soft tissue, and the other half is present in bone. Less than 1% of the total body Mg^{2+} is circulating in blood (Elin 1994). In healthy individuals, plasma Mg^{2+} levels are maintained in a narrow range (0.7–1.1 mmol/l; Konrad et al. 2004) by the balance between intestinal absorption, renal excretion, and dynamic exchange with the stored Mg^{2+} in bone. Several regulatory processes control these Mg^{2+} transport processes in kidney and intestine to account for variations in dietary Mg^{2+} content (Quamme and de Rouffignac 2000). In analogy with renal and intestinal Ca^{2+} transport, Mg^{2+} is transported via a paracellular and a transcellular pathway. First, the various tissues will be discussed where Mg^{2+} is transported and the pathway(s) involved in this process.

Localization of epithelial Mg^{2+} transport

Gastrointestinal tract

Gastrointestinal Mg^{2+} absorption in healthy adults is balanced by the renal excretion of Mg^{2+} into the urine. The fate of intestinal Mg^{2+} in human volunteers was quantified by Fine and coworkers (1991). They measured net Mg^{2+} absorption after ingestion of a standard meal supplemented with various amounts of Mg-acetate. The relationship between absorption and intake was curvilinear, which could be fitted perfectly to the sum of an unsaturable linear component and a saturable component. Saturation kinetics of the transcellular transport system are explained by the limited transport capacity of active transport. This suggests that intestinal Mg^{2+} absorption is mainly transcellular (forming the saturable component of the sum) in conditions of low Mg^{2+} intake and that the relative role of paracellular absorption (forming the unsaturable ~7% of the intake) increases with increasing Mg^{2+} intake.

However, the data provided limited information on the localization of the uptake process within the gastrointestinal tract. In contrast to the kidney, functional data on the distribution of Mg^{2+} absorption in the stomach and gut is relatively scarce. In ruminating animals, most of the required Mg^{2+} is absorbed from the forestomachs by active, transcellular mechanisms (Schweigel and Martens 2000; Schweigel et al. 2006). However, in other mammals, the localization of the Mg^{2+} absorption in the gastrointestinal tract is less clear. The main reason for this lack of data is the absence of a suitable isotope of Mg^{2+}, as $^{28}Mg^{2+}$ has low specific activity and a short half-life (Avioli and Berman 1966; Dai et al. 2001). It was, however, shown that colectomy in rats and humans results in decreased urinary Mg^{2+} excretion, normal plasma Mg^{2+} levels, and decreased bone Mg^{2+} content, suggesting an important role for this intestinal segment in Mg^{2+} absorption (Croner et al. 2000; Fagan and Phelan 2001). Kayne and Lee have reviewed the available literature and suggested a prominent role for the distal segments of the small intestine, in particular the ileum and colon in Mg^{2+} absorption (Kayne and Lee 1993). However, they indicated that this information is mainly derived from isolated segments and may not adequately reflect absorption. Schweigel and Martens studied the sites of intestinal Mg^{2+} absorption in various animals and concluded that in dogs, cats, and pigs, Mg^{2+} is predominantly absorbed from the ileum and colon, whereas in horses and probably rabbits, Mg^{2+} is mainly absorbed from the small intestine (Schweigel and Martens 2000).

Intestinal Ca^{2+} absorption was not affected by increased Mg^{2+} intake in humans (Fine et al. 1991), suggesting that these cations utilize distinct, noncompeting uptake mechanisms. Recently, Groenestege and coworkers confirmed this in animal studies. Ca^{2+} absorption, as measured by $^{45}Ca^{2+}$ uptake in the blood, was unaffected by the Mg^{2+} content of the food (Groenestege et al. 2006). However, both in humans and mice, the renal handling of Mg^{2+} and Ca^{2+} pointed to competitive reabsorption of the two divalent cations, suggesting a common transport pathway (Fine et al. 1991; Groenestege et al. 2006). In addition, the distribution of the Ca^{2+} and Mg^{2+} transport proteins further pinpoints the localization of transepithelial absorption of divalents (Groenestege et al. 2006). Transepithelial absorption of Ca^{2+} takes place predominantly in duodenum and colon as illustrated by the robust expression of the epithelial Ca^{2+} channel TRPV6 in these particular intestinal segments. However, the luminal Mg^{2+} channel TRPM6 was mainly expressed in colon while being virtually absent in duodenum (Groenestege et al. 2006). Therefore, active Mg^{2+} and Ca^{2+} absorption occurs in the distal part of the intestine, whereas in duodenum only active Ca^{2+} absorption takes place.

Kidney

Fine-tuning of the Mg^{2+} balance mainly resides within the kidney (Dai et al. 2001; Quamme 1997). Approximately 80% of the total plasma Mg^{2+} is filtered by the glomerulus. Along the nephron 95% of this filtrate is being reabsorbed. Of the ultrafiltrable Mg^{2+}, 5–15% is reabsorbed by passive transport in the PT. Remarkably, Lelievre-Pegorier et al. reported that the permeability for Mg^{2+} of the PT dramatically changes during development. The immature kidney of the newborn can absorb up to 70% of the filtered Mg^{2+} in this segment (de Rouffignac and Quamme 1994; Lelievre-Pegorier et al. 1983). Later in development (early childhood), PT reabsorbs only 5–15% of the filtered Mg^{2+}, whereas the fractional reabsorption of Na^{+} and Ca^{2+} remains high (70%). The cTAL plays a major role in the determination of Mg^{2+} excretion, as it accounts for approximately 70% of Mg^{2+} reabsorption (Mandon et al. 1993; Shareghi and Agus 1982), whereas the medullary segment (mTAL) is Mg^{2+} impermeable. Transport in the cTAL is passive and paracellular in nature, driven by the lumen-positive transepithelial potential difference (Mandon et al. 1993). Processes that affect the transepithelial voltage (i.e., rate of Na^{+}-K^{+}-Cl^{-} cotransport) or alter the permeability of the paracellular pathway will therefore alter Mg^{2+} reabsorption in this segment. Finally, 10–15% of the Mg^{2+} that is filtered at the glomerulus will be delivered distally from TAL. The remaining reabsorption takes place in DCT with no evidence for significant Mg^{2+} transport in CNT or CD. Mg^{2+} reabsorption in DCT is mediated by the active transcellular pathway and plays a pivotal role in determining the final urinary Mg^{2+} excretion (Dai et al. 2001).

Interestingly, Wilson et al. identified a mutation in mitochondrial DNA resulting in a cluster of metabolic effects including hypomagnesemia due to renal Mg^{2+} wasting. They suggested that impaired mitochondrial function and consequent lower ATP production in DCT has significant effects on the Mg^{2+} transport capacities of this highly energy-consuming nephron segment (Bastin et al. 1987; Wilson et al. 2004). Given an average Mg^{2+} intake, less than 5% of the filtered Mg^{2+} appears in the urine.

Bone

Several studies demonstrated a positive correlation between dietary Mg^{2+} intake and bone density as indicated by increased bone loss in the situation of deficient Mg^{2+} intake (New et al. 2000; Tucker et al. 1999). In a rat model, this effect was already observed even in situations of moderate Mg^{2+} deficiency as demonstrated by Rude et al. Here, only a 50% reduction of the advised dietary Mg^{2+} intake resulted in bone Mg^{2+} deficiency, although no significant change in plasma Mg^{2+} was observed (Rude et al. 2006). This suggests that bone operates as a Mg^{2+} storage in situations of low Mg^{2+} supply. The increase in trabecular bone mineral content and bone volume normally observed in rats between 3 and 6 months of age was reduced in the group with dietary Mg^{2+} restriction. Furthermore, osteoclast number was also significantly increased with Mg^{2+} depletion. Dietary Mg^{2+} deficiency was accompanied by low $1,25(OH)_2D_3$ and high PTH levels in plasma and elevated tumor necrosis factor α ($TNF\alpha$) and interleukin 1β (IL-1β) levels in osteoclasts and megakaryocytes, suggesting that these may contribute to bone loss. (Rude et al. 2006; Vidal et al. 2006). However, PTH levels in humans with hypomagnesemia and hypocalcemia have been either low or even undetectable (Agus 1999; Chase and Slatopolsky 1974; Rude and Gruber 2004; Rude et al. 1978).

Similar results were obtained in dogs that were fed a Mg^{2+}-free diet for 4–6 months (Freitag et al. 1979). This effect might be due to impaired PTH production or secretion. The short-term restoration of plasma PTH levels on Mg^{2+} supplementation suggests that the latter is mainly affected (Agus 1999). In addition, Mg^{2+} depletion diminished the cyclic AMP (cAMP) production during PTH administration in isolated bone in line with an organ resistance to PTH (Chase and Slatopolsky 1974; Freitag et al. 1979; Rude et al. 1976). As described, several studies illustrated the important role of Mg^{2+} in bone formation and integrity. In contrast, in which manner bone affects the Mg^{2+} balance is mechanistically not clear. Mg^{2+} exists abundantly in bone (0.5–1%, Rude et al. 2006), suggesting significant Mg^{2+} transport in this tissue. However, the Mg^{2+} entry pathway in bone and the molecular mechanism of release is poorly understood. With the current insight of the molecular identity of the players involved in epithelial Mg^{2+} transport, more information on these mechanisms should become available.

(Patho)physiology of Mg^{2+} influx in kidney and intestine

The genetic basis of a number of inherited renal Mg^{2+} wasting disorders has been elucidated in recent years. This yielded the identification of novel proteins involved in epithelial Mg^{2+} transport (Table 3). Furthermore, the various inherited diseases described to date often affect distinct nephron segments and lead to variable phenotypic presentations. As a whole, this novel information has significantly boosted our understanding of epithelial Mg^{2+} handling.

Table 3 Inherited disorders of Mg^{2+} homeostasis

Disorder	Inheritance	OMIM	Affected protein	Reference(s)
Hypomagnesemia with hypercalciuria and nephrocalcinosis (HHN)	AR	248250	Claudin-16, claudin-19	Konrad et al. 2006; Simon et al. 1999
Dominant isolated hypomagnesemia with hypocalciuria (IDH)	AD	154020	γ-Subunit Na^+, K^+-ATPase	Meij et al. 2000
Dominant isolated hypomagnesemia with hypocalciuria	AD	–	ND	
Hypomagnesemia with normocalciuria	AR	–	ND	
Autosomal dominant hypoparathyroidism (ADH)	AD, AR	146200	CaSR, Activating mutations	Lovlie et al. 1996
Familial hypomagnesemia/ neonatal severe hyperparathyroidism	AR	239200	CaSR, Inactivating mutations	Pollak et al. 1993
Familial hypocalciuric hypercalcemia (FHH/ or HHC1) and neonatal severe hyperparathyroidism (NSHPT)		145980		
Gitelman's syndrome (GS)	AR	263800	NCC	Simon et al. 1996
Hypomagnesemia, hypertension and hypercholesterolemia	M	500005	MTTI	Wilson et al. 2004
Familial hypomagnesemia with secondary hypocalcemia (HSH)	AR	602014	TRPM6	Schlingmann et al. 2002; Walder et al. 2002

AR, autosomal-recessive; AD, autosomal-dominant; NCC, Na^+/Cl^- cotransporter; MTTI, mitochondrial transfer RNA; TRPM6, transient receptor potential cation channel subfamily M member 6; CaSR, Ca^{2+} sensing receptor; ND, not determined. See text for explanation

Hypomagnesemia with hypercalciuria and nephrocalcinosis

Hypomagnesemia with hypercalciuria and nephrocalcinosis (HHN, OMIM #248250) is mainly characterized by hypomagnesemia with inappropriately high urinary Mg^{2+} excretion. It is an autosomal-recessive disorder that is further characterized by hypercalciuria often leading to nephrocalcinosis (Manz et al. 1978; Milazzo et al. 1981; Nicholson et al. 1995; Praga et al. 1995; Richard and Freycon 1992; Torralbo et al. 1995). Furthermore, several patients with this disorder displayed ocular disorders (Torralbo et al. 1995). Renal transplantation corrected the abnormal Mg^{2+} and Ca^{2+} handling and normalizes plasma Mg^{2+} and Ca^{2+} concentrations (Praga et al. 1995), further substantiating the role of the kidney in the etiology of the disease. Using whole genome analysis in 12 kindreds with recessive renal hypomagnesemia, Simon et al. demonstrated linkage to a segment at chromosome 3q27 and subsequently identified the responsible gene, which was called *PCLN-1* or *CLDN-16* (Simon et al. 1999). Subsequently, several other groups demonstrated patients with mutations in this gene (Muller et al. 2003; Weber et al. 2000, 2001b). The *PCLN-1* gene encodes a protein of 305 amino acids with 4 TMs and intracellular amino- and carboxy-termini named paracellin-1 (Simon et al. 1999).

Paracellin-1 shows sequence and structural similarity to members of the claudin family and was therefore renamed claudin-16. Most claudins have an amino-terminus of only 6 to 7 amino acids (Morita et al. 1999). However, claudin-16 contains a cytoplasmic amino-terminus of 73 amino acids. Northern blot analysis demonstrated that claudin-16 is exclu-

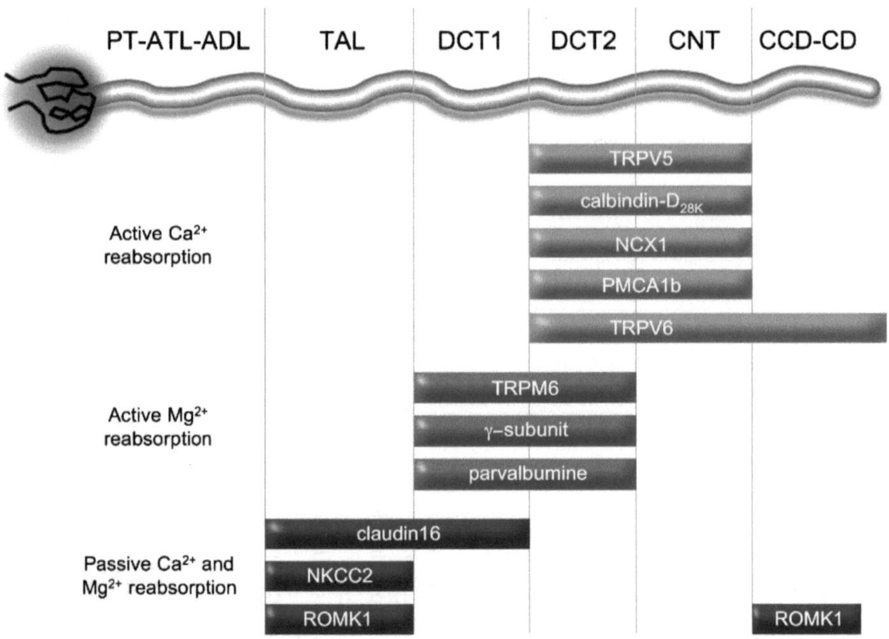

Fig. 3 Localization of transport proteins involved in active and passive Ca^{2+} and Mg^{2+} reabsorption along the nephron. *TRPV5* transient receptor potential cation channel subfamily V member 5, *TRPV6* transient receptor potential cation channel subfamily V member 6, *NCX1* Na^+/Ca^{2+} exchanger type 1, *PMCA1b* ATP-dependent Ca^{2+}-ATPase type 1b, *TRPM6* transient receptor potential cation channel subfamily M member 6, *γ-subunit* γ-subunit of the Na^+/K^+ ATPase, *NKCC2* Na^+-K^+-$2Cl^-$ cotransporter type 2 *ROMK1* renal outer medullary K^+ channel

sively expressed in kidney (Simon et al. 1999). Here, its expression was localized to TAL and DCT (Figs. 1 and 3). Significant colocalization between claudin-16 and occludin was demonstrated by confocal microscopy, indicating that this protein is a component of tight junctions (Schneeberger and Lynch 2004; Simon et al. 1999). This was corroborated by Muller et al. who identified a novel homozygous mutation in the *CLDN-16* gene (T233R) in two families (Muller et al. 2003). They showed that the mutation results in an activation of a PDZ-domain binding motif in claudin-16, disabling the association of claudin-16 with the tight junction protein ZO1. The mutant claudin-16 was no longer localized to tight junctions in kidney epithelial cells, but instead accumulated in lysosomes. Furthermore, these patients displayed serious childhood hypercalciuria, in contrast to more classic symptoms as hypomagnesemia with hypercalciuria (Muller et al. 2003). Thus, mutations at different sites in the claudin gene may lead to particular clinical phenotypes with a distinct prognosis.

Konrad and coworkers recently described several families with a similar renal phenotype as patients with *CLDN16* mutations but the affected individuals also showed severe visual impairment (Konrad et al. 2006). The *CLDN16* gene in these patients was not mutated pointing to a distinct affected gene. Genome-wide linkage and subsequent mutation analyses pointed to disease causative mutations in *CLDN19*, the gene encoding claudin-19. Claudin-19 is a tight junction protein (Lee et al. 2006b) expressed in the eye and in the kidney where it prominently colocalizes with claudin-16 in TAL (Konrad et al. 2006). On the whole, this shows that at least two distinct tight junction proteins of the claudin family are involved in Mg^{2+} homeostasis and explains HHN with ocular defects in several families with unaffected *CLDN16*.

Dominant isolated hypomagnesemia with hypocalciuria

Geven et al. described two unrelated families in which hypomagnesemia due to renal Mg^{2+} wasting was inherited as an autosomal dominant trait (Geven et al. 1987a) (OMIM #154020). Mg^{2+} infusions performed in two patients showed a reduced renal transport maximum for Mg^{2+} and Ca^{2+}. Subsequently, Meij et al. performed a genome-wide linkage study in these families (Meij et al. 1999). Linkage was demonstrated to locus 11q23. Detailed haplotype analyses suggested the existence of a single, hypomagnesemia-causing mutation in both families. In a follow-up study Meij et al. identified a putative dominant-negative mutation in the *FXYD2* gene encoding the Na^+,K^+-ATPase γ-subunit (γ-subunit Na^+,K^+-ATPase) (Meij et al. 2000). It was shown that the mutant γ-subunit accumulates in the cytoplasm, whereas the wild-type protein localizes predominantly to the plasma membrane (Meij et al. 2000).

The γ-subunit of Na^+,K^+-ATPase was originally cloned by Kim et al. and encodes a small (58 amino acids), type I membrane protein (Kim et al. 1997). The protein is localized in DCT, the main site of active renal Mg^{2+} reabsorption. Two individuals with an 11q23.3-qter deletion including FXYD2 had normal plasma Mg^{2+} levels, showing that the hypomagnesemia results from the presence of mutant γ-subunit Na^+,K^+-ATPase, rather than from haploinsufficiency, consistent with a dominant-negative inheritance pattern (Meij et al. 2000). This was corroborated in mice lacking the FXYD2 gene, which show no disturbances in the Mg^{2+} balance (Jones et al. 2005).

The Na^+-K^+-ATPase complex consists of three subunits. The catalytic α-subunit hydrolyzes ATP and transports the cations, whereas the β- and γ-subunits function in the membrane insertion of the α-subunit and modulate its transport properties (Levenson 1994). There is some controversy whether the presence of the mutant γ-subunit results in the impaired trafficking of the entire Na^+-K^+-ATPase complex to the plasma membrane or the affected subunit (with normal membrane insertion of the α- and β-subunits) (Meij et al. 2000; Pu et al. 2002). Furthermore, the precise cellular mechanism of decreased Mg^{2+} reabsorption and the hypocalciuria remains to be further refined. Kantorovich et al. described a family with a similar phenotype as patients with mutated γ-subunit Na^+,K^+-ATPase, but without linkage to the 11q23 locus. Genetic analysis suggested an autosomal dominant inheritance, although a X-linked dominant inheritance could not be excluded (Kantorovich et al. 2002). This shows that dominant inheritance of hypomagnesemia can be caused by a gene other than *FXYD2* and points to the existence of another protein involved in Mg^{2+} homeostasis.

Autosomal-recessive hypomagnesemia with normocalciuria

Another variant of hypomagnesemia that is more consistent with isolated renal Mg^{2+} loss with autosomal-recessive inheritance was described (Geven et al. 1987b; Meij et al. 2002). Unlike several other Mg^{2+} wasting diseases, no abnormalities in plasma or urine Ca^{2+} concentrations were observed in these patients. However, despite the hypomagnesemia, Mg^{2+} excretion in the urine was in the normal range, indicating a renal defect in these patients. Meij et al. excluded linkage to the FXYD2 locus and mutations in any other previously reported loci associated with hypomagnesemia, indicating a distinct disease (Meij et al. 2003). Therefore, this suggests that a second so-far-unidentified genetic cause for hypomagnesemia exists. Identification of the etiology of this disease should provide novel insight into renal Mg^{2+} handling and maintenance of the Mg^{2+} balance.

Disorders associated with abnormal extracellular Mg^{2+}/Ca^{2+} sensing

Malfunctioning of CaSR is frequently associated with disturbed Mg^{2+} handling (OMIM 239200 and 146200). The CaSR is sensitive to both Ca^{2+} and Mg^{2+}, although the relative affinity for Ca^{2+} is higher (Brown 1991). In the parathyroid glands, CaSR senses the plasma levels of these divalent ions to regulate the PTH production and secretion. Both activating and inactivating mutations in the CaSR have been identified. Activating mutations are dominant and lead to hypoparathyroidism (Pollak et al. 1994), which is mainly characterized by hypocalcemia and hypercalciuria. However, hypomagnesemia is observed in up to about 50% of affected individuals (Okazaki et al. 1999; Pearce et al. 1996). This is explained by a shift of the set point of the receptor, resulting in decreased PTH secretion by the parathyroid glands and inhibition of divalent cation reabsorption in the kidney (Brown and MacLeod 2001). On the other hand, patients with inactivating CaSR mutations display hypocalciuric hypercalcemia (Pollak et al. 1993). Furthermore, affected individuals also show a tendency toward hypermagnesemia (Marx et al. 1981). Inactivation of both alleles of the CaSR gene requires parathyroidectomy early in life to prevent fatal consequences of the severe hyperparathyroidism (Pollak et al. 1993).

Gitelman's syndrome

In Gitelman's syndrome, hypomagnesemia is accompanied by hypocalciuria, hypokalemia, and metabolic alkalosis (OMIM #263800). The low urinary Ca^{2+} excretion is the biochemical parameter to distinguish between Gitelman's and Bartter's syndrome, genetically distinct tubular transport disorders that share a hypokalemic metabolic alkalosis (Knoers et al. 2003). Dissociation of renal Ca^{2+} from Mg^{2+} transport together with exaggerated natriuresis after furosemide treatment in these patients indicated the presence of a defect in DCT rather than in TAL. Furthermore, hydrochlorothiazide administration had almost no effect in patients with Gitelman's syndrome compared to controls (Colussi et al. 1997).

The molecular explanation of Gitelman's syndrome was provided by the demonstration of linkage of the disease to the Na^+-Cl^- cotransporter (NCC) gene on 16q13 (Pollak et al. 1996; Simon et al. 1996). This cotransporter is the target of thiazide diuretics, one of the major classes of agents used in the treatment of hypertension (Gamba et al. 1993), and expressed in the apical membrane of DCT (Bachmann et al. 1995; Loffing et al. 2001; Plotkin et al. 1996). The correlation between Gitelman's syndrome and NCC was confirmed in a mouse model (Loffing et al. 2004; Morris et al. 2006; Schultheis et al. 1998) that shows all the characteristics of the disease. More than 100 different, putative loss-of-function mutations in the NCC encoding gene (*SLC12A3*) have been identified in Gitelman's patients (Knoers et al. 2003). Functional expression studies and results of immunocytochemistry in *Xenopus laevis* oocytes showed that most disease-causing NCC mutants are impaired in their trafficking to the plasma membrane (de Jong et al. 2002).

The mechanisms of hypocalciuria and hypomagnesemia in Gitelman's syndrome remain unclear. It has been suggested that inactivating mutations of NCC cause hypocalciuria by the same mechanisms as postulated for chronic thiazide administration (Nijenhuis et al. 2005). Hypomagnesemia has been suggested to be associated with hypokalemia, a hypothesis disputed by studies in NCC knockout mice that develop severe hypocalciuria and hypomagnesemia despite the absence of hypokalemia or alkalosis (Knoers et al. 2003), although this might be dependent on the amount of potassium in the diet (Morris et al. 2006). An alternative hypothesis to explain hypomagnesemia in Gitelman's syndrome is based on the observation in rats that blockage of NCC by chronic thiazide treatment results in an increased rate of apoptosis in DCT cells (Loffing et al. 1996). Furthermore, remodeling of DCT has been observed in mouse models of Gitelman's syndrome, with a significant reduction in parvalbumin positive (DCT1) tubules (Loffing et al. 2004). It is, therefore, possible that reduced number of DCT cells in Gitelman's syndrome compromises Mg^{2+} reabsorption in this nephron segment resulting in hypomagnesemia. This is in line with recent data from Nijenhuis et al. who showed a significant reduction of the expression of the epithelial Mg^{2+} channel TRPM6 during chronic thiazide treatment (Nijenhuis et al. 2005). However, this latter study does not discriminate whether the DCT cells are reduced in number or that the cells display lower expression of proteins involved in transepithelial Mg^{2+} transport. This latter would be another explanation for the observed hypomagnesemia during thiazide treatment or in Gitelman's patients.

Hypomagnesemia with secondary hypocalcemia

Familial hypomagnesemia with secondary hypocalcemia (HSH) is an autosomal recessive disease due to defective intestinal and renal Mg^{2+} (re)absorption (OMIM #602014). The disease can be fatal or results in permanent neuronal damage if untreated. As the passive component of intestinal Mg^{2+} transport is not affected, the disease can be controlled with high oral Mg^{2+} supplements.

The mechanisms leading to hypocalcemia are not completely understood, but might be related to Mg^{2+}-dependent effects on PTH secretion or PTH resistance as described in "Disorders associated with abnormal extracellular Mg^{2+}/Ca^{2+} sensing". However, plasma Ca^{2+} and PTH levels are usually restored by administration of high doses of Mg^{2+}. In addition to the observed decreased intestinal Mg^{2+} absorption in HSH, there may be a renal leak, deteriorating the efficacy of oral Mg^{2+} supplements to normalize plasma Mg^{2+} or the symptoms of the hypomagnesemia (Konrad et al. 2004; Matzkin et al. 1989). It was postulated that the renal defect was due to impaired Mg^{2+} reabsorption in DCT (Cole and Quamme 2000). HSH was initially postulated to be a X-linked recessive disorder based on its initial predominant occurrence in males (Skyberg et al. 1967; Stromme et al. 1969; Vainsel et al. 1970), and on the basis of a case of X-autosome translocation t(9;X) (Chery et al. 1994; Walder et al. 1997). The proposed X-linked inheritance of this disorder was later questioned, as an autosomal recessive inheritance was suggested (Garty et al. 1983; Hennekam and Donckerwolcke 1983; Pronicka and Gruszczynska 1991; Walder et al. 1997).

Finally, Walder et al. convincingly demonstrated that the disorder is, in fact, autosomal recessive and is determined by a mutation in a gene located on 9q12-q22.2 (Walder et al. 1997). The disease segregated with a single affected haplotype in three inbred Bedouin kindreds from Israel, suggesting that hypomagnesemia is caused by a common ancestral mutation. Walder et al. suggested that a likely candidate gene for HSH would be a receptor or ion channel involved in the absorption of intestinal Mg^{2+} (Walder et al. 1997). Indeed, using positional cloning, Schlingmann et al. and Walder et al. identified HSH causing mutations in a gene within this region encoding a novel member of the TRP superfamily of cation channels, TRPM6 (Schlingmann et al. 2002; Walder et al. 2002).

Subsequently, many novel mutations in several families with HSH were elucidated, further confirming the role of TRPM6 in this disease (Schlingmann et al. 2005). TRPM6 (also known as CHAK2) is a member of the transient receptor potential melastatin (TRPM) cation channel family. Remarkably, TRPM6 comprises both an ion channel domain and a protein kinase and will be described in detail in section "TRPM6".

Molecular mechanism of epithelial Mg^{2+} transport

Elucidation of the genetic basis of a number of diseases that involve disturbances of the Mg^{2+} balance has increased our understanding of epithelial Mg^{2+} transport. Some of the molecular players that are mutated in certain pathological states affect mainly paracellular transport of Mg^{2+}, whereas other diseases are caused by disturbances in transcellular transport. In this section our current knowledge and some hypothesis are summarized about the molecular mechanisms of these Mg^{2+} transport pathways.

Paracellular Mg^{2+} transport

Paracellular Mg^{2+} reabsorption accounts for a significant fraction of the total Mg^{2+} (re)absorption in intestine and kidney. In TAL paracellular Mg^{2+} absorption is driven by the lumen-positive transepithelial voltage (Mandon et al. 1993). This transepithelial potential is mainly created by the luminal K^+ conductance of the ROMK channel and drives the positively charged Mg^{2+} and Ca^{2+} ions from the lumen through the paracellular pathway into the interstitium (Fig. 1).

The high paracellular Mg^{2+} and Ca^{2+} conductance contrasts with the low water permeability of this segment (Kokko 1974). As described in "Hypomagnesemia with hypercalciuria and nephrocalcinosis", the identification of genetic defects in claudin-16 revealed that this protein governs the divalent ion conductance of the tight junction complex (Simon et al. 1999). Wild-type claudin-16 resides on the plasma membrane in HeLa, MDCK, and LLC-PK1 cells. In these latter two cell types it is confined to tight junctions where it colocalizes and associates with ZO-1 (Hou et al. 2005; Muller et al. 2003). Disease-causing mutations in claudin-16 can lead to the intracellular retention of this protein or affect its capacity to facilitate paracellular Mg^{2+} transport (Kausalya et al. 2006; Muller et al. 2006). Kausalya et al. reported that several claudin-16 mutants are retained in the endoplasmic reticulum, where they undergo proteasomal degradation or accumulate in the Golgi complex (Kausalya et al. 2006). In addition, two mutants were delivered to lysosomes, one via clathrin-mediated endocytosis following transport to the cell surface and the other without appearing on the plasma membrane (Kausalya et al. 2006). This indicates that claudin-16 mutations have distinct cellular consequences. Cell surface localization of some of these mutants was rescued by inhibiting endocytosis (Muller et al. 2006) or using compounds acting as pharmacological chaperones (Kausalya et al. 2006). These compounds include glycerol, dimethylsulfoxide, thapsigargin, curcumin, and 4-phenylbutyrate (4-PBA) and have been postulated to aid in the correct folding of transmembrane proteins (Ulloa-Aguirre et al. 2004), including mutant forms of the cystic fibrosis transmembrane regulator (CFTR) (Egan et al. 2002, 2004; Sato et al. 1996), the V_2 vasopressin receptor (Robben et al. 2006; Tan et al. 2003), and aquaporin 2 (Tamarappoo et al. 1999). However, these chaperones did not restore paracellular Mg^{2+} permeability, suggesting that this claudin-16 mutant has additional defects on paracellular Mg^{2+} transport capacities (Kausalya et al. 2006). Remarkably, in LLC-PK1 cells, the G121R claudin-16 mutant was localized in tight junctions even without pharmacological intervention, further supporting a transport defect of this mutant protein (Hou et al. 2005).

There are currently two models explaining the molecular mechanism by which claudin-16 mediates paracellular Mg^{2+} transport. Upon identification of mutations, it was initially postulated that claudin-16 directly promotes paracellular Mg^{2+} and Ca^{2+} movement by creating a selective paracellular conductance for these divalent cations, allowing paracellular fluxes of Mg^{2+} and Ca^{2+} down their electrochemical gradients (Simon et al. 1999; Wong and Goodenough 1999). This hypothesis was supported by studies on two other claudins (claudin-4 and 14), showing that these proteins also influence ion selectivity by forming a charge-selective passage through the tight-junction barrier (Colegio et al. 2002; Van Itallie et al. 2001).

A recent report by Hou et al., however, suggested that in LLC-PK1 epithelial cells claudin-16 modulates the ion selectivity of the tight junction by significantly and selectively increasing the permeability of Na^+ (with no effects on Cl^-) and generating a high permeability ratio of Na^+ to Cl^- (Hou et al. 2005). Remarkably, Mg^{2+} flux across cell monolayers showed a far less-pronounced change (compared to monovalent cations) following claudin-16 expression, suggesting that this protein does not form Mg^{2+}-selective paracellular channels (Hou et al. 2005). Therefore, a second hypothesis was postulated to explain the phenotype of patients with defects in the claudin-16 gene by causing a reduction in driving force for transepithelial Mg^{2+}/Ca^{2+} movement without directly affecting a divalent-selective shunt (Hou et al. 2005).

Early microperfusion studies (Greger and Schlatter 1983) have indicated that the transepithelial voltage is lumen positive (+3–10 mV) when TAL is perfused with isotonic solutions. This is mainly due to apical ROMK1-mediated membrane K^+ secretion. However, in conditions where the lumen content in TAL is hypotonic compared to the interstitial fluid (as is the physiological condition in the TAL) the transepithelial potential in TAL increases to as much as +30 mV (lumen positive) (Greger 1981; Rocha and Kokko 1973). This results from the back-flow of Na^+ from the interstitium to the lumen down its concentration gradient via the paracellular pathway. Claudin-16 dysfunction in TAL, with a concomitant loss of cation selectivity, could therefore contribute to a reduction of the lumen-positive potential that constitutes the driving force for the reabsorption of Mg^{2+} and Ca^{2+} (Hou et al. 2005). Future studies that address the transmembrane potential in the TAL in the presence and absence of functional claudin-16 are needed to distinguish between these two models.

Physiological regulation of paracellular Mg^{2+} transport

Paracellular Mg^{2+} movement is affected by two distinct mechanisms. First, all factors that affect the transepithelial potential also control the amount of Mg^{2+} transported via the paracellular pathway, for reasons described above. Second, the expression of claudins is subject to regulation (reviewed in Balkovetz 2006). So far, little information is available on the molecular regulation of claudin-16, which seems the most relevant claudin for epithelial Mg^{2+} transport. However, the role of other claudins in paracellular Mg^{2+} movement, i.e., in the intestine, cannot be excluded. In vitro analysis of the human claudin-16 (using luciferase reporter vectors) indicated promoter activity in renal cell lines only (Efrati et al. 2005). Interestingly, a high ambient Mg^{2+} concentration ($[Mg^{2+}]$) increased, whereas low $[Mg^{2+}]$ reduced the promoter activity (Efrati et al. 2005). Furthermore, $1,25(OH)_2D_3$ decreased claudin-16 promoter activity, and this action appeared to be mediated through the single peroxisome-proliferator-responsive element (PPRE) within the promoter region (Efrati et al. 2005). It is currently unclear to what extent these findings correlate with a physiological regulation of claudin-16 expression and/or paracellular Mg^{2+} transport by the circulating Mg^{2+} and $1,25(OH)_2D_3$ concentration.

Transcellular Mg^{2+} transport

In analogy with active Ca^{2+} (re)absorption (as described in "Transcellular Ca^{2+} transport"), the process of transcellular Mg^{2+} transport is envisaged by a series of sequential steps. This transport starts with the entry into the epithelial cell through the luminal epithelial Mg^{2+} channel TRPM6. This step is driven by a favorable transmembrane potential. Subsequently, Mg^{2+} diffuses through the cytosol to reach the basolateral plasma membrane. Here, Mg^{2+} is actively extruded into the interstitium that is in contact with the blood compartment. It was postulated that luminal Mg^{2+} entry forms the rate-limiting step (Dai et al. 2001; Groenestege et al. 2006; Voets et al. 2004b) and therefore the major site of regulation (Fig. 1).

Mg^{2+} influx

In contrast to the steep chemical gradient for Ca^{2+}, mammalian cells lack a significant chemical gradient driving Mg^{2+} influx, since $[Mg^{2+}]_i$ is typically in the submillimolar range (Grubbs 2002; Romani and Maguire 2002; Wolf et al. 2003). Consequently, the negative membrane potential primarily drives the movement of Mg^{2+} into the cell. Until recently, the molecular nature of the luminal Mg^{2+} influx pathway remained elusive. Goytain and Quamme attempted to elucidate the characteristics of the epithelial Mg^{2+} transporter and identified a number of proteins with similarity to bacterial Mg^{2+} transporting proteins. Moreover, with the recent elucidation of TRPM6, a novel TRP channel required for renal and intestinal Mg^{2+} absorption, and TRPM7, a ubiquitously expressed cellular Mg^{2+} channel, our understanding of Mg^{2+} influx pathways has significantly advanced.

MagT1, SLC41A1, SLC41A2, and ACDP2 Two approaches have recently been applied to identify proteins with Mg^{2+}-transporting capacities. First, eukaryotic proteins were cloned based on homology with the prokaryote CorA Mg^{2+} transport protein family. A Mg^{2+} transport protein family, termed Mrs2, was isolated from yeast, mouse, and human mitochondria (Bui et al. 1999; Zsurka et al. 2001). However, Mrs2 does not seem to operate as a cellular Mg^{2+} influx channel, but is merely involved in Mg^{2+} transport in mitochondria (Kolisek et al. 2003). In addition, homology searches with prokaryotic Mg^{2+} transport proteins yielded the elucidation of members of the solute carrier 41 (SLC41) family as novel Mg^{2+} transporters (Goytain and Quamme 2005b; Goytain and Quamme 2005c; Wabakken et al. 2003). The second approach to find novel Mg^{2+} transport proteins was performed by Goytain et al. who screened for genes that are upregulated by low extracellular Mg^{2+} in a DCT-like cell line and in mouse kidney. They identified SLC41A1 (Goytain and Quamme 2005b), a Mg^{2+} transporter called MagT (Goytain and Quamme 2005d), and the ancient conserved domain protein 2 (ADCP2) (Goytain and Quamme 2005a). SLC41 proteins consist of 10 putative TMs, and SLC41A1 has a predicted molecular mass of 56 kDa (Wabakken et al. 2003). Goytain et al. showed that expression of members of the SLC41 family, including SLC41A1 and SLC41A2, in *Xenopus* oocytes resulted in currents that are dependent on extracellular Mg^{2+} (Goytain and Quamme 2005b, c). Both SLC41A1 and SLC41A2 also transport a variety of other divalent cations. Expression of SLC41A1 resulted in transport of a range of divalent cations: Mg^{2+}, Sr^{2+}, Zn^{2+}, Cu^{2+}, Fe^{2+}, Co^{2+}, Ba^{2+}, and Cd^{2+}. The divalent cations Ca^{2+}, Mn^{2+}, and Ni^{2+} and the trivalent ion Gd^{3+} did not induce currents nor did they inhibit Mg^{2+} transport, whereas La^{3+} abolished Mg^{2+} uptake (Goytain and Quamme 2005b). In addition to Mg^{2+}, cells overexpressing SLC41A2 transported a range of other divalent cations: Ba^{2+}, Ni^{2+}, Co^{2+}, Fe^{2+}, or Mn^{2+}, but not Ca^{2+}, Zn^{2+}, or Cu^{2+}. Mg^{2+} transport was inhibited by high concentrations of Ca^{2+} (Goytain and Quamme 2005c).

MagT comprises 335 amino acids with a relative molecular mass of approximately 38 kDa. Hydropathy profile analysis suggested that MagT1 is an integral membrane protein containing five hydrophobic transmembrane-spanning (TM) α-helical regions, the first of which is likely cleaved to form the final product with four TM domains (Goytain and Quamme 2005d). MagT proteins form a novel family without major similarity to other transport proteins. This family consists of two members, MagT1 and MagT2, of which MagT1 is selective for Mg^{2+}, whereas MagT2 is also permeable for other divalent cations including Fe^{2+}, Cu^{2+}, and Mn^{2+} (Goytain and Quamme 2005d).

The ancient conserved domain gene, *ACDP2*, is responsive to Mg^{2+} and encodes a protein of 874 amino acids with a postulated 4 TM topology (Wang et al. 2004a). ACDP2 is part of a family of four homologous ubiquitous proteins with homology to the microbial CorC protein, which is involved in bacterial Co^{2+} resistance (Gibson et al. 1991; Wang et al. 2003). The cellular localization of the endogenous ACDP proteins is unclear, although plasma membrane localization was suggested for ACDP1 in neurons (Wang et al. 2004a). However, overexpressed ACDP1 in HeLa cells primarily localized to the nucleus (Wang et al. 2003). When expressed in oocytes, ACDP2 mediated saturable Mg^{2+} uptake with a Michaelis constant of 0.56 mM (Goytain and Quamme 2005a). ACDP2 has a low substrate selectivity as it transports a range of divalent cations: Mg^{2+}, Co^{2+}, Mn^{2+}, Sr^{2+}, Ba^{2+}, Cu^{2+}, and Fe^{2+}. The cations Ca^{2+}, Cd^{2+}, Zn^{2+}, and Ni^{2+} did not induce currents (Goytain and Quamme 2005a).

Although a role for these novel proteins in maintaining the Mg^{2+} balance was postulated by Goytain et al., several lines of evidence remain untouched. The ubiquitous tissue distribution does not support a specialized role in epithelial Mg^{2+} transport. Likewise, the localization of these transporters in kidney and intestine remains unknown, as well as the cellular distribution of these proteins. Finally, the phenotype of patients with mutations in TRPM6 showed that neither of these putative Mg^{2+} transporters is able to compensate the physiological role of TRPM6.

TRPM6 The TRPM6 gene contains 39 exons encoding a protein of 2,022 amino acids with a calculated molecular mass of 234 kDa. (Clapham 2003; Schlingmann et al. 2002; Voets et al. 2004b; Walder et al. 2002). TRPM6 shares the highest (approximately 50%) sequence homology with TRPM7, which was identified by Runnels et al. in a screen for phospholipase C-associated proteins (Nadler et al. 2001; Runnels et al. 2001). TRPM6 and TRPM7 share the predicted topology of 6 TM domains with a putative pore region between TM5 and TM6. However, unlike other members of the TRP family, TRPM6 and TRPM7 contain long carboxyl-terminal domains with similarity to the atypical α-kinases (Runnels et al. 2001; Schlingmann et al. 2002; Walder et al. 2002). The combination of channel and enzyme domains in TRPM6 and TRPM7, also known as chanzymes, is unique among known proteins and raises intriguing questions concerning the physiological role(s) of these chanzymes (Montell 2003). TRPM6 has a restricted expression pattern and is predominantly present in absorbing epithelia, whereas TRPM7 is ubiquitously expressed and implicated in cellular Mg^{2+} homeostasis (Groenestege et al. 2006; Nadler et al. 2001; Schlingmann et al. 2002; Schmitz et al. 2003; Voets et al. 2004b; Walder et al. 2002). In kidney, immunohistochemical studies using NCC as a marker of DCT showed that the localization of TRPM6 was restricted to this segment (Voets et al. 2004b; Fig. 3).

Importantly, this localization strongly supports a function in transcellular Mg^{2+} reabsorption, which is restricted to DCT. The localization in the apical domain of DCT cells is further in line with TRPM6 as the gatekeeper of Mg^{2+} influx. In small intestine, TRPM6-positive signal was detected in absorptive epithelial cells by in situ hybridization and immunohistochemistry (Schlingmann et al. 2002; Voets et al. 2004b). In these cells, TRPM6 was localized along the brush-border membrane (Voets et al. 2004b). Groenestege et al. recently addressed the relative expression of TRPM6 and TRPM7 (Groenestege et al. 2006). They showed that TRPM6 is expressed predominantly in kidney, lung, and intestine, whereas TRPM7 is distributed ubiquitously. Several other tissues including spleen, heart, brain, and liver, were virtually negative for TRPM6. In intestine, highest TRPM6 mRNA levels were measured in cecum and colon, in line with a predominant localization of the active component of intestinal Mg^{2+} absorption in these latter segments.

Voets and coworkers expressed full-length TRPM6 in HEK293 cells to functionally characterize this novel ion channel. They showed that TRPM6-transfected HEK293 cells perfused with an extracellular solution containing 1 mM Mg^{2+} or Ca^{2+} exhibit characteristic outwardly rectifying currents upon establishment of the whole-cell configuration (Voets et al. 2004b). These results were similar to those obtained upon overexpression of TRPM7 (Nadler et al. 2001; Runnels et al. 2001, 2002). In contrast, other groups reported that TRPM6 alone does not yield any additional currents compared to mock-transfected cells (Chubanov et al. 2004; Schmitz et al. 2005). Chubanov and coworkers suggested that association with TRPM7 is a prerequisite for proper plasma membrane localization of TRPM6, and therefore for functional activity (Chubanov et al. 2004). The S141L TRPM6 missense mutation that causes HSH abrogated the oligomeric assembly of TRPM6 and therefore provides a molecular explanation for this disease. This further demonstrates a role for the amino-terminus in the oligomerization of TRPM6. Furthermore, during the procedure to isolate full-length TRPM6, the human TRPM6 gene was found to encode for multiple mRNA isoforms (Chubanov et al. 2004). Therefore, it seems likely that subtle differences are present between the various TRPM6 constructs used in the literature. This notion is further supported by Li and coworkers who characterized homomeric and heteromeric TRPM6 and TRPM7. This latter group used the TRPM6 construct employed in the study by Voets et al. and confirmed that expression of TRPM6 alone yields functional channels in both HEK293 and CHOK1 cells, which exhibit lower endogenous TRPM7 expression than HEK203 cells (Li et al. 2006).

At physiological membrane potentials, significant inward currents were observed in TRPM6-expressing HEK293 and CHOK1 cells with all tested divalent cations as the sole charge carrier, including Ca^{2+} and Mg^{2+}. The linkage of mutations in TRPM6 with the HSH phenotype indicated that this current is essential and perhaps sufficient for epithelial Mg^{2+} uptake. It is possible that the TRPM6-mediated Mg^{2+} inward current is more pronounced in native DCT and intestinal cells as a result of specific cofactors, such as intracellular Mg^{2+} buffers, that are missing in overexpression systems used to date. Voets et al. evaluated the effect of $[Mg^{2+}]_i$ on TRPM6 activity, using flash photolysis of the photolabile Mg^{2+} chelator DM-nitrophen to rapidly alter the $[Mg^{2+}]$ in a spatially uniform manner (Voets et al. 2004b). The TRPM6-mediated current was significantly inhibited by increased levels of $[Mg^{2+}]$ (K_D ~0.5 mM) indicating that the channel is tightly regulated by $[Mg^{2+}]_i$. Similarly, TRPM7 channel activity is strongly reduced by millimolar concentrations of Mg^{2+}-ATP (Hermosura et al. 2002; Nadler et al. 2001). Kozak and Cahalan demonstrated that TRPM7 inhibition is mediated by intracellular Mg^{2+} rather than ATP (Kozak and Cahalan 2003).

Importantly, data obtained by Voets et al. indicated that the TRPM6 pore has a higher affinity for Mg^{2+} than for Ca^{2+} (Voets et al. 2004b). This is physiologically important as micropuncture studies have shown that the luminal concentration of free Mg^{2+} in DCT ranges from 0.2 to 0.7 mM (Dai et al. 2001), whereas the luminal Ca^{2+} concentration is in the millimolar range. Therefore, luminal Mg^{2+} influx should exhibit a higher affinity for Mg^{2+} than for Ca^{2+} to ascertain specific divalent cation transport in DCT. TRPM6 uniquely fulfills this role as all known Ca^{2+}-permeable channels, including members of the TRP superfamily, generally display a 10 to 1,000 times lower affinity for Mg^{2+} than for Ca^{2+} (Voets et al. 2004b).

TRPM7 TRPM7 is the protein with the highest homology to TRPM6 and was cloned independently by two groups (Nadler et al. 2001; Runnels et al. 2001). Similar to TRPM6, TRPM7 forms a cation channel conducting both Mg^{2+} and Ca^{2+} ions (Monteilh-Zoller et al. 2003), exhibits constitutive activity, and is both an ion channel and a kinase (Nadler et al. 2001). TRPM7 activity is regulated by the free $[Mg^{2+}]_i$ (Matsushita et al. 2005; Schmitz et al. 2003; Takezawa et al. 2004). However, the regulation by Mg^{2+}-nucleotides including ATP proposed by Nadler et al. has been challenged by Cahalan and colleagues, who suggested that the inhibitory effect of Mg^{2+}-nucleotides can be explained by free Mg^{2+} alone (Kozak and Cahalan 2003). A recent report by Demeuse et al. suggested that TRPM6 is regulated by both free Mg^{2+} and by Mg^{2+}-nucleotides and this dual mode of regulation could be modulated by its kinase domain (Demeuse et al. 2006). Furthermore, modulation of TRPM7 activity by various factors including PIP_2 (Runnels et al. 2002) and PKA (Takezawa et al. 2004) has been suggested.

Recent studies have been performed to address the similarities and differences in the physiological function and electrophysiological properties of the structurally analogous channels TRPM6 and TRPM7. Particular emphasis has initially been placed on the function of the atypical protein α-kinase domain located in the carboxyl-terminus. α-Kinases are a recently discovered family of proteins that have low sequence homology to conventional protein kinases (Drennan and Ryazanov 2004). Both the TRPM7 and TRPM6 α-kinase domain display autophosphorylation activity when expressed in bacteria (Ryazanova et al. 2004). The TRPM7 α-kinase is specific for ATP and cannot use GTP as a substrate, requires Mg^{2+} or Mn^{2+} for optimal activity, and has been shown to phosphorylate several substrates such as myelin basic protein and histone H3 on serine and threonine residues in vitro (Ryazanova et al. 2004). Dorovkov and Ryazanov identified annexin 1 as a novel substrate for the TRPM7 α-kinase. TRPM7 α-kinase phosphorylated annexin 1 at a conserved serine residue (Ser5) located within the amino-terminal amphipathic α-helix of annexin 1 (Dorovkov and Ryazanov 2004). This region plays a crucial role in the interaction of annexin 1 with other proteins, lipids, and phosphatidylinositides, suggesting that TRPM7 modulates the function of annexin 1. In addition, Clark et al. demonstrated that TRPM7 phosphorylates myosin IIA heavy chain. Association of TRPM7 with myosin IIA was regulated by Ca^{2+} entering the cell via TRPM7, and both activation of TRPM7 and inhibition of myosin II resulted in actomyosin remodeling (Clark et al. 2006).

It is currently unclear to what extent the substrate-specificity of TRPM6 and TRPM7 are conserved. Future studies should address the question whether the α-kinase domain, present in TRPM6, has specific cellular targets that might modulate ion channel activity or transepithelial Mg^{2+} transport and, therefore, the Mg^{2+} balance. In a comprehensive study aiming to compare the functional properties of TRPM6 and TRPM7, Li and coworkers convincingly demonstrated that TRPM6 alone forms functional channels with biophysical properties distinct from TRPM7 or TRPM6/TRPM7 hetero-oligomers (Li et al. 2006). Heterologous expression of TRPM6 produced functional channels with a divalent permeability profile, pH sensitivity and unitary conductance that were distinct from those of TRPM7 channels or from cells heterologously coexpressing TRPM6 and TRPM7 (called TRPM6/7 currents). The relative permeability for Ni^{2+} over Ca^{2+} was significantly higher for TRPM7 than for TRPM6 or TRPM6/7. The relative permeability for other cations, including Ba^{2+}, Mg^{2+}, Ca^{2+}, Mn^{2+}, Sr^{2+}, and Cd^{2+} of TRPM6 was similar to TRPM7. Furthermore, both TRPM6 and TRPM7 currents were increased upon extracellular acidification. However, the relative increase of currents at low pH compared to currents at pH 7.4 was much greater for TRPM7 than for TRPM6.

Li et al. also measured the single channel conductance of TRPM6, TRPM6/7, and TRPM7 in the outside-out configuration and demonstrated that these conditions yielded distinct unitary conductances (Li et al. 2006). TRPM7 showed a unitary conductance of approximately 40 pS, whereas TRPM6 channels reveal higher current amplitudes at all measured voltages, displaying a unitary conductance of approximately 84 pS. Coexpressing TRPM6 with TRPM7 in a 1:1 ratio yielded a single channel conductance of about 57 pS. When TRPM6 and TRPM7 are coexpressed oligomeric channels may be formed at random yielding a mixture of single channel conductances. Indeed, in patches that contained two or more channels, single channel events indicative of distinct amplitudes were observed. These amplitudes matched with the conductances of TRPM6 or TRPM7, but also a single novel conductance with an intermediate value was obtained, suggesting that heteromeric TRPM6/7 channels have a preferred configuration with a distinct single channel conductance. Only one type of channels was observed when TRPM6 or TRPM7 were expressed alone.

Finally, Li et al. also identified 2-aminoethoxydiphenyl borate (2-APB) as a novel pharmacological tool to distinguish between TRPM6 and TRPM7 currents, as micromolar levels of this compound maximally increased TRPM6, but significantly inhibited TRPM7 channel activities. In contrast, millimolar concentrations of 2-APB potentiated TRPM6/7 and TRPM7 channel activities.

Not only the biophysical properties, but also the physiological roles of TRPM6 and TRPM7 are distinct. TRPM6 and TRPM7 have different expression patterns, with TRPM6 present mainly in absorbing epithelia, whereas TRPM7 is ubiquitously expressed (Groenestege et al. 2006). TRPM7 has been implicated in several processes including anoxic neuronal death (Aarts et al. 2003), regulation of actomyosin contractility and cell adhesion

(Clark et al. 2006), influx of toxic divalent metals (Monteilh-Zoller et al. 2003), and maintenance of cellular Mg^{2+} homeostasis (Schmitz et al. 2003). However, the physiological role of TRPM6 seems to be more specialized as the gatekeeper of epithelial Mg^{2+} influx (Schlingmann et al. 2002; Voets et al. 2004b; Walder et al. 2002). Furthermore, several lines of evidence suggest that both channels are functionally nonredundant. Deletion of the TRPM7 gene in chicken DT40 B-lymphocytes is lethal (Nadler et al. 2001) and TRPM7 downregulation using siRNA impairs viability of human neuroblastoma cells. The viability of TRPM7-deficient DT40 cells cannot be complemented by heterologously expressed TRPM6 (Schmitz et al. 2005). Similarly, mutations in TRPM6 result in HSH despite the ubiquitous presence of TRPM7 (Schlingmann et al. 2002; Walder et al. 2002).

Intracellular Mg^{2+} transport

The second step of transcellular Mg^{2+} transport is the diffusion of cytosolic Mg^{2+} from the luminal entry site toward the basolateral membrane. Importantly, $[Ca^{2+}]_i$ is typically in the 0.1 μM range, which the cell maintains for signaling purposes, while $[Mg^{2+}]_i$ with estimated values of 0.8 mM is substantially higher (Grubbs 2002; Romani and Maguire 2002; Wolf et al. 2003). This has considerable consequences for buffering of $[Mg^{2+}]_i$ with respect to the process of epithelial Mg^{2+} transport. Transcellular Ca^{2+} transport requires a massive Ca^{2+} buffering capacity to maintain low $[Ca^{2+}]_i$ during large luminal to basolateral transport rates. Given the much higher $[Mg^{2+}]_i$, this requirement seems less strict for Mg^{2+}. Thus, the question is whether the molecular mechanism of transcellular Mg^{2+} transport indeed requires intracellular Mg^{2+} buffers? If so, which proteins could fulfill this buffering function? Certainly, the physiological role of putative Mg^{2+} buffers needs to be calculated using mathematical models for epithelial Mg^{2+} transport and experimentally determined, i.e., using animal models lacking certain Mg^{2+} buffering proteins. However, the strong regulation of TRPM6 by $[Mg^{2+}]_i$ points to the importance of a significant local Mg^{2+} buffering capacity (Voets et al. 2004b). Therefore, we postulate that specific Mg^{2+}-binding proteins play an important role in the process of transcellular Mg^{2+} transport in general and in particular in the Mg^{2+} influx step.

It is interesting to mention that parvalbumin and calbindins in addition to Ca^{2+} also bind Mg^{2+}(Eberhard and Erne 1994; Yang et al. 2002) with dissociation constants that are in the same order of magnitude as the respective cellular concentrations for these two ions. Importantly, TRPM6 colocalizes with parvalbumin in DCT1 and with calbindin-D_{28K} in DCT2 (Voets et al. 2004b; Fig. 3). Therefore, these proteins are candidates to fulfill the role of intracellular Mg^{2+} buffers in Mg^{2+} reabsorption in DCT; however, other buffers including members of the S100 family (Gribenko and Makhatadze 1998) cannot be excluded. A role of calbindins and parvalbumin in neuronal Ca^{2+} signaling has been studied using mouse knockout models, but possible effects on the Mg^{2+} balance in these mice have not yet been addressed (Caillard et al. 2000; Servais et al. 2005).

Mg^{2+} extrusion mechanisms

To date, little experimental data are available on the extrusion of Mg^{2+} across the basolateral membrane. The chemical gradient for Mg^{2+} across this plasma membrane is negligibly small. However, the negative membrane potential dictates the participation of primary (ATP consuming Mg^{2+}-pump) or secondary (Mg^{2+} efflux-coupled to Na^+ influx) active transport processes governing the exit of Mg^{2+}. At present, experimental data only support the involvement of the latter transport mechanism. Schweigel et al. addressed the Mg^{2+} efflux step of epithelial cells of the bovine forestomachs. This organ fulfills the Mg^{2+} absorption in ruminate animals (cow, sheep, goat, etc.) similar to intestinal Mg^{2+} absorption in monogastric animals and humans (Schweigel and Martens 2000). Mg^{2+} efflux of isolated rumen epithelial cells required the presence of extracellular Na^+ (Schweigel et al. 2006). The activation of the Mg^{2+} efflux by extracellular Na^+ followed a simple Michaelis-Menten relationship with a K_m of 24 mM, in line with observations in other cell types (Gunther and Vormann 1985). Previously, it was shown that inhibition of the Na^+-K^+-ATPase by ouabain reduces transcellular Mg^{2+} transport across isolated sheep rumen epithelia by 90% (Martens and Harmeyer 1978). Therefore, a Na^+-linked mechanism was suggested utilizing the electrochemical gradient of Na^+ (generated by Na^+-K^+-ATPase) to extrude Mg^{2+} via a Na^+/Mg^{2+} exchanger. Na^+/Mg^{2+} exchange has been proposed to regulate $[Mg^{2+}]_i$ of a variety of other cells types including human red blood cells (Feray and Garay 1986), suggesting that the Na^+/Mg^{2+} exchanger has a more ubiquitous expression. Schweigel et al. generated monoclonal antibodies using material from red blood cells as an antigen. This yielded an antibody that blocks Na^+/Mg^{2+} exchange in rumen epithelial cells. Importantly, this antibody recognized a product that is upregulated by high extracellular Mg^{2+} and detected a protein of approximately 70 kDa in lysates of bovine rumen epithelial cells and porcine red blood cells. This size is distinct from the size of the Na/Ca^{2+} exchanger, suggesting the presence of a specific Mg^{2+} efflux system. However, the molecular identity of this protein remains unknown.

Regulation of epithelial Mg^{2+} transport

Initially, Mg^{2+} transport was generally described as functionally characterized using electrophysiological methods employing isolated tissues or immortalized nonpolarized cells (Dai et al. 2001). Quamme and coworkers have performed several experiments using nonpolarized immortalized cells with a partial DCT phenotype. Their investigations on the regulation of Mg^{2+} influx have mainly relied on fluorescence measurements of $[Mg^{2+}]_i$ (Dai et al. 2001). To this end, cells were depleted for Mg^{2+} by incubation in Mg^{2+}-deficient medium for 16 h and subsequent exposure of the cells to Mg^{2+}-containing medium to measure the rate of Mg^{2+} influx. They suggested that Mg^{2+} entry into cultured DCT-like cells is mediated by a specific and regulated Mg^{2+} channel (Dai et al. 2001). However, electrophysiological measurements have not been performed, making it difficult to compare the Mg^{2+} influx properties of these cells with measured TRPM6 currents. In addition, several studies addressed the regulation of epithelial Mg^{2+} (re)absorption using micropuncture and microperfusion methods. Finally, the recent identification of TRPM6 allowed for studies performed at the molecular level, yielding novel insight into the regulation of epithelial Mg^{2+} transport. In the following we will present an overview of the regulation of epithelial Mg^{2+} transport and its possible molecular rationalization obtained using these various methods.

Diet-dependent Mg^{2+} (re)absorption

Mg^{2+} (re)absorption in the intestine and kidney is load dependent. In intestine, dietary Mg^{2+} content has been shown to affect the relative contribution of transcellular and paracellular Mg^{2+} transport. The fraction of transcellular Mg^{2+} absorption is saturable whereas paracellular Mg^{2+} absorption is not (Fine et al. 1991). In addition, Groenestege et al. have recently addressed the effect of the dietary Mg^{2+} content on the expression of Mg^{2+} transporters in intestine and kidney (Groenestege et al. 2006). Expression levels of TRPM6 mRNA in colon were upregulated by the Mg^{2+}-enriched diet, whereas Mg^{2+} restriction did not significantly affect TRPM6 mRNA expression levels. This suggests that mice can increase their transcellular Mg^{2+} absorption capacity when fed a Mg^{2+}-enriched diet (Groenestege et al. 2006). In contrast, in the same study an increased TRPM6 expression level upon dietary Mg^{2+} restriction was demonstrated in the kidney. Therefore, it was suggested that an excess of Mg^{2+} absorption as a result of high dietary Mg^{2+} intake together with TRPM6 upregulation in colon can be corrected by the kidney. Indeed, although the kidney normally excretes only 2–4% of the filtered Mg^{2+}, it is capable of increasing fractional excretion to nearly 100% in the face of increased plasma Mg^{2+} levels (Sutton and Domrongkitchaiporn 1993). Furthermore, the unaltered expression levels of TRPM6 mRNA in colon during Mg^{2+} restriction indicates that the Mg^{2+} absorptive capacity is sufficient to obtain maximal transcellular Mg^{2+} transport.

The load dependence of Mg^{2+} reabsorption in the kidney has been investigated by microperfusion (Massry et al. 1969; Quamme and Dirks 1980). Mg^{2+} absorption in PT is not saturable and increases linear with the luminal Mg^{2+} concentration or the delivered load, in line with the paracellular Mg^{2+} transport. In the loop of Henle, similar results were obtained. Interestingly, even in conditions when Mg^{2+} is absent from the luminal fluid, back-flow of Mg^{2+} from the blood side into the lumen was not detectable, indicating that paracellular Mg^{2+} transport in the TAL is strictly unidirectional (Quamme and Dirks 1980). Finally, the load dependence of Mg^{2+} absorption in DCT was measured. Although the absolute amount of Mg^{2+} that was absorbed in DCT increased with the enlarged load, the relative amount strongly decreased. This suggests that the Mg^{2+} absorptive capacity of DCT has a maximum. Interestingly, the fractional Mg^{2+} reabsorption in DCT decreased with hypermagnesemia. This is in line with the results of Groenestege et al. demonstrating an inverse relation between plasma Mg^{2+} levels and TRPM6 expression in DCT (Groenestege et al. 2006). Furthermore, it was postulated that elevated extracellular Mg^{2+} or Ca^{2+} inhibits fractional Mg^{2+} reabsorption in DCT through activation of the CaSR (Bapty et al. 1998a; Bapty et al. 1998b).

Hormonal control of epithelial Mg^{2+} transport

Although the significance of a strict Mg^{2+} balance is clearly reflected by the severity of pathology associated with hypomagnesemia and hypermagnesemia, a specific "magnesiotropic" hormone has not been identified (Kelepouris and Agus 1998). Several hormones, including PTH, calcitonin, $1,25(OH)_2D_3$, insulin, glucagons, antidiuretic hormone, aldosterone, and sex steroids have been reported to influence the Mg^{2+} balance (Bailly et al. 1984; Dai et al. 1999, 2001; Elalouf et al. 1983, 1984; Harris et al. 1979). PTH stimulated renal Mg^{2+} reabsorption in parathyroidectomized animals (Bailly et al. 1985; Harris et al. 1979). This stimulation was localized to TAL and DCT. In addition, Dai and Quamme showed that PTH enhances Mg^{2+} influx in immortalized DCT-like cells (Dai et al. 1999). This effect was accompanied by increased cAMP values, suggesting that PTH acts via PKA. Other cellular stimulations that are associated with increases in cytosolic cAMP levels, including prostaglandin E_2, vasopressin, glucagon, and insulin treatment, also enhanced Mg^{2+} influx in these cells (Dai et al. 2001). It was further shown that the signaling pathway underlying the effect of PTH on Mg^{2+} influx in immortalized DCT-like cells also involves PKC, as both inhibitors for PKA and PKC pathways blocked the PTH-mediated increase in Mg^{2+} influx (Dai et al. 2001). In contrast, PTH had no effect on TRPM6 and TRPM7 expression level in kidney, as the expression of these channels was not affected by parathyroidectomy alone or parathyroidectomy with subsequent pharmacological PTH supplementation (Groenestege et al. 2006). Similarly, $1,25(OH)_2D_3$

enhanced the influx of Mg^{2+} in a mouse DCT cell line (Ritchie et al. 2001), but this calciotropic hormone did not upregulate renal TRPM6 expression levels (Groenestege et al. 2006). The expression of TRPM6 was identical in wild-type mice, 25-hydroxyvitamin D_3-1α-hydroxylase knockout mice (showing no detectable $1,25(OH)_2D_3$ synthesis), and in the same knockout mice supplemented with $1,25(OH)_2D_3$. In addition, Karbach showed that cellular Mg^{2+} transport in rat colon is not responsive to $1,25(OH)_2D_3$ (Karbach 1989a, b). In contrast to $1,25(OH)_2D_3$ and PTH, estrogen displayed significant effects on the expression level of TRPM6 in kidney. In ovariectomized rats, the renal TRPM6 mRNA level was significantly reduced and subsequently normalized by 17β-estradiol (17β-E2) supplementation (Groenestege et al. 2006). These findings indicate a role for estrogens in Mg^{2+} homeostasis via regulation of TRPM6. Postmenopausal estrogen loss is associated with hypermagnesuria, which is corrected after estrogen substitution therapy (McNair et al. 1984). This finding could be explained by estrogen-mediated enhancement of renal TRPM6 expression resulting in increased Mg^{2+} reabsorption. The stimulatory effect of 17β-E2 could be due to enhanced transcriptional activity or mRNA stabilization. Thus far, sequence analysis indicated 17β-E2-responsive elements in the putative promoter sequence of human and mouse TRPM6.

Tacrolimus

Tacrolimus (also FK-506 or Fujimycin, tradename Prograf) was discovered as a novel immunosuppressant in 1984 from the fermentation broth of the bacteria *Streptomyces tsukubaensis* (Kino et al. 1987; Wallemacq and Reding 1993). Tacrolimus interacts with the immunophilin FKBP-12 (FK506 binding protein) (Liu et al. 1991). This complex binds and inhibits calcineurin, resulting in decreased T cell receptor signal transduction (Liu et al. 1991). Tacrolimus is mainly used as immunosuppressive drug to reduce the risk of organ rejection after transplant. Hypomagnesemia is a significant side effect of tacrolimus, even at relatively low doses (Lote et al. 2000; Nijenhuis et al. 2004). The effect of tacrolimus on plasma $[Mg^{2+}]$ is mediated via decreased renal tubular Mg^{2+} reabsorption (Lote et al. 2000; Nijenhuis et al. 2004). The underlying principle for the inappropriate hypermagnesuria, however, was unknown. Nijenhuis et al. have recently addressed the effect of tacrolimus on renal TRPM6 expression in male Wistar rats. These rats received tacrolimus by oral gavage for 7 days. Analysis of serum and urine samples showed a robust hypomagnesemia in the tacrolimus treated group (Nijenhuis et al. 2004). In line with a defect in renal Mg^{2+} reabsorption, a significant increase in urinary Mg^{2+} excretion was observed. Interestingly, these effects were accompanied by a significant reduced TRPM6 expression in kidney. The authors further addressed whether this downregulation reflects a general nephrotoxic effect on DCT, or a more specific effect on TRPM6 expression (Nijenhuis et al. 2004). The expression of several proteins with marked expression in DCT, including kallikrein (Zolotnitskaya and Satlin 1999) and NCC, did not significantly differ between controls and the tacrolimus-treated group. This indicates that no overt tacrolimus-mediated nephrotoxicity is present in DCT. Furthermore, no signs of a general deleterious effect of tacrolimus were detected, glomerular filtration rate was unaffected and enzymuria was not increased (Nijenhuis et al. 2004). Together, these factors suggest that tacrolimus has a direct effect on the expression of TRPM6 via an unknown regulatory pathway.

Thiazides

Thiazide diuretics are among the most commonly prescribed drugs, particularly in the treatment of hypertension. These compounds inhibit NCC present in the apical membrane of DCT to enhance renal Na^+ excretion (Gamba et al. 1993). Besides this diuretic effect, thiazides are known to cause hypocalciuria and hypomagnesemia. Several features of chronic thiazide treatment are mimicked in NCC knockout mice, which form an animal model for Gitelman's syndrome (Loffing et al. 2004; Schultheis et al. 1998). Intriguingly, the molecular mechanisms responsible for hypomagnesemia during thiazide administration and Gitelman's syndrome have remained elusive. Thiazides have opposing effects on the Mg^{2+} and Ca^{2+} balance. Therefore, direct inhibitory actions on active Mg^{2+} absorption in DCT have been proposed, as Ca^{2+} and Mg^{2+} are generally mutually reabsorbed in other segments of the nephron.

Identification of TRPM6 provides a powerful new tool to study the mechanism of active Mg^{2+} transport at the molecular level. A recent study by Voets et al., demonstrating complete colocalization of NCC with TRPM6, provided the first indication of a role of TRPM6 in thiazide-mediated hypomagnesemia (Voets et al. 2004b). Subsequently, Nijenhuis et al. addressed the mechanism of thiazide-induced hypomagnesemia. They showed that, although a single dose of thiazides resulted in a clear natriuresis within 24 h after administration, urinary Mg^{2+} excretion remained unaltered, which contradicts the hypothesis that Mg^{2+} reabsorption is directly inhibited by thiazide-mediated reduction in Na^+-Cl^- influx in DCT (Nijenhuis et al. 2005). Furthermore, the renal expression level of TRPM6 in NCC knockout mice, and in mice chronically receiving thiazides, was analyzed. Renal TRPM6 mRNA expression was significantly reduced in NCC knockout mice compared to control littermates. Furthermore, immunohistochemical analysis revealed that TRPM6 protein abundance along the apical membrane of the DCT is profoundly decreased in these mice. Similarly, renal TRPM6 expression was diminished in thiazide-treated animals. NCC expression was enhanced in these animals, illustrating that TRPM6 downregulation is a specific nondeleterious effect (Nijenhuis et al. 2005). On the whole, this demonstrates that chronic application of thiazide diuretics results in specific downregulation of renal TRPM6, resulting in inappropriate high renal Mg^{2+} excretion and hypomagnesemia.

Acid–base balance

Metabolic acidosis induces hypermagnesuria in several animal models (Nijenhuis et al. 2006; Shapiro et al. 1987; Wong et al. 1986) and in humans (Ariceta et al. 2004; Blumberg et al. 1998). Micropuncture experiments indicated that this effect was localized beyond the late PT (Wong et al. 1986). Likewise, metabolic alkalosis has a renal Mg^{2+} sparing effect and results in hypermagnesemia.

The role of TRPM6 in the acid–base effects on Mg^{2+} balance was recently addressed (Nijenhuis et al. 2006). NH_4Cl-induced chronic metabolic acidosis decreased renal TRPM6 abundance accompanied by increased Mg^{2+} excretion and hypomagnesemia. Conversely, chronic metabolic alkalosis increased TRPM6 expression as well as renal Mg^{2+} reabsorption, resulting in hypermagnesemia (Nijenhuis et al. 2006). These data indicate that regulation of TRPM6 explains the effects of acid–base status on renal Mg^{2+} handling. However, other studies support the existence of additional mechanisms to explain acid–base-mediated effects on epithelial Mg^{2+} transport.

First, an alkaline extracellular pH was shown to enhance Mg^{2+} uptake in immortalized mouse DCT cells, and conversely a low pH diminished this uptake. This effect was immediate and did not require preincubation of the cells with low or high pH (Dai et al. 1997). Therefore, pH-mediated effects on TRPM6 expression seem unlikely at this timescale. Furthermore, Li et al. recently indicated that TRPM6 currents are increased upon a shift of the extracellular medium pH from 7.0 to lower values with a half-maximal activation at pH 4.3 (Li et al. 2006). At present, it is unclear how these findings reconcile with the reduced epithelial Mg^{2+} transport in acidosis. Furthermore, it is possible that the acidosis-mediated downregulation of TRPM6 provides the determining factor for Mg^{2+} transport, as the proton-dependence of TRPM6 channel activity cannot explain the hypermagnesuria during acidosis. Interestingly, inhibition of carbonic anhydrase by acetazolamide displayed a Mg^{2+}-sparing effect in mice (Nijenhuis et al. 2006) and humans (Sutton and Walker 1980) via an unknown mechanism. Treatment with this drug resulted in acidosis. However, the urine pH was alkaline, in contrast to situations with NH_4Cl-induced acidosis. Importantly, TRPM6 expression was significantly diminished during chronic acetazolamide treatment (Nijenhuis et al. 2006). This suggests that chronic metabolic acidosis results in TRPM6 downregulation, irrespective of the luminal pH.

Downregulation of TRPM6 does not support, however, the stimulated active Mg^{2+} reabsorption as a likely explanation for the decreased Mg^{2+} excretion during acetazolamide exposure. This treatment was associated with a larger urine volume and Na^+ excretion, which has been shown to result in increased mRNA and protein abundance (Attmane-Elakeb et al. 1998) and activity (Kwon et al. 2003) of the Na^+-K^+-$2Cl^-$ cotransporter (NKCC2) in the TAL. Therefore, it was postulated that these additive effects of acetazolamide enhance passive Mg^{2+} reabsorption in TAL, and that this dominates the effect of TRPM6 downregulation in DCT (Nijenhuis et al. 2006).

Other factors affecting epithelial Mg^{2+} transport

Although our understanding of epithelial Mg^{2+} transport has greatly improved in recent years, information concerning the molecular regulation of this process is still relatively scarce. It is anticipated that the molecular mechanism of multiple factors related to changes in the Mg^{2+} balance will be elucidated in the near future. For instance, it is currently unclear why hypomagnesemia is observed so frequently in the clinical setting, as this occurs in up to 12% of hospitalized patients, a number which rises to about 60% in the intensive care setting (Aglio et al. 1991; Agus 1999). Furthermore, disturbances in the Mg^{2+} balance are associated with diabetes mellitus. Urinary Mg^{2+} excretion in diabetic adolescents is significantly higher than in healthy persons, with higher excretion in boys than in girls (Driziene et al. 2005). In addition, it has been suggested that Mg^{2+} intake may be inversely related to the risk of hypertension and type 2 diabetes mellitus and that decreased cellular and plasma Mg^{2+} concentration is related to impaired insulin efficacy (He et al. 2006; Huerta et al. 2005; Kao et al. 1999; Paolisso and Barbagallo 1997).

Although several studies point to an association of diabetes with renal Mg^{2+} wasting, the etiology of the hypomagnesemia is largely unknown. In a recent study Lee et al. investigated the effect of streptozotocin-induced diabetes on the expression of claudin-16 and TRPM6 (Lee et al. 2006a). They showed that diabetic rats have a significant increase in the fractional excretion of Mg^{2+} and Ca^{2+}, but not of Na^+. Remarkably, a significant increase in mRNA levels of TRPM6 was observed. No change was found in claudin-16 mRNA or protein levels. Furthermore, several transport proteins including TRPV5, TRPV6, calbindin-D_{28K}, and also NCC were upregulated. Insulin administration completely corrected the hyperglycemia-associated hypercalciuria and hypermagnesuria, and normalized the augment of Ca^{2+} transporter and TRPM6 abundance (Lee et al. 2006a). These findings suggest that the increased

TRPM6 expression might reflect an adaptation to the higher Mg^{2+} load present in the lumen of DCT. Furthermore, this could indicate a compensatory mechanism for the diabetes-induced Mg^{2+} wasting, which was mainly localized to TAL (Garland 1992; Garland et al. 1991). The etiology of this transport defect in TAL is unknown and could be related to several factors including the level of hyperglycemia (Djurhuus et al. 2000), insulin concentration, or tubular alterations in osmolarity, pH, or membrane potential.

Mutual regulation of epithelial Ca^{2+} and Mg^{2+} transport

A coupling of the Ca^{2+} and Mg^{2+} balance is observed in certain pathological conditions and animal models (Groenestege et al. 2006; Hebert et al. 1997; Simon et al. 1999). For several situations, there is consensus in the field on the molecular mechanism that links transport of these divalent ions, whereas in some conditions the reason for this coupling remains unknown.

To explain the HHN phenotype, it was proposed that claudin-16 either controls the Ca^{2+} and Mg^{2+} permeability of the paracellular pathway in TAL, or the driving force for the reabsorption of both ions. Therefore, mutations in claudin-16 affect both Ca^{2+} and Mg^{2+} reabsorption, although the effect is more prominent for Mg^{2+}. In addition, mutations in the CaSR are associated with disturbed Ca^{2+} and Mg^{2+} handling. Mutations in the CaSR resulted in a lower set point for plasma Ca^{2+} and Mg^{2+} to activate the receptor. Consequently, renal Ca^{2+} and Mg^{2+} reabsorption and PTH secretion are suppressed, resulting in inappropriately low plasma PTH levels, and increased Ca^{2+} and Mg^{2+} excretion.

However, in other conditions Ca^{2+} and Mg^{2+} reabsorption are oppositely affected. Patients with mutations in TRPM6, the γ-subunit of the Na^+- K^+-ATPase, or NCC all exhibit hypermagnesuria, while renal Ca^{2+} excretion is reduced. The expression of these particular Na^+, Ca^{2+} and Mg^{2+} transporting proteins is restricted to DCT, suggesting a direct regulatory connection in this nephron segment.

The nature of this interaction between transcellular Ca^{2+} and Mg^{2+} pathways in the distal part of the nephron, however, is still unclear. There is limited overlap in expression between the Ca^{2+} transport proteins and TRPM6, γ-subunit of the Na^+- K^+-ATPase, or NCC (Fig. 3). Furthermore, it is interesting that hypocalcemia in HSH patients can be corrected only by supplementation of the diet with high amounts of Mg^{2+}, probably linked to restoration of PTH secretion and efficacy (Konrad and Weber 2003). Similarly, alterations of Ca^{2+} and Mg^{2+} excretion in Gitelman's syndrome have been attributed to distinct mechanisms. Renal Mg^{2+} loss in patients with chronic thiazide treatment, with Gitelman's syndrome, and in NCC knockout mice is most likely due to reduced TRPM6 expression, whereas the increased Ca^{2+} reabsorption is mediated by adaptive mechanisms in the PT to compensate for the hypovolemia resulting from reduced or abolished NCC function (Nijenhuis et al. 2005). Furthermore, mutations in the γ-subunit of Na^+-K^+-ATPase are the cause of dominant isolated hypomagnesemia with hypocalciuria. It was proposed that the mutated γ-subunit impairs the activity of the Na^+,K^+-ATPase, resulting in reduced $[K^+]_i$, increased $[Na^+]_i$ or depolarization of the plasma membrane (Meij et al. 2000). This might subsequently lead to reduced Mg^{2+} influx through the apical TRPM6 channel, resulting in Mg^{2+} wasting. However, the molecular mechanism of the decreased Mg^{2+} reabsorption and the associated hypocalciuria remains to be further substantiated in this disorder.

On the whole, many diseases show disturbances in both Ca^{2+} and Mg^{2+} balance. In some cases, there is an explanation for the mutual disorder in divalent renal handling, but in the majority of the diseases, the origin of this coupling is still unclear. Particularly, the limited segmental overlap between the Mg^{2+} transport (DCT1-DCT2) and Ca^{2+} transport (DCT2-CNT) machinery suggests that additional mechanisms might be involved in the kidney.

Future perspective

In the last decade, significant advances were made in the field of epithelial Ca^{2+} and Mg^{2+} (re)absorption. The identification of the proteins mediating this transport, including proteins involved in paracellular Ca^{2+}/Mg^{2+} transport (claudin-16), active Ca^{2+} transport (TRPV5/6 and novel channel associated proteins), and active Mg^{2+} transport (TRPM6) has provided novel insight and means to study the molecular aspects of divalent ion transport. Several questions on the molecular mechanisms of divalent ion transport remain unknown. One example is the unknown etiology of hypomagnesemia in several diseases, suggesting that novel molecular players involved in epithelial Mg^{2+} transport remain to be elucidated. Although the identification of TRPM6 provided a first view on the luminal Mg^{2+} influx pathway, molecular data explaining the diffusion and basolateral extrusion of Mg^{2+} are still elusive (Fig. 1). Furthermore, in view of life-threatening consequences of large deviations in the plasma $[Mg^{2+}]$, it is surprising how little we know about the maintenance of the Mg^{2+} balance. In conclusion, the timely area of epithelial Ca^{2+} and Mg^{2+} transport is very dynamic and will likely remain so for the years to come. The large number of recent novel developments and the ones to be expected in the near future will further increase our understanding of epithelial ion homeostasis and provide new insights in the diagnoses and management of corresponding diseases.

Acknowledgements. S.F.G. was supported by an EMBO long-term fellowship (ALTF 700–2005). The authors were further supported by grants of the Dutch Organization of Scientific Research (ZonMw 016.006.001, ZonMW 9120.6110, NWO-ALW 814.02.001), the Stomach Liver Intestine foundation (MWO 03–19), Human Frontiers Science Program (RGP32/2004), and the Dutch Kidney foundation (C02.2030, C03.6017, C05.2134).

References

Aarts M, Iihara K, Wei WL, Xiong ZG, Arundine M, Cerwinski W, MacDonald JF, Tymianski M (2003) A key role for TRPM7 channels in anoxic neuronal death. Cell 115:863–877

Abeele FV, Shuba Y, Roudbaraki M, Lemonnier L, Vanoverberghe K, Mariot P, Skryma R, Prevarskaya N (2003) Store-operated Ca^{2+} channels in prostate cancer epithelial cells: function, regulation, and role in carcinogenesis. Cell Calcium 33:357–373

Aglio LS, Stanford GG, Maddi R, Boyd JL 3rd, Nussbaum S, Chernow B (1991) Hypomagnesemia is common following cardiac surgery. J Cardiothorac Vasc Anesth 5:201–208

Agus ZS (1999) Hypomagnesemia. J Am Soc Nephrol 10:1616–1622

Airaksinen MS, Eilers J, Garaschuk O, Thoenen H, Konnerth A, Meyer M (1997a) Ataxia and altered dendritic calcium signaling in mice carrying a targeted null mutation of the calbindin D28K gene. Proc Natl Acad Sci USA 94:1488–1493

Airaksinen MS, Thoenen H, Meyer M (1997b) Vulnerability of midbrain dopaminergic neurons in calbindin-D28k-deficient mice: lack of evidence for a neuroprotective role of endogenous calbindin in MPTP-treated and weaver mice. Eur J Neurosci 9:120–127

Ariceta G, Vallo A, Rodriguez-Soriano J (2004) Acidosis increases magnesiuria in children with distal renal tubular acidosis. Pediatr Nephrol 19:1367–1370

Arking DE, Krebsova A, Macek M Sr, Macek M Jr, Arking A, Mian IS, Fried L, Hamosh A, Dey S, McIntosh I, Dietz HC (2002) Association of human aging with a functional variant of klotho. Proc Natl Acad Sci USA 99:856–861

Arking DE, Becker DM, Yanek LR, Fallin D, Judge DP, Moy TF, Becker LC, Dietz HC (2003) KLOTHO allele status and the risk of early-onset occult coronary artery disease. Am J Hum Genet 72:1154–1161

Arking DE, Atzmon G, Arking A, Barzilai N, Dietz HC (2005) Association between a functional variant of the KLOTHO gene and high-density lipoprotein cholesterol, blood pressure, stroke, and longevity. Circ Res 96:412–418

Arniges M, Fernandez-Fernandez JM, Albrecht N, Schaefer M, Valverde MA (2006) Human TRPV4 channel splice variants revealed a key role of ankyrin domains in multimerization and trafficking. J Biol Chem 281:1580–1586

Arnold DB, Heintz N (1997) A calcium responsive element that regulates expression of two calcium binding proteins in Purkinje cells. Proc Natl Acad Sci USA 94:8842–8847

Ashby MC, Tepikin AV (2002) Polarized calcium and calmodulin signaling in secretory epithelia. Physiol Rev 82:701–734

Attmane-Elakeb A, Mount DB, Sibella V, Vernimmen C, Hebert SC, Bichara M (1998) Stimulation by in vivo and in vitro metabolic acidosis of expression of rBSC-1, the Na^{+}-K^{+}($NH4^{+}$)-2Cl– cotransporter of the rat medullary thick ascending limb. J Biol Chem 273:33681–33691

Avioli LV, Berman M (1966) Mg28 kinetics in man. J Appl Physiol 21:1688–1694

Bachmann S, Velazquez H, Obermuller N, Reilly RF, Moser D, Ellison DH (1995) Expression of the thiazide-sensitive Na-Cl cotransporter by rabbit distal convoluted tubule cells. J Clin Invest 96:2510–2514

Bacskai BJ, Friedman PA (1990) Activation of latent Ca^{2+} channels in renal epithelial cells by parathyroid hormone. Nature 347:388–391

Bailly C, Roinel N, Amiel C (1984) PTH-like glucagon stimulation of Ca and Mg reabsorption in Henle's loop of the rat. Am J Physiol 246:F205–F212

Bailly C, Roinel N, Amiel C (1985) Stimulation by glucagon and PTH of Ca and Mg reabsorption in the superficial distal tubule of the rat kidney. Pflugers Arch 403:28–34

Bailly C, Imbert-Teboul M, Roinel N, Amiel C (1990) Isoproterenol increases Ca, Mg, and NaCl reabsorption in mouse thick ascending limb. Am J Physiol 258:F1224–F1231

Baker PF, Blaustein MP, Hodgkin AL, Steinhardt RA (1969) The influence of calcium on sodium efflux in squid axons. J Physiol 200:431–458

Balkovetz DF (2006) Claudins at the gate: determinants of renal epithelial tight junction paracellular permeability. Am J Physiol Renal Physiol 290:F572–F579

Balsan S, Garabedian M, Larchet M, Gorski AM, Cournot G, Tau C, Bourdeau A, Silve C, Ricour C (1986) Long-term nocturnal calcium infusions can cure rickets and promote normal mineralization in hereditary resistance to 1,25-dihydroxyvitamin D. J Clin Invest 77:1661–1667

Bapty BW, Dai LJ, Ritchie G, Canaff L, Hendy GN, Quamme GA (1998a) Mg^{2+}/Ca^{2+} sensing inhibits hormone-stimulated Mg^{2+} uptake in mouse distal convoluted tubule cells. Am J Physiol 275:F353–F360

Bapty BW, Dai LJ, Ritchie G, Jirik F, Canaff L, Hendy GN, Quamme GA (1998b) Extracellular Mg^{2+}- and Ca^{2+}-sensing in mouse distal convoluted tubule cells. Kidney Int 53:583–592

Barley NF, Howard A, O'Callaghan D, Legon S, Walters JR (2001) Epithelial calcium transporter expression in human duodenum. Am J Physiol Gastrointest Liver Physiol 280:G285–G290

Barski JJ, Hartmann J, Rose CR, Hoebeek F, Morl K, Noll-Hussong M, De Zeeuw CI, Konnerth A, Meyer M (2003) Calbindin in cerebellar Purkinje cells is a critical determinant of the precision of motor coordination. J Neurosci 23:3469–3477

Bastin J, Cambon N, Thompson M, Lowry OH, Burch HB (1987) Change in energy reserves in different segments of the nephron during brief ischemia. Kidney Int 31:1239–1247

Baumann JM (1998) Stone prevention: why so little progress? Urol Res 26:77–81

Bawden JW (1989) Calcium transport during mineralization. Anat Rec 224:226–233

Beck L, Karaplis AC, Amizuka N, Hewson AS, Ozawa H, Tenenhouse HS (1998) Targeted inactivation of Npt2 in mice leads to severe renal phosphate wasting, hypercalciuria, and skeletal abnormalities. Proc Natl Acad Sci USA 95:5372–5377

Belkacemi L, Simoneau L, Lafond J (2002) Calcium-binding proteins: distribution and implication in mammalian placenta. Endocrine 19:57–64

Belkacemi L, Gariepy G, Mounier C, Simoneau L, Lafond J (2003) Expression of calbindin-D28K (CaBP28K) in trophoblasts from human term placenta. Biol Reprod 68:1943–1950

Belkacemi L, Gariepy G, Mounier C, Simoneau L, Lafond J (2004) Calbindin-D9K (CaBP9K) localization and levels of expression in trophoblast cells from human term placenta. Cell Tissue Res 315:107–117

Belkacemi L, Bedard I, Simoneau L, Lafond J (2005) Calcium channels, transporters and exchangers in placenta: a review. Cell Calcium 37:1–8

Benais-Pont G, Punn A, Flores-Maldonado C, Eckert J, Raposo G, Fleming TP, Cereijido M, Balda MS, Matter K (2003) Identification of a tight junction-associated guanine nucleotide exchange factor that activates Rho and regulates paracellular permeability. J Cell Biol 160:729–740

Bindels RJ, Hartog A, Timmermans J, Van Os CH (1991) Active Ca^{2+} transport in primary cultures of rabbit kidney CCD: stimulation by 1,25-dihydroxyvitamin D3 and PTH. Am J Physiol 261:F799–807

Bindels RJ, Ramakers PL, Dempster JA, Hartog A, van Os CH (1992) Role of Na^+/Ca^{2+} exchange in transcellular Ca^{2+} transport across primary cultures of rabbit kidney collecting system. Pflugers Arch 420:566–572

Bindels RJ, Dempster JA, Ramakers PL, Willems PH, van Os CH (1993) Effect of protein kinase C activation and down-regulation on active calcium transport. Kidney Int 43:295–300

Bindels RJ, Hartog A, Abrahamse SL, Van Os CH (1994) Effects of pH on apical calcium entry and active calcium transport in rabbit cortical collecting system. Am J Physiol 266:F620–F627

Biner HL, Arpin-Bott MP, Loffing J, Wang X, Knepper M, Hebert SC, Kaissling B (2002) Human cortical distal nephron: distribution of electrolyte and water transport pathways. J Am Soc Nephrol 13:836–847

Birkenfeld J, Kartmann B, Anliker B, Ono K, Schlotcke B, Betz H, Roth D (2003) Characterization of zetin 1/rBSPRY, a novel binding partner of 14-3-3 proteins. Biochem Biophys Res Commun 302:526–533

Blaustein MP, Lederer WJ (1999) Sodium/calcium exchange: its physiological implications. Physiol Rev 79:763–854

Blaustein MP, Santiago EM (1977) Effects of internal and external cations and of ATP on sodium-calcium and calcium-calcium exchange in squid axons. Biophys J 20:79–111

Blaustein MP, Juhaszova M, Golovina VA, Church PJ, Stanley EF (2002) Na/Ca exchanger and PMCA localization in neurons and astrocytes: functional implications. Ann NY Acad Sci 976:356–366

Blumberg D, Bonetti A, Jacomella V, Capillo S, Truttmann AC, Luthy CM, Colombo JP, Bianchetti MG (1998) Free circulating magnesium and renal magnesium handling during acute metabolic acidosis in humans. Am J Nephrol 18:233–236

Bodding M, Wissenbach U, Flockerzi V (2002) The recombinant human TRPV6 channel functions as Ca^{2+} sensor in human embryonic kidney and rat basophilic leukemia cells. J Biol Chem 277:36656–36664

Borden KL (1998) RING fingers and B-boxes: zinc-binding protein-protein interaction domains. Biochem Cell Biol 76:351–358

Borghi L, Schianchi T, Meschi T, Guerra A, Allegri F, Maggiore U, Novarini A (2002) Comparison of two diets for the prevention of recurrent stones in idiopathic hypercalciuria. N Engl J Med 346:77–84

Borke JL, Minami J, Verma A, Penniston JT, Kumar R (1987) Monoclonal antibodies to human erythrocyte membrane Ca^{2+}-Mg^{2+} adenosine triphosphatase pump recognize an epitope in the basolateral membrane of human kidney distal tubule cells. J Clin Invest 80:1225–1231

Borke JL, Caride A, Verma AK, Penniston JT, Kumar R (1989) Plasma membrane calcium pump and 28-kDa calcium binding protein in cells of rat kidney distal tubules. Am J Physiol 257:F842–F849

Bouillon R, Van Cromphaut S, Carmeliet G (2003) Intestinal calcium absorption: molecular vitamin D mediated mechanisms. J Cell Biochem 88:332–339

Boulay G (2002) Ca^{2+}-calmodulin regulates receptor-operated Ca^{2+} entry activity of TRPC6 in HEK-293 cells. Cell Calcium 32:201–207

Bourdeau JE, Burg MB (1979) Voltage dependence of calcium transport in the thick ascending limb of Henle's loop. Am J Physiol 236:F357–F364

Bourdeau JE, Burg MB (1980) Effect of PTH on calcium transport across the cortical thick ascending limb of Henle's loop. Am J Physiol 239:F121–F126

Bourdeau JE, Langman CB, Bouillon R (1987) Parathyroid hormone-stimulated calcium absorption in cTAL from vitamin D-deficient rabbits. Kidney Int 31:913–917

Bowl MR, Nesbit MA, Harding B, Levy E, Jefferson A, Volpi E, Rizzoti K, Lovell-Badge R, Schlessinger D, Whyte MP, Thakker RV (2005) An interstitial deletion-insertion involving chromosomes 2p25.3 and Xq27.1, near SOX3, causes X-linked recessive hypoparathyroidism. J Clin Invest 115:2822–2831

Bronner F (2003) Mechanisms of intestinal calcium absorption. J Cell Biochem 88:387–393

Bronner F, Pansu D (1999) Nutritional aspects of calcium absorption. J Nutr 129:9–12

Bronner F, Stein WD (1988) CaBPr facilitates intracellular diffusion for Ca pumping in distal convoluted tubule. Am J Physiol 255:F558–F562

Bronner F, Pansu D, Stein WD (1986) An analysis of intestinal calcium transport across the rat intestine. Am J Physiol 250:G561–G569

Brown EM (1991) Extracellular Ca^{2+} sensing, regulation of parathyroid cell function, and role of Ca^{2+} and other ions as extracellular (first) messengers. Physiol Rev 71:371–411

Brown EM, MacLeod RJ (2001) Extracellular calcium sensing and extracellular calcium signaling. Physiol Rev 81:239–297

Brown EM, Gamba G, Riccardi D, Lombardi M, Butters R, Kifor O, Sun A, Hediger MA, Lytton J, Hebert SC (1993) Cloning and characterization of an extracellular Ca^{2+}-sensing receptor from bovine parathyroid. Nature 366:575–580

Brown PS, Wang E, Aroeti B, Chapin SJ, Mostov KE, Dunn KW (2000) Definition of distinct compartments in polarized Madin-Darby canine kidney (MDCK) cells for membrane-volume sorting, polarized sorting and apical recycling. Traffic 1:124–140

Brunette MG (1988) Calcium transport through the placenta. Can J Physiol Pharmacol 66:1261–1269

Bui DM, Gregan J, Jarosch E, Ragnini A, Schweyen RJ (1999) The bacterial magnesium transporter CorA can functionally substitute for its putative homologue Mrs2p in the yeast inner mitochondrial membrane. J Biol Chem 274:20438–20443

Bulger RE, Tisher CC, Myers CH, Trump BF (1967) Human renal ultrastructure. II. The thin limb of Henle's loop and the interstitium in healthy individuals. Lab Invest 16:124–141

Bushinsky DA, Kittaka MK, Weisinger JR, Langman CB, Favus MJ (1989) Effects of chronic metabolic alkalosis on Ca^{2+}, PTH and 1,25(OH)2D3 in the rat. Am J Physiol 257:E578–E582

Cai X, Lytton J (2004) Molecular cloning of a sixth member of the K^+-dependent Na^+/Ca^{2+} exchanger gene family, NCKX6. J Biol Chem 279:5867–5876

Caillard O, Moreno H, Schwaller B, Llano I, Celio MR, Marty A (2000) Role of the calcium-binding protein parvalbumin in short-term synaptic plasticity. Proc Natl Acad Sci USA 97:13372–13377

Canzanello VJ, Bodvarsson M, Kraut JA, Johns CA, Slatopolsky E, Madias NE (1990) Effect of chronic respiratory acidosis on urinary calcium excretion in the dog. Kidney Int 38:409–416

Carafoli E (1985) The homeostasis of calcium in heart cells. J Mol Cell Cardiol 17:203–212

Carani C, Qin K, Simoni M, Faustini-Fustini M, Serpente S, Boyd J, Korach KS, Simpson ER (1997) Effect of testosterone and estradiol in a man with aromatase deficiency. N Engl J Med 337:91–95

Caride AJ, Chini EN, Homma S, Penniston JT, Dousa TP (1998) mRNA encoding four isoforms of the plasma membrane calcium pump and their variants in rat kidney and nephron segments. J Lab Clin Med 132:149–156

Casanova JE, Wang X, Kumar R, Bhartur SG, Navarre J, Woodrum JE, Altschuler Y, Ray GS, Goldenring JR (1999) Association of Rab25 and Rab11a with the apical recycling system of polarized Madin-Darby canine kidney cells. Mol Biol Cell 10:47–61

Caterina MJ, Julius D (2001) The vanilloid receptor: a molecular gateway to the pain pathway. Annu Rev Neurosci 24:487–517

Chang Q, Gyftogianni E, van de Graaf SF, Hoefs S, Weidema FA, Bindels RJ, Hoenderop JG (2004) Molecular determinants in TRPV5 channel assembly. J Biol Chem 279:54304–54311

Chang Q, Hoefs S, van der Kemp AW, Topala CN, Bindels RJ, Hoenderop JG (2005) The beta-glucuronidase klotho hydrolyzes and activates the TRPV5 channel. Science 310:490–493

Chantret I, Barbat A, Dussaulx E, Brattain MG, Zweibaum A (1988) Epithelial polarity, villin expression, and enterocytic differentiation of cultured human colon carcinoma cells: a survey of twenty cell lines. Cancer Res 48:1936–1942

Chase LR, Slatopolsky E (1974) Secretion and metabolic efficacy of parathyroid hormone in patients with severe hypomagnesemia. J Clin Endocrinol Metab 38:363–371

Chen H, Hewison M, Hu B, Adams JS (2003) Heterogeneous nuclear ribonucleoprotein (hnRNP) binding to hormone response elements: a cause of vitamin D resistance. Proc Natl Acad Sci USA 100:6109–6114

Chery M, Biancalana V, Philippe C, Malpuech G, Carla H, Gilgenkrantz S, Mandel JL, Hanauer A (1994) Hypomagnesemia with secondary hypocalcemia in a female with balanced X;9 translocation: mapping of the Xp22 chromosome breakpoint. Hum Genet 93:587–591

Chubanov V, Waldegger S, Mederos y Schnitzler M, Vitzthum H, Sassen MC, Seyberth HW, Konrad M, Gudermann T (2004) Disruption of TRPM6/TRPM7 complex formation by a mutation in the TRPM6 gene causes hypomagnesemia with secondary hypocalcemia. Proc Natl Acad Sci USA 101:2894–2899

Clapham DE (2003) TRP channels as cellular sensors. Nature 426:517–524

Clark K, Langeslag M, van Leeuwen B, Ran L, Ryazanov AG, Figdor CG, Moolenaar WH, Jalink K, van Leeuwen FN (2006) TRPM7, a novel regulator of actomyosin contractility and cell adhesion. EMBO J 25:290–301

Cole DE, Quamme GA (2000) Inherited disorders of renal magnesium handling. J Am Soc Nephrol 11:1937–1947

Colegio OR, Van Itallie CM, McCrea HJ, Rahner C, Anderson JM (2002) Claudins create charge-selective channels in the paracellular pathway between epithelial cells. Am J Physiol Cell Physiol 283:C142–C147

Colussi G, Rombola G, Brunati C, De Ferrari ME (1997) Abnormal reabsorption of Na^+/Cl– by the thiazide-inhibitable transporter of the distal convoluted tubule in Gitelman's syndrome. Am J Nephrol 17:103–111

Costanzo LS, Windhager EE (1978) Calcium and sodium transport by the distal convoluted tubule of the rat. Am J Physiol 235:F492–506

Costanzo LS, Windhager EE, Ellison DH (2000) Calcium and sodium transport by the distal convoluted tubule of the rat. J Am Soc Nephrol 11:1562–1580

Cotter AA, Cashman KD (2006) Effect of 17beta-oestradiol on transepithelial calcium transport in human intestinal-like Caco-2 cells and its interactions with 1,25-dihydroxycholecalciferol and 9-cis retinoic acid. Eur J Nutr 45:234–241

Crayen ML, Thoenes W (1978) Architecture and cell structures in the distal nephron of the rat kidney. Cytobiologie 17:197–211

Croner R, Schwille PO, Erben RG, Gepp H, Stahr K, Rumenapf G, Parth R, Scheuerlein H (2000) Effects of partial and total colectomy on mineral and acid-base homoeostasis in the rat: magnesium deficiency, hyperphosphaturia and osteopathy, in the presence of high serum 1,25-dihydroxyvitamin D but normal parathyroid hormone. Clin Sci (Lond) 98:649–659

Dai LJ, Friedman PA, Quamme GA (1997) Acid-base changes alter Mg^{2+} uptake in mouse distal convoluted tubule cells. Am J Physiol 272:F759–F766

Dai LJ, Ritchie G, Bapty BW, Kerstan D, Quamme GA (1999) Insulin stimulates Mg^{2+} uptake in mouse distal convoluted tubule cells. Am J Physiol 277:F907–F913

Dai LJ, Ritchie G, Kerstan D, Kang HS, Cole DE, Quamme GA (2001) Magnesium transport in the renal distal convoluted tubule. Physiol Rev 81:51–84

Dardenne O, Prud'homme J, Arabian A, Glorieux FH, St-Arnaud R (2001) Targeted inactivation of the 25-hydroxyvitamin D3-1(alpha)-hydroxylase gene (CYP27B1) creates an animal model of pseudovitamin D-deficiency rickets. Endocrinology 142:3135–3141

Dardenne O, Prud'homme J, Hacking SA, Glorieux FH, St-Arnaud R (2003a) Correction of the abnormal mineral ion homeostasis with a high-calcium, high-phosphorus, high-lactose diet rescues the PDDR phenotype of mice deficient for the 25-hydroxyvitamin D-1alpha-hydroxylase (CYP27B1). Bone 32:332–340

Dardenne O, Prudhomme J, Hacking SA, Glorieux FH, St-Arnaud R (2003b) Rescue of the pseudo-vitamin D deficiency rickets phenotype of CYP27B1-deficient mice by treatment with 1,25-dihydroxyvitamin D3: biochemical, histomorphometric, and biomechanical analyses. J Bone Miner Res 18:637–643

de Jong JC, van der Vliet A, van den Heuvel LP, Willems PH, Knoers NV, Bindels RJ (2002) Functional expression of mutations in the human NaCl cotransporter: evidence for impaired routing mechanisms in Gitelman's syndrome. J Am Soc Nephrol 13:1442–1448

de Rouffignac C, Quamme G (1994) Renal magnesium handling and its hormonal control. Physiol Rev 74:305–322

DeLuca HF (2004) Overview of general physiologic features and functions of vitamin D. Am J Clin Nutr 80:1689S–1696S

Demeuse P, Penner R, Fleig A (2006) TRPM7 channel is regulated by magnesium nucleotides via its kinase domain. J Gen Physiol 127:421–434

Di Stefano A, Wittner M, Nitschke R, Braitsch R, Greger R, Bailly C, Amiel C, Elalouf JM, Roinel N, de Rouffignac C (1989) Effects of glucagon on Na^+, Cl–, K^+, Mg^{2+} and Ca^{2+} transports in cortical and medullary thick ascending limbs of mouse kidney. Pflugers Arch 414:640–646

Di Stefano A, Wittner M, Nitschke R, Braitsch R, Greger R, Bailly C, Amiel C, Roinel N, de Rouffignac C (1990) Effects of parathyroid hormone and calcitonin on Na^+, Cl–, K^+, Mg^{2+} and Ca^{2+} transport in cortical and medullary thick ascending limbs of mouse kidney. Pflugers Arch 417:161–167

DiPolo R (1979) Calcium influx in internally dialyzed squid giant axons. J Gen Physiol 73:91–113

Dirks JH, Cirksena WJ, Berliner RW (1966) Micropuncture study of the effect of various diuretics on sodium reabsorption by the proximal tubules of the dog. J Clin Invest 45:1875–1885

Dixon LB, Pellizzon MA, Jawad AF, Tershakovec AM (2005) Calcium and dairy intake and measures of obesity in hyper- and normocholesterolemic children. Obes Res 13:1727–1738

Djurhuus MS, Skott P, Vaag A, Hother-Nielsen O, Andersen P, Parving HH, Klitgaard NA (2000) Hyperglycaemia enhances renal magnesium excretion in type 1 diabetic patients. Scand J Clin Lab Invest 60:403–409

Dobnig H, Turner RT (1997) The effects of programmed administration of human parathyroid hormone fragment (1–34) on bone histomorphometry and serum chemistry in rats. Endocrinology 138:4607–4612

Dodier Y, Banderali U, Klein H, Topalak O, Dafi O, Simoes M, Bernatchez G, Sauve R, Parent L (2004) Outer pore topology of the ECaC-TRPV5 channel by cysteine scan mutagenesis. J Biol Chem 279:6853–6862

Donier E, Rugiero F, Okuse K, Wood JN (2005) Annexin II light chain p11 promotes functional expression of acid-sensing ion channel ASIC1a. J Biol Chem 280:38666–38672

Dorovkov MV, Ryazanov AG (2004) Phosphorylation of annexin I by TRPM7 channel-kinase. J Biol Chem 279:50643–50646

Doucet A, Katz AI (1982) High-affinity Ca-Mg-ATPase along the rabbit nephron. Am J Physiol 242:F346–F352

Doyle DA, Morais Cabral J, Pfuetzner RA, Kuo A, Gulbis JM, Cohen SL, Chait BT, MacKinnon R (1998) The structure of the potassium channel: molecular basis of K^+ conduction and selectivity. Science 280:69–77

Drennan D, Ryazanov AG (2004) Alpha-kinases: analysis of the family and comparison with conventional protein kinases. Prog Biophys Mol Biol 85:1–32

Drenth JP, te Morsche RH, Smink R, Bonifacino JS, Jansen JB (2003) Germline mutations in PRKCSH are associated with autosomal dominant polycystic liver disease. Nat Genet 33:345–347

Drenth JP, Martina JA, Te Morsche RH, Jansen JB, Bonifacino JS (2004) Molecular characterization of hepatocystin, the protein that is defective in autosomal dominant polycystic liver disease. Gastroenterology 126:1819–1827

Driziene Z, Stakisaitis D, Balsiene J (2005) Magnesium urinary excretion in diabetic adolescents. Acta Medica (Hradec Kralove) 48:157–161

Eberhard M, Erne P (1994) Calcium and magnesium binding to rat parvalbumin. Eur J Biochem 222:21–26

Ebnet K, Aurrand-Lions M, Kuhn A, Kiefer F, Butz S, Zander K, Meyer zu Brickwedde MK, Suzuki A, Imhof BA, Vestweber D (2003) The junctional adhesion molecule (JAM) family members JAM-2 and JAM-3 associate with the cell polarity protein PAR-3: a possible role for JAMs in endothelial cell polarity. J Cell Sci 116:3879–3891

Edwards BR, Baer PG, Sutton RA, Dirks JH (1973) Micropuncture study of diuretic effects on sodium and calcium reabsorption in the dog nephron. J Clin Invest 52:2418–2427

Efrati E, Arsentiev-Rozenfeld J, Zelikovic I (2005) The human paracellin-1 gene (hPCLN-1): renal epithelial cell-specific expression and regulation. Am J Physiol Renal Physiol 288:F272–F283

Egan ME, Glockner-Pagel J, Ambrose C, Cahill PA, Pappoe L, Balamuth N, Cho E, Canny S, Wagner CA, Geibel J, Caplan MJ (2002) Calcium-pump inhibitors induce functional surface expression of Delta F508-CFTR protein in cystic fibrosis epithelial cells. Nat Med 8:485–492

Egan ME, Pearson M, Weiner SA, Rajendran V, Rubin D, Glockner-Pagel J, Canny S, Du K, Lukacs GL, Caplan MJ (2004) Curcumin, a major constituent of turmeric, corrects cystic fibrosis defects. Science 304:600–602

Elalouf JM, Roinel N, de Rouffignac C (1983) Stimulation by human calcitonin of electrolyte transport in distal tubules of rat kidney. Pflugers Arch 399:111–118

Elalouf JM, Roinel N, de Rouffignac C (1984) Effects of antidiuretic hormone on electrolyte reabsorption and secretion in distal tubules of rat kidney. Pflugers Arch 401:167–173

Elin RJ (1994) Magnesium: the fifth but forgotten electrolyte. Am J Clin Pathol 102:616–622

Ellison DH (2000) Divalent cation transport by the distal nephron: insights from Bartter's and Gitelman's syndromes. Am J Physiol Renal Physiol 279:F616–F625

Embark HM, Setiawan I, Poppendieck S, van de Graaf SF, Boehmer C, Palmada M, Wieder T, Gerstberger R, Cohen P, Yun CC, Bindels RJ, Lang F (2004) Regulation of the epithelial Ca^{2+} channel TRPV5 by the NHE regulating factor NHERF2 and the serum and glucocorticoid inducible kinase isoforms SGK1 and SGK3 expressed in Xenopus oocytes. Cell Physiol Biochem 14:203–212

Erler I, Hirnet D, Wissenbach U, Flockerzi V, Niemeyer BA (2004) Ca^{2+}-selective transient receptor potential V channel architecture and function require a specific ankyrin repeat. J Biol Chem 279:34456–34463

Fagan C, Phelan D (2001) Severe convulsant hypomagnesaemia and short bowel syndrome. Anaesth Intensive Care 29:281–283

Faulk WP, McIntyre JA (1983) Immunological studies of human trophoblast: markers, subsets and functions. Immunol Rev 75:139–175

Feher JJ (1983) Facilitated calcium diffusion by intestinal calcium-binding protein. Am J Physiol 244:C303–C307

Feher JJ, Wasserman RH (1979) Calcium absorption and intestinal calcium-binding protein: quantitative relationship. Am J Physiol 236:E556–E561

Feher JJ, Fullmer CS, Wasserman RH (1992) Role of facilitated diffusion of calcium by calbindin in intestinal calcium absorption. Am J Physiol 262:C517–C526

Feray JC, Garay R (1986) An Na^+-stimulated Mg^{2+}-transport system in human red blood cells. Biochim Biophys Acta 856:76–84

Fine KD, Santa Ana CA, Porter JL, Fordtran JS (1991) Intestinal absorption of magnesium from food and supplements. J Clin Invest 88:396–402

Fixemer T, Wissenbach U, Flockerzi V, Bonkhoff H (2003) Expression of the Ca^{2+}-selective cation channel TRPV6 in human prostate cancer: a novel prognostic marker for tumor progression. Oncogene 22:7858–7861

Flatman PW (1984) Magnesium transport across cell membranes. J Membr Biol 80:1–14

Fleet JC, Wood RJ (1999) Specific 1,25(OH)2D3-mediated regulation of transcellular calcium transport in Caco-2 cells. Am J Physiol 276:G958–G964

Fleet JC, Eksir F, Hance KW, Wood RJ (2002) Vitamin D-inducible calcium transport and gene expression in three Caco-2 cell lines. Am J Physiol Gastrointest Liver Physiol 283:G618–G625

Flik G, Schoenmakers TJ, Groot JA, van Os CH, Wendelaar Bonga SE (1990) Calcium absorption by fish intestine: the involvement of ATP- and sodium-dependent calcium extrusion mechanisms. J Membr Biol 113:13–22

Fraser DR, Kodicek E (1970) Unique biosynthesis by kidney of a biological active vitamin D metabolite. Nature 228:764–766

Fraser DR, Kodicek E (1973) Regulation of 25-hydroxycholecalciferol-1-hydroxylase activity in kidney by parathyroid hormone. Nat New Biol 241:163–166

Freitag JJ, Martin KJ, Conrades MB, Bellorin-Font E, Teitelbaum S, Klahr S, Slatopolsky E (1979) Evidence for skeletal resistance to parathyroid hormone in magnesium deficiency. Studies in isolated perfused bone. J Clin Invest 64:1238–1244

Frick KK, Bushinsky DA (2003) Molecular mechanisms of primary hypercalciuria. J Am Soc Nephrol 14:1082–1095

Friedman PA (1988) Basal and hormone-activated calcium absorption in mouse renal thick ascending limbs. Am J Physiol 254:F62–70

Friedman PA (1998) Codependence of renal calcium and sodium transport. Annu Rev Physiol 60:179–197

Friedman PA (1999) Calcium transport in the kidney. Curr Opin Nephrol Hypertens 8:589–595

Friedman PA, Bushinsky DA (1999) Diuretic effects on calcium metabolism. Semin Nephrol 19:551–556

Friedman PA, Gesek FA (1993) Vitamin D3 accelerates PTH-dependent calcium transport in distal convoluted tubule cells. Am J Physiol 265:F300–F308

Friedman PA, Gesek FA (1994) Hormone-responsive Ca^{2+} entry in distal convoluted tubules. J Am Soc Nephrol 4:1396–1404

Friedman PA, Gesek FA (1995a) Cellular calcium transport in renal epithelia: measurement, mechanisms, and regulation. Physiol Rev 75:429–471

Friedman PA, Gesek FA (1995b) Stimulation of calcium transport by amiloride in mouse distal convoluted tubule cells. Kidney Int 48:1427–1434

Friedman PA, Coutermarsh BA, Kennedy SM, Gesek FA (1996) Parathyroid hormone stimulation of calcium transport is mediated by dual signaling mechanisms involving protein kinase A and protein kinase C. Endocrinology 137:13–20

Fujioka Y, Komeda M, Matsuoka S (2000) Stoichiometry of Na^{+}-Ca^{2+} exchange in inside-out patches excised from guinea-pig ventricular myocytes. J Physiol 523:339–351

Fukuoka H, Satoh K (1982) Characterization of the three calcium binding proteins in the human placenta. Nippon Naibunpi Gakkai Zasshi 58:662–678

Furuse M, Furuse K, Sasaki H, Tsukita S (2001) Conversion of zonulae occludentes from tight to leaky strand type by introducing claudin-2 into Madin-Darby canine kidney I cells. J Cell Biol 153:263–272

Gamba G, Saltzberg SN, Lombardi M, Miyanoshita A, Lytton J, Hediger MA, Brenner BM, Hebert SC (1993) Primary structure and functional expression of a cDNA encoding the thiazide-sensitive, electroneutral sodium-chloride cotransporter. Proc Natl Acad Sci USA 90:2749–2753

Garcia NH, Ramsey CR, Knox FG (1998) Understanding the role of paracellular transport in the proximal tubule. News Physiol Sci 13:38–43

Garfield N, Karaplis AC (2001) Genetics and animal models of hypoparathyroidism. Trends Endocrinol Metab 12:288–294

Garland HO (1992) New experimental data on the relationship between diabetes mellitus and magnesium. Magnes Res 5:193–202

Garland HO, Harris PJ, Morgan TO (1991) Calcium transport in the proximal convoluted tubule and loop of Henle of rats made diabetic with streptozotocin. J Endocrinol 131:373–380

Garty R, Alkalay A, Bernheim JL (1983) Parathyroid hormone secretion and responsiveness to parathyroid hormone in primary hypomagnesemia. Isr J Med Sci 19:345–348

Gennari C, Agnusdei D, Nardi P, Civitelli R (1990) Estrogen preserves a normal intestinal responsiveness to 1,25-dihydroxyvitamin D3 in oophorectomized women. J Clin Endocrinol Metab 71:1288–1293

Gentili C, Morelli S, de Boland AR (2003) Characterization of PTH/PTHrP receptor in rat duodenum: effects of ageing. J Cell Biochem 88:1157–1167

Gerke V, Creutz CE, Moss SE (2005) Annexins: linking Ca^{2+} signalling to membrane dynamics. Nat Rev Mol Cell Biol 6:449–461

Gesek FA, Friedman PA (1992a) Mechanism of calcium transport stimulated by chlorothiazide in mouse distal convoluted tubule cells. J Clin Invest 90:429–438

Gesek FA, Friedman PA (1992b) On the mechanism of parathyroid hormone stimulation of calcium uptake by mouse distal convoluted tubule cells. J Clin Invest 90:749–758

Geven WB, Monnens LA, Willems HL, Buijs WC, ter Haar BG (1987a) Renal magnesium wasting in two families with autosomal dominant inheritance. Kidney Int 31:1140–1144

Geven WB, Monnens LA, Willems JL, Buijs W, Hamel CJ (1987b) Isolated autosomal recessive renal magnesium loss in two sisters. Clin Genet 32:398–402

Gibson MM, Bagga DA, Miller CG, Maguire ME (1991) Magnesium transport in Salmonella typhimurium: the influence of new mutations conferring Co^{2+} resistance on the CorA Mg^{2+} transport system. Mol Microbiol 5:2753–2762

Girard C, Tinel N, Terrenoire C, Romey G, Lazdunski M, Borsotto M (2002) p11, an annexin II subunit, an auxiliary protein associated with the background K^+ channel, TASK-1. EMBO J 21:4439–4448

Gisler SM, Stagljar I, Traebert M, Bacic D, Biber J, Murer H (2001) Interaction of the type IIa Na/Pi cotransporter with PDZ proteins. J Biol Chem 276:9206–9213

Gitelman HJ, Graham JB, Welt LG (1966) A new familial disorder characterized by hypokalemia and hypomagnesemia. Trans Assoc Am Physicians 79:221–235

Giuliano AR, Wood RJ (1991) Vitamin D-regulated calcium transport in Caco-2 cells: unique in vitro model. Am J Physiol 260:G207–G212

Giuliano AR, Franceschi RT, Wood RJ (1991) Characterization of the vitamin D receptor from the Caco-2 human colon carcinoma cell line: effect of cellular differentiation. Arch Biochem Biophys 285:261–269

Gkika D, Mahieu F, Nilius B, Hoenderop JG, Bindels RJ (2004) 80K-H as a new Ca^{2+} sensor regulating the activity of the epithelial Ca^{2+} channel transient receptor potential cation channel V5 (TRPV5). J Biol Chem 279:26351–26357

Gkika D, Hsu YJ, van der Kemp AW, Christakos S, Bindels RJ, Hoenderop JG (2006) Critical role of the epithelial Ca^{2+} channel TRPV5 in active Ca^{2+} reabsorption as revealed by TRPV5/calbindin-D28K knockout mice. J Am Soc Nephrol 17:3020–3027

Goh KC, Lim YP, Ong SH, Siak CB, Cao X, Tan YH, Guy GR (1996) Identification of p90, a prominent tyrosine-phosphorylated protein in fibroblast growth factor-stimulated cells, as 80K-H. J Biol Chem 271:5832–5838

Goltzman D, Miao D, Panda DK, Hendy GN (2004) Effects of calcium and of the Vitamin D system on skeletal and calcium homeostasis: lessons from genetic models. J Steroid Biochem Mol Biol 89–90:485–489

Gonzalez-Mariscal L, Betanzos A, Nava P, Jaramillo BE (2003) Tight junction proteins. Prog Biophys Mol Biol 81:1–44

Gonzalez-Perrett S, Batelli M, Kim K, Essafi M, Timpanaro G, Moltabetti N, Reisin IL, Arnaout MA, Cantiello HF (2002) Voltage dependence and pH regulation of human polycystin-2-mediated cation channel activity. J Biol Chem 277:24959–24966

Goodenough DA (1999) Plugging the leaks. Proc Natl Acad Sci USA 96:319–321

Gopalakrishnan S, Dunn KW, Marrs JA (2002) Rac1, but not RhoA, signaling protects epithelial adherens junction assembly during ATP depletion. Am J Physiol Cell Physiol 283:C261–C272

Goytain A, Quamme GA (2005a) Functional characterization of ACDP2 (ancient conserved domain protein), a divalent metal transporter. Physiol Genomics 22:382–389

Goytain A, Quamme GA (2005b) Functional characterization of human SLC41A1, a Mg^{2+} transporter with similarity to prokaryotic MgtE Mg^{2+} transporters. Physiol Genomics 21:337–342

Goytain A, Quamme GA (2005c) Functional characterization of the human solute carrier, SLC41A2. Biochem Biophys Res Commun 330:701–705

Goytain A, Quamme GA (2005d) Identification and characterization of a novel mammalian Mg^{2+} transporter with channel-like properties. BMC Genomics 6:48

Greeb J, Shull GE (1989) Molecular cloning of a third isoform of the calmodulin-sensitive plasma membrane Ca^{2+}-transporting ATPase that is expressed predominantly in brain and skeletal muscle. J Biol Chem 264:18569–18576

Greer FR, Krebs NF (2006) Optimizing bone health and calcium intakes of infants, children, and adolescents. Pediatrics 117:578–585

Greger R (1981) Cation selectivity of the isolated perfused cortical thick ascending limb of Henle's loop of rabbit kidney. Pflugers Arch 390:30–37

Greger R, Schlatter E (1983) Properties of the lumen membrane of the cortical thick ascending limb of Henle's loop of rabbit kidney. Pflugers Arch 396:315–324

Greger R, Lang F, Oberleithner H (1978) Distal site of calcium reabsorption in the rat nephron. Pflugers Arch 374:153–157

Gribenko AV, Makhatadze GI (1998) Oligomerization and divalent ion binding properties of the S100P protein: a Ca^{2+}/Mg^{2+}-switch model. J Mol Biol 283:679–694

Groenestege WM, Hoenderop JG, van den Heuvel L, Knoers N, Bindels RJ (2006) The epithelial Mg^{2+} channel transient receptor potential melastatin 6 is regulated by dietary Mg^{2+} content and estrogens. J Am Soc Nephrol 17:1035–1043

Grubbs RD (2002) Intracellular magnesium and magnesium buffering. Biometals 15:251–259

Gunther T, Vormann J (1985) Mg^{2+} efflux is accomplished by an amiloride-sensitive Na^{+}/Mg^{2+} antiport. Biochem Biophys Res Commun 130:540–545

Hamelin E, Theriault C, Laroche G, Parent JL (2005) The intracellular trafficking of the G protein-coupled receptor TPβ depends on a direct interaction with Rab11. J Biol Chem 280:36195–36205

Harada S, Rodan GA (2003) Control of osteoblast function and regulation of bone mass. Nature 423:349–355

Harris CA, Burnatowska MA, Seely JF, Sutton RA, Quamme GA, Dirks JH (1979) Effects of parathyroid hormone on electrolyte transport in the hamster nephron. Am J Physiol 236:F342–F348

Harteneck C (2003) Proteins modulating TRP channel function. Cell Calcium 33:303–310

He K, Liu K, Daviglus ML, Morris SJ, Loria CM, Van Horn L, Jacobs DR Jr, Savage PJ (2006) Magnesium intake and incidence of metabolic syndrome among young adults. Circulation 113:1675–1682

Heaney RP, Recker RR, Stegman MR, Moy AJ (1989) Calcium absorption in women: relationships to calcium intake, estrogen status, and age. J Bone Miner Res 4:469–475

Hebert SC, Brown EM, Harris HW (1997) Role of the Ca^{2+}-sensing receptor in divalent mineral ion homeostasis. J Exp Biol 200:295–302

Hellwig N, Albrecht N, Harteneck C, Schultz G, Schaefer M (2005) Homo- and heteromeric assembly of TRPV channel subunits. J Cell Sci 118:917–928

Hennekam RC, Donckerwolcke RA (1983) Primary hypomagnesaemia, an autosomal recessive inherited disease? Lancet 1:927

Hermosura MC, Monteilh-Zoller MK, Scharenberg AM, Penner R, Fleig A (2002) Dissociation of the store-operated calcium current I(CRAC) and the Mg-nucleotide-regulated metal ion current MagNuM. J Physiol 539:445–458

Hildmann B, Schmidt A, Murer H (1982) Ca^{2+}-transport across basal-lateral plasma membranes from rat small intestinal epithelial cells. J Membr Biol 65:55–62

Hirase T, Staddon JM, Saitou M, Ando-Akatsuka Y, Itoh M, Furuse M, Fujimoto K, Tsukita S, Rubin LL (1997) Occludin as a possible determinant of tight junction permeability in endothelial cells. J Cell Sci 110:1603–1613

Hirnet D, Olausson J, Fecher-Trost C, Bodding M, Nastainczyk W, Wissenbach U, Flockerzi V, Freichel M (2003) The TRPV6 gene, cDNA and protein. Cell Calcium 33:509–518

Hodgkinson CP, Mander A, Sale GJ (2005) Identification of 80K-H as a protein involved in GLUT4 vesicle trafficking. Biochem J 388:785–793

Hodsman AB, Hanley DA, Ettinger MP, Bolognese MA, Fox J, Metcalfe AJ, Lindsay R (2003) Efficacy and safety of human parathyroid hormone-(1–84) in increasing bone mineral density in postmenopausal osteoporosis. J Clin Endocrinol Metab 88:5212–5220

Hoenderop JG, Hartog A, Willems PH, Bindels RJ (1998) Adenosine-stimulated Ca^{2+} reabsorption is mediated by apical A1 receptors in rabbit cortical collecting system. Am J Physiol 274:F736–F743

Hoenderop JG, De Pont JJ, Bindels RJ, Willems PH (1999a) Hormone-stimulated Ca^{2+} reabsorption in rabbit kidney cortical collecting system is cAMP-independent and involves a phorbol ester-insensitive PKC isotype. Kidney Int 55:225–233

Hoenderop JG, Vaandrager AB, Dijkink L, Smolenski A, Gambaryan S, Lohmann SM, de Jonge HR, Willems PH, Bindels RJ (1999b) Atrial natriuretic peptide-stimulated Ca^{2+} reabsorption in rabbit kidney requires membrane-targeted, cGMP-dependent protein kinase type II. Proc Natl Acad Sci USA 96:6084–6089

Hoenderop JG, van der Kemp AW, Hartog A, van de Graaf SF, van Os CH, Willems PH, Bindels RJ (1999c) Molecular identification of the apical Ca^{2+} channel in 1, 25-dihydroxyvitamin D3-responsive epithelia. J Biol Chem 274:8375–8378

Hoenderop JG, Hartog A, Stuiver M, Doucet A, Willems PH, Bindels RJ (2000) Localization of the epithelial Ca^{2+} channel in rabbit kidney and intestine. J Am Soc Nephrol 11:1171–1178

Hoenderop JG, Muller D, Van Der Kemp AW, Hartog A, Suzuki M, Ishibashi K, Imai M, Sweep F, Willems PH, Van Os CH, Bindels RJ (2001a) Calcitriol controls the epithelial calcium channel in kidney. J Am Soc Nephrol 12:1342–1349

Hoenderop JG, Dardenne O, Van Abel M, Van Der Kemp AW, Van Os CH, St -Arnaud R, Bindels RJ (2002) Modulation of renal Ca^{2+} transport protein genes by dietary Ca^{2+} and 1,25-dihydroxyvitamin D3 in 25-hydroxyvitamin D3-1alpha-hydroxylase knockout mice. FASEB J 16:1398–1406

Hoenderop JG, van Leeuwen JP, van der Eerden BC, Kersten FF, van der Kemp AW, Merillat AM, Waarsing JH, Rossier BC, Vallon V, Hummler E, Bindels RJ (2003a) Renal Ca^{2+} wasting, hyperabsorption, and reduced bone thickness in mice lacking TRPV5. J Clin Invest 112:1906–1914

Hoenderop JG, Voets T, Hoefs S, Weidema F, Prenen J, Nilius B, Bindels RJ (2003b) Homo- and heterotetrameric architecture of the epithelial Ca^{2+} channels TRPV5 and TRPV6. EMBO J 22:776–785

Hoenderop JG, Chon H, Gkika D, Bluyssen HA, Holstege FC, St-Arnaud R, Braam B, Bindels RJ (2004) Regulation of gene expression by dietary Ca^{2+} in kidneys of 25-hydroxyvitamin D3-1α-hydroxylase knockout mice. Kidney Int 65:531–539

Hoenderop JG, Nilius B, Bindels RJ (2005) Calcium absorption across epithelia. Physiol Rev 85:373–422

Hoenderop JGJ, Vennekens R, Müller D, Prenen J, Droogmans G, Bindels RJM, Nilius B (2001b) Function and expression of the epithelial Ca^{2+} channel family: comparison of the epithelial Ca^{2+} channel 1 and 2. J Physiol (Lond) 537:747–761

Hou J, Paul DL, Goodenough DA (2005) Paracellin-1 and the modulation of ion selectivity of tight junctions. J Cell Sci 118:5109–5118

Huerta MG, Roemmich JN, Kington ML, Bovbjerg VE, Weltman AL, Holmes VF, Patrie JT, Rogol AD, Nadler JL (2005) Magnesium deficiency is associated with insulin resistance in obese children. Diabetes Care 28:1175–1181

Hughes MR, Malloy PJ, Kieback DG, Kesterson RA, Pike JW, Feldman D, O'Malley BW (1988) Point mutations in the human vitamin D receptor gene associated with hypocalcemic rickets. Science 242:1702–1705

Huybers S, Naber TH, Bindels RJ, Hoenderop JG (2006) Prednisolone-induced Ca^{2+} malabsorption is caused by diminished expression of the epithelial Ca^{2+} channel TRPV6. Am J Physiol Gastrointest Liver Physiol (in press)

Imai M (1978) Calcium transport across the rabbit thick ascending limb of Henle's loop perfused in vitro. Pflugers Arch 374:255–263

Imai M (1981) Effects of parathyroid hormone and N6,O2′-dibutyryl cyclic AMP on Ca^{2+} transport across the rabbit distal nephron segments perfused in vitro. Pflugers Arch 390:145–151

Imura A, Iwano A, Tohyama O, Tsuji Y, Nozaki K, Hashimoto N, Fujimori T, Nabeshima Y (2004) Secreted Klotho protein in sera and CSF: implication for post-translational cleavage in release of Klotho protein from cell membrane. FEBS Lett 565:143–147

Jones DH, Li TY, Arystarkhova E, Barr KJ, Wetzel RK, Peng J, Markham K, Sweadner KJ, Fong GH, Kidder GM (2005) Na,K-ATPase from mice lacking the gamma subunit (FXYD2) exhibits altered Na^+ affinity and decreased thermal stability. J Biol Chem 280:19003–19011

Jones G, Strugnell SA, DeLuca HF (1998) Current understanding of the molecular actions of vitamin D. Physiol Rev 78:1193–1231

Jonsson KB, Zahradnik R, Larsson T, White KE, Sugimoto T, Imanishi Y, Yamamoto T, Hampson G, Koshiyama H, Ljunggren O, Oba K, Yang IM, Miyauchi A, Econs MJ, Lavigne J, Juppner H (2003) Fibroblast growth factor 23 in oncogenic osteomalacia and X-linked hypophosphatemia. N Engl J Med 348:1656–1663

Juppner H, Abou-Samra AB, Freeman M, Kong XF, Schipani E, Richards J, Kolakowski LF Jr, Hock J, Potts JT Jr, Kronenberg HM, et al (1991) A G protein-linked receptor for parathyroid hormone and parathyroid hormone-related peptide. Science 254:1024–1026

Kahle KT, Macgregor GG, Wilson FH, Van Hoek AN, Brown D, Ardito T, Kashgarian M, Giebisch G, Hebert SC, Boulpaep EL, Lifton RP (2004) Paracellular Cl– permeability is regulated by WNK4 kinase: insight into normal physiology and hypertension. Proc Natl Acad Sci USA 101:14877–14882

Kaissling B (1982) Structural aspects of adaptive changes in renal electrolyte excretion. Am J Physiol 243:F211–F226

Kanai M, Goke M, Tsunekawa S, Podolsky DK (1997) Signal transduction pathway of human fibroblast growth factor receptor 3. Identification of a novel 66-kDa phosphoprotein. J Biol Chem 272:6621–6628

Kang TM, Hilgemann DW (2004) Multiple transport modes of the cardiac Na^+/Ca^{2+} exchanger. Nature 427:544–548

Kantorovich V, Adams JS, Gaines JE, Guo X, Pandian MR, Cohn DH, Rude RK (2002) Genetic heterogeneity in familial renal magnesium wasting. J Clin Endocrinol Metab 87:612–617

Kao WH, Folsom AR, Nieto FJ, Mo JP, Watson RL, Brancati FL (1999) Serum and dietary magnesium and the risk for type 2 diabetes mellitus: the Atherosclerosis Risk in Communities Study. Arch Intern Med 159:2151–2159

Karbach U (1989a) Cellular-mediated and diffusive magnesium transport across the descending colon of the rat. Gastroenterology 96:1282–1289

Karbach U (1989b) Magnesium transport across colon ascendens of the rat. Dig Dis Sci 34:1825–1831

Karppanen H, Karppanen P, Mervaala E (2005) Why and how to implement sodium, potassium, calcium, and magnesium changes in food items and diets? J Hum Hypertens 19(Suppl 3):S10–S19

Kausalya PJ, Amasheh S, Gunzel D, Wurps H, Muller D, Fromm M, Hunziker W (2006) Disease-associated mutations affect intracellular traffic and paracellular Mg^{2+} transport function of Claudin-16. J Clin Invest 116:878–891

Kawano K, Ogata N, Chiano M, Molloy H, Kleyn P, Spector TD, Uchida M, Hosoi T, Suzuki T, Orimo H, Inoue S, Nabeshima Y, Nakamura K, Kuro-o M, Kawaguchi H (2002) Klotho gene polymorphisms associated with bone density of aged postmenopausal women. J Bone Miner Res 17:1744–1751

Kayne LH, Lee DB (1993) Intestinal magnesium absorption. Miner Electrolyte Metab 19:210–217

Keeton TP, Burk SE, Shull GE (1993) Alternative splicing of exons encoding the calmodulin-binding domains and C termini of plasma membrane Ca^{2+}-ATPase isoforms 1, 2, 3, and 4. J Biol Chem 268:2740–2748

Kelepouris E, Agus ZS (1998) Hypomagnesemia: renal magnesium handling. Semin Nephrol 18:58–73

Kikuchi K, Kikuchi T, Ghishan FK (1988) Characterization of calcium transport by basolateral membrane vesicles of human small intestine. Am J Physiol 255:G482–G489

Kim JW, Lee Y, Lee IA, Kang HB, Choe YK, Choe IS (1997) Cloning and expression of human cDNA encoding Na^+, K^+-ATPase gamma-subunit. Biochim Biophys Acta 1350:133–135

Kino T, Hatanaka H, Hashimoto M, Nishiyama M, Goto T, Okuhara M, Kohsaka M, Aoki H, Imanaka H (1987) FK-506, a novel immunosuppressant isolated from a Streptomyces. I. Fermentation, isolation, and physico-chemical and biological characteristics. J Antibiot (Tokyo) 40:1249–1255

Kip SN, Strehler EE (2003) Characterization of PMCA isoforms and their contribution to transcellular Ca^{2+} flux in MDCK cells. Am J Physiol Renal Physiol 284:F122–F132

Kip SN, Strehler EE (2004) Vitamin D3 upregulates plasma membrane Ca^{2+}-ATPase expression and potentiates apico-basal Ca^{2+} flux in MDCK cells. Am J Physiol Renal Physiol 286:F363–F369

Kitanaka S, Takeyama K, Murayama A, Sato T, Okumura K, Nogami M, Hasegawa Y, Niimi H, Yanagisawa J, Tanaka T, Kato S (1998) Inactivating mutations in the 25-hydroxyvitamin D3 1alpha-hydroxylase gene in patients with pseudovitamin D-deficiency rickets. N Engl J Med 338:653–661

Knoers NV, de Jong JC, Meij IC, Van Den Heuvel LP, Bindels RJ (2003) Genetic renal disorders with hypomagnesemia and hypocalciuria. J Nephrol 16:293–296

Kofuji P, Lederer WJ, Schulze DH (1994) Mutually exclusive and cassette exons underlie alternatively spliced isoforms of the Na/Ca exchanger. J Biol Chem 269:5145–5149

Kokko JP (1974) Membrane characteristics governing salt and water transport in the loop of Henle. Fed Proc 33:25–30

Kolisek M, Zsurka G, Samaj J, Weghuber J, Schweyen RJ, Schweigel M (2003) Mrs2p is an essential component of the major electrophoretic Mg^{2+} influx system in mitochondria. EMBO J 22:1235–1244

Konrad M, Weber S (2003) Recent advances in molecular genetics of hereditary magnesium-losing disorders. J Am Soc Nephrol 14:249–260

Konrad M, Schlingmann KP, Gudermann T (2004) Insights into the molecular nature of magnesium homeostasis. Am J Physiol Renal Physiol 286:F599–605

Konrad M, Schaller A, Seelow D, Pandey AV, Waldegger S, Lesslauer A, Vitzthum H, Suzuki Y, Luk JM, Becker C, Schlingmann KP, Schmid M, Rodriguez-Soriano J, Ariceta G, Cano F, Enriquez R, Juppner H, Bakkaloglu SA, Hediger MA, Gallati S, Neuhauss SC, Nurnberg P, Weber S (2006) Mutations in the tight-junction gene claudin 19 (CLDN19) are associated with renal magnesium wasting, renal failure, and severe ocular involvement. Am J Hum Genet 79:949–957

Koster HP, Hartog A, Van Os CH, Bindels RJ (1995) Calbindin-D28K facilitates cytosolic calcium diffusion without interfering with calcium signaling. Cell Calcium 18:187–196

Koushik SV, Wang J, Rogers R, Moskophidis D, Lambert NA, Creazzo TL, Conway SJ (2001) Targeted inactivation of the sodium-calcium exchanger (NCX1) results in the lack of a heartbeat and abnormal myofibrillar organization. FASEB J 15:1209–1211

Kozak JA, Cahalan MD (2003) MIC channels are inhibited by internal divalent cations but not ATP. Biophys J 84:922–927

Kozel PJ, Friedman RA, Erway LC, Yamoah EN, Liu LH, Riddle T, Duffy JJ, Doetschman T, Miller ML, Cardell EL, Shull GE (1998) Balance and hearing deficits in mice with a null mutation in the gene encoding plasma membrane Ca^{2+}-ATPase isoform 2. J Biol Chem 273:18693–18696

Krieger NS, Frick KK, Bushinsky DA (2004) Mechanism of acid-induced bone resorption. Curr Opin Nephrol Hypertens 13:423–436

Kuro-o M, Matsumura Y, Aizawa H, Kawaguchi H, Suga T, Utsugi T, Ohyama Y, Kurabayashi M, Kaname T, Kume E, Iwasaki H, Iida A, Shiraki-Iida T, Nishikawa S, Nagai R, Nabeshima YI (1997) Mutation of the mouse klotho gene leads to a syndrome resembling ageing. Nature 390:45–51

Kurosu H, Yamamoto M, Clark JD, Pastor JV, Nandi A, Gurnani P, McGuinness OP, Chikuda H, Yamaguchi M, Kawaguchi H, Shimomura I, Takayama Y, Herz J, Kahn CR, Rosenblatt KP, Kuro-o M (2005) Suppression of aging in mice by the hormone Klotho. Science 309:1829–1833

Kutuzova GD, Akhter S, Christakos S, Vanhooke J, Kimmel-Jehan C, Deluca HF (2006) Calbindin-D9K knockout mice are indistinguishable from wild-type mice in phenotype and serum calcium level. Proc Natl Acad Sci USA 103:12377–12381

Kwon TH, Nielsen J, Kim YH, Knepper MA, Frokiaer J, Nielsen S (2003) Regulation of sodium transporters in the thick ascending limb of rat kidney: response to angiotensin II. Am J Physiol Renal Physiol 285:F152–F165

Lafond J, Leclerc M, Brunette MG (1991) Characterization of calcium transport by basal plasma membranes from human placental syncytiotrophoblast. J Cell Physiol 148:17–23

Lamberg BA, Kuhlback B (1959) Effect of chlorothiazide and hydrochlorothiazide on the excretion of calcium in urine. Scand J Clin Lab Invest 11:351–357

Lambers TT, Weidema AF, Nilius B, Hoenderop JG, Bindels RJ (2004) Regulation of the mouse epithelial Ca^{2+} channel TRPV6 by the Ca^{2+}-sensor calmodulin. J Biol Chem 279:28855–28861

Lambers TT, Bindels RJ, Hoenderop JG (2006a) Coordinated control of renal Ca^{2+} handling. Kidney Int 69:650–654

Lambers TT, Mahieu F, Oancea E, Hoofd L, De Lange F, Mensenkamp AR, Voets T, Nilius B, Clapham DE, Hoenderop JG, Bindels RJ (2006b) Calbindin-D28K dynamically controls TRPV5-mediated Ca^{2+} transport. EMBO J 25:2978–2988

Lambers TT, Oancea E, De Groot T, Topala CN, Hoenderop JG, Bindels RJ (2006c) Extracellular pH dynamically controls cell surface delivery of functional TRPV5 channels. Mol Cell Biol (in press)

Lamprecht G, Weinman EJ, Yun CH (1998) The role of NHERF and E3KARP in the cAMP-mediated inhibition of NHE3. J Biol Chem 273:29972–29978

Larsson D, Nemere I (2002) Vectorial transcellular calcium transport in intestine: integration of current models. J Biomed Biotechnol 2:117–119

Lee CT, Shang S, Lai LW, Yong KC, Lien YH (2004) Effect of thiazide on renal gene expression of apical calcium channels and calbindins. Am J Physiol Renal Physiol 287:F1164–F1170

Lee CT, Lien YH, Lai LW, Chen JB, Lin CR, Chen HC (2006a) Increased renal calcium and magnesium transporter abundance in streptozotocin-induced diabetes mellitus. Kidney Int 69:1786–1791

Lee J, Cha SK, Sun TJ, Huang CL (2005) PIP2 activates TRPV5 and releases its inhibition by intracellular Mg^{2+}. J Gen Physiol 126:439–451

Lee K, Brown D, Urena P, Ardaillou N, Ardaillou R, Deeds J, Segre GV (1996) Localization of parathyroid hormone/parathyroid hormone-related peptide receptor mRNA in kidney. Am J Physiol 270:F186–F191

Lee NP, Tong MK, Leung PP, Chan VW, Leung S, Tam PC, Chan KW, Lee KF, Yeung WS, Luk JM (2006b) Kidney claudin-19: localization in distal tubules and collecting ducts and dysregulation in polycystic renal disease. FEBS Lett 580:923–931

Lee SL, Yu AS, Lytton J (1994) Tissue-specific expression of Na^+-Ca^{2+} exchanger isoforms. J Biol Chem 269:14849–14852

Lelievre-Pegorier M, Merlet-Benichou C, Roinel N, de Rouffignac C (1983) Developmental pattern of water and electrolyte transport in rat superficial nephrons. Am J Physiol 245:F15–21

Lemann J Jr, Bushinsky DA, Hamm LL (2003) Bone buffering of acid and base in humans. Am J Physiol Renal Physiol 285:F811–F832

Levenson R (1994) Isoforms of the Na,K-ATPase: family members in search of function. Rev Physiol Biochem Pharmacol 123:1–45

Levitan IB (1999) It is calmodulin after all! Mediator of the calcium modulation of multiple ion channels. Neuron 22:645–648

Li J, Dai Z, Jana D, Callaway DJ, Bu Z (2005) Ezrin controls the macromolecular complexes formed between an adapter protein Na^+/H^+ exchanger regulatory factor and the cystic fibrosis transmembrane conductance regulator. J Biol Chem 280:37634–37643

Li M, Jiang J, Yue L (2006) Functional characterization of homo- and heteromeric channel kinases TRPM6 and TRPM7. J Gen Physiol 127:525–537

Li XF, Kraev AS, Lytton J (2002) Molecular cloning of a fourth member of the potassium-dependent sodium-calcium exchanger gene family, NCKX4. J Biol Chem 277:48410–48417

Li YC, Bolt MJ, Cao LP, Sitrin MD (2001) Effects of vitamin D receptor inactivation on the expression of calbindins and calcium metabolism. Am J Physiol Endocrinol Metab 281:E558–E564

Li YM, Mitsuhashi T, Wojciechowicz D, Shimizu N, Li J, Stitt A, He C, Banerjee D, Vlassara H (1996) Molecular identity and cellular distribution of advanced glycation endproduct receptors: relationship of p60 to OST-48 and p90 to 80K-H membrane proteins. Proc Natl Acad Sci USA 93:11047–11052

Li Z, Matsuoka S, Hryshko LV, Nicoll DA, Bersohn MM, Burke EP, Lifton RP, Philipson KD (1994) Cloning of the NCX2 isoform of the plasma membrane Na^{+}-Ca^{2+} exchanger. J Biol Chem 269:17434–17439

Liedtke CM, Yun CH, Kyle N, Wang D (2002) Protein kinase C epsilon-dependent regulation of cystic fibrosis transmembrane regulator involves binding to a receptor for activated C kinase (RACK1) and RACK1 binding to Na^{+}/H^{+} exchange regulatory factor. J Biol Chem 277:22925–22933

Linck B, Qiu Z, He Z, Tong Q, Hilgemann DW, Philipson KD (1998) Functional comparison of the three isoforms of the Na^{+}/Ca^{2+} exchanger (NCX1, NCX2, NCX3). Am J Physiol 274:C415–C423

Liu J, Farmer JD Jr, Lane WS, Friedman J, Weissman I, Schreiber SL (1991) Calcineurin is a common target of cyclophilin-cyclosporin A and FKBP-FK506 complexes. Cell 66:807–815

Loffing J, Loffing-Cueni D, Hegyi I, Kaplan MR, Hebert SC, Le Hir M, Kaissling B (1996) Thiazide treatment of rats provokes apoptosis in distal tubule cells. Kidney Int 50:1180–1190

Loffing J, Loffing-Cueni D, Valderrabano V, Klausli L, Hebert SC, Rossier BC, Hoenderop JG, Bindels RJ, Kaissling B (2001) Distribution of transcellular calcium and sodium transport pathways along mouse distal nephron. Am J Physiol Renal Physiol 281:F1021–F1027

Loffing J, Vallon V, Loffing-Cueni D, Aregger F, Richter K, Pietri L, Bloch-Faure M, Hoenderop JG, Shull GE, Meneton P, Kaissling B (2004) Altered renal distal tubule structure and renal Na^{+} and Ca^{2+} handling in a mouse model for Gitelman's syndrome. J Am Soc Nephrol 15:2276–2288

Lorentzon M, Swanson C, Eriksson AL, Mellstrom D, Ohlsson C (2006) Polymorphisms in the aromatase gene predict areal BMD as a result of affected cortical bone size: the GOOD study. J Bone Miner Res 21:332–339

Lote CJ, Thewles A, Wood JA, Zafar T (2000) The hypomagnesaemic action of FK506: urinary excretion of magnesium and calcium and the role of parathyroid hormone. Clin Sci (Lond) 99:285–292

Lovlie R, Eiken HG, Sorheim JI, Boman H (1996) The Ca^{2+}-sensing receptor gene (PCAR1) mutation T151M in isolated autosomal dominant hypoparathyroidism. Hum Genet 98:129–133

Magosci M, Yamaki M, Penniston JT, Dousa TP (1992) Localization of mRNAs coding for isozymes of plasma membrane Ca^{2+}-ATPase pump in rat kidney. Am J Physiol 263:F7–14

Magyar CE, White KE, Rojas R, Apodaca G, Friedman PA (2002) Plasma membrane Ca^{2+}-ATPase and NCX1 Na^{+}/Ca^{2+} exchanger expression in distal convoluted tubule cells. Am J Physiol Renal Physiol 283:F29–F40

Mahon MJ, Donowitz M, Yun CC, Segre GV (2002) Na^{+}/H^{+} exchanger regulatory factor 2 directs parathyroid hormone 1 receptor signalling. Nature 417:858–861

Malloy PJ, Eccleshall TR, Gross C, Van Maldergem L, Bouillon R, Feldman D (1997) Hereditary vitamin D resistant rickets caused by a novel mutation in the vitamin D receptor that results in decreased affinity for hormone and cellular hyporesponsiveness. J Clin Invest 99:297–304

Mandon B, Siga E, Roinel N, de Rouffignac C (1993) Ca^{2+}, Mg^{2+} and K^{+} transport in the cortical and medullary thick ascending limb of the rat nephron: influence of transepithelial voltage. Pflugers Arch 424:558–560

Mannstadt M, Juppner H, Gardella TJ (1999) Receptors for PTH and PTHrP: their biological importance and functional properties. Am J Physiol 277:F665–F675

Manz F, Scharer K, Janka P, Lombeck J (1978) Renal magnesium wasting, incomplete tubular acidosis, hypercalciuria and nephrocalcinosis in siblings. Eur J Pediatr 128:67–79

Margolis DS, Kim D, Szivek JA, Lai LW, Lien YH (2006) Functionally improved bone in Calbindin-D28K knockout mice. Bone 39:477–484

Martens H, Harmeyer J (1978) Magnesium transport by isolated rumen epithelium of sheep. Res Vet Sci 24:161–168

Martin-Padura I, Lostaglio S, Schneemann M, Williams L, Romano M, Fruscella P, Panzeri C, Stoppacciaro A, Ruco L, Villa A, Simmons D, Dejana E (1998) Junctional adhesion molecule, a novel member of the immunoglobulin superfamily that distributes at intercellular junctions and modulates monocyte transmigration. J Cell Biol 142:117–127

Martini LA, Wood RJ (2000) Should dietary calcium and protein be restricted in patients with nephrolithiasis? Nutr Rev 58:111–117

Marx SJ, Attie MF, Levine MA, Spiegel AM, Downs RW Jr, Lasker RD (1981) The hypocalciuric or benign variant of familial hypercalcemia: clinical and biochemical features in fifteen kindreds. Medicine (Baltimore) 60:397–412

Massry SG, Coburn JW, Kleeman CR (1969) Renal handling of magnesium in the dog. Am J Physiol 216:1460–1467

Matsushita M, Kozak JA, Shimizu Y, McLachlin DT, Yamaguchi H, Wei FY, Tomizawa K, Matsui H, Chait BT, Cahalan MD, Nairn AC (2005) Channel function is dissociated from the intrinsic kinase activity and autophosphorylation of TRPM7/ChaK1. J Biol Chem 280:20793–20803

Matzkin H, Lotan D, Boichis H (1989) Primary hypomagnesemia with a probable double magnesium transport defect. Nephron 52:83–86

McCarthy KM, Skare IB, Stankewich MC, Furuse M, Tsukita S, Rogers RA, Lynch RD, Schneeberger EE (1996) Occludin is a functional component of the tight junction. J Cell Sci 109:2287–2298

McNair P, Christiansen C, Transbol I (1984) Effect of menopause and estrogen substitutional therapy on magnesium metabolism. Miner Electrolyte Metab 10:84–87

Means AR, Dedman JR (1980) Calmodulin—an intracellular calcium receptor. Nature 285:73–77

Meij IC, Saar K, van den Heuvel LP, Nuernberg G, Vollmer M, Hildebrandt F, Reis A, Monnens LA, Knoers NV (1999) Hereditary isolated renal magnesium loss maps to chromosome 11q23. Am J Hum Genet 64:180–188

Meij IC, Koenderink JB, van Bokhoven H, Assink KF, Tiel Groenestege W, de Pont JJ, Bindels RJ, Monnens LA, van den Heuvel LP, Knoers NV (2000) Dominant isolated renal magnesium loss is caused by misrouting of the Na^+,K^+-ATPase gamma-subunit. Nat Genet 26:265–266

Meij IC, van den Heuvel LP, Knoers NV (2002) Genetic disorders of magnesium homeostasis. Biometals 15:297–307

Meij IC, van den Heuvel LP, Hemmes S, van der Vliet WA, Willems JL, Monnens LA, Knoers NV (2003) Exclusion of mutations in FXYD2, CLDN16 and SLC12A3 in two families with primary renal Mg^{2+} loss. Nephrol Dial Transplant 18:512–516

Meyer MB, Watanuki M, Kim S, Shevde NK, Pike JW (2006) The human Trpv6 distal promoter contains multiple vitamin D receptor binding sites that mediate activation by 1,25-dihydroxyvitamin D3 in intestinal cells. Mol Endocrinol 20:1447–1461

Milazzo SC, Ahern MJ, Cleland LG, Henderson DR (1981) Calcium pyrophosphate dihydrate deposition disease and familial hypomagnesemia. J Rheumatol 8:767–771

Miller LA, Stapleton FB (1989) Urinary volume in children with urolithiasis. J Urol 141:918–920

Moe OW (2006) Kidney stones: pathophysiology and medical management. Lancet 367:333–344

Monteilh-Zoller MK, Hermosura MC, Nadler MJ, Scharenberg AM, Penner R, Fleig A (2003) TRPM7 provides an ion channel mechanism for cellular entry of trace metal ions. J Gen Physiol 121:49–60

Montell C (2003) Mg^{2+} homeostasis: the Mg^{2+}nificent TRPM chanzymes. Curr Biol 13:R799–801

Montell C, Birnbaumer L, Flockerzi V, Bindels RJ, Bruford EA, Caterina MJ, Clapham DE, Harteneck C, Heller S, Julius D, Kojima I, Mori Y, Penner R, Prawitt D, Scharenberg AM, Schultz G, Shimizu N, Zhu MX (2002) A unified nomenclature for the superfamily of TRP cation channels. Mol Cell 9:229–231

Moreau R, Daoud G, Bernatchez R, Simoneau L, Masse A, Lafond J (2002a) Calcium uptake and calcium transporter expression by trophoblast cells from human term placenta. Biochim Biophys Acta 1564:325–332

Moreau R, Hamel A, Daoud G, Simoneau L, Lafond J (2002b) Expression of calcium channels along the differentiation of cultured trophoblast cells from human term placenta. Biol Reprod 67:1473–1479

Moreau R, Daoud G, Masse A, Simoneau L, Lafond J (2003a) Expression and role of calcium-ATPase pump and sodium-calcium exchanger in differentiated trophoblasts from human term placenta. Mol Reprod Dev 65:283–288

Moreau R, Simoneau L, Lafond J (2003b) Calcium fluxes in human trophoblast (BeWo) cells: calcium channels, calcium-ATPase, and sodium-calcium exchanger expression. Mol Reprod Dev 64:189–198

Morita K, Furuse M, Fujimoto K, Tsukita S (1999) Claudin multigene family encoding four-transmembrane domain protein components of tight junction strands. Proc Natl Acad Sci USA 96:511–516

Morris RG, Hoorn EJ, Knepper MA (2006) Hypokalemia in a mouse model of Gitelman's syndrome. Am J Physiol Renal Physiol 290:F1416–F1420

Moseley JM, Kubota M, Diefenbach-Jagger H, Wettenhall RE, Kemp BE, Suva LJ, Rodda CP, Ebeling PR, Hudson PJ, Zajac JD, et al (1987) Parathyroid hormone-related protein purified from a human lung cancer cell line. Proc Natl Acad Sci USA 84:5048–5052

Muller D, Hoenderop JG, Meij IC, van den Heuvel LP, Knoers NV, den Hollander AI, Eggert P, Garcia-Nieto V, Claverie-Martin F, Bindels RJ (2000) Molecular cloning, tissue distribution, and chromosomal mapping of the human epithelial Ca^{2+} channel (ECAC1). Genomics 67:48–53

Muller D, Kausalya PJ, Claverie-Martin F, Meij IC, Eggert P, Garcia-Nieto V, Hunziker W (2003) A novel claudin 16 mutation associated with childhood hypercalciuria abolishes binding to ZO-1 and results in lysosomal mistargeting. Am J Hum Genet 73:1293–1301

Muller D, Kausalya PJ, Meij IC, Hunziker W (2006) Familial hypomagnesemia with hypercalciuria and nephrocalcinosis: blocking endocytosis restores surface expression of a novel Claudin-16 mutant that lacks the entire C-terminal cytosolic tail. Hum Mol Genet 15:1049–1058
Muslin AJ, Tanner JW, Allen PM, Shaw AS (1996) Interaction of 14-3-3 with signaling proteins is mediated by the recognition of phosphoserine. Cell 84:889–897
Nadler MJ, Hermosura MC, Inabe K, Perraud AL, Zhu Q, Stokes AJ, Kurosaki T, Kinet JP, Penner R, Scharenberg AM, Fleig A (2001) LTRPC7 is a Mg.ATP-regulated divalent cation channel required for cell viability. Nature 411:590–595
Nakasaki Y, Iwamoto T, Hanada H, Imagawa T, Shigekawa M (1993) Cloning of the rat aortic smooth muscle Na^+/Ca^{2+} exchanger and tissue-specific expression of isoforms. J Biochem (Tokyo) 114:528–534
Neer RM (1975) The evolutionary significance of vitamin D, skin pigment, and ultraviolet light. Am J Phys Anthropol 43:409–416
Nemere I, Norman AW (1986) Parathyroid hormone stimulates calcium transport in perfused duodena from normal chicks: comparison with the rapid (transcaltachic) effect of 1,25-dihydroxyvitamin D3. Endocrinology 119:1406–1408
Nemere I, Norman AW (1990) Transcaltachia, vesicular calcium transport, and microtubule-associated calbindin-D28K: emerging views of 1,25-dihydroxyvitamin D3-mediated intestinal calcium absorption. Miner Electrolyte Metab 16:109–114
Nemere I, Szego CM (1981) Early actions of parathyroid hormone and 1,25-dihydroxycholecalciferol on isolated epithelial cells from rat intestine. II. Analyses of additivity, contribution of calcium, and modulatory influence of indomethacin. Endocrinology 109:2180–2187
Nemere I, Leathers V, Norman AW (1986) 1,25-Dihydroxyvitamin D3-mediated intestinal calcium transport. Biochemical identification of lysosomes containing calcium and calcium-binding protein (calbindin-D28K). J Biol Chem 261:16106–16114
New SA, Robins SP, Campbell MK, Martin JC, Garton MJ, Bolton-Smith C, Grubb DA, Lee SJ, Reid DM (2000) Dietary influences on bone mass and bone metabolism: further evidence of a positive link between fruit and vegetable consumption and bone health? Am J Clin Nutr 71:142–151
Ng RC, Peraino RA, Suki WN (1982) Divalent cation transport in isolated tubules. Kidney Int 22:492–497
Nicholson JC, Jones CL, Powell HR, Walker RG, McCredie DA (1995) Familial hypomagnesaemia—hypercalciuria leading to end-stage renal failure. Pediatr Nephrol 9:74–76
Nicoll DA, Quednau BD, Qui Z, Xia YR, Lusis AJ, Philipson KD (1996) Cloning of a third mammalian Na^+-Ca^{2+} exchanger, NCX3. J Biol Chem 271:24914–24921
Niemeyer BA, Bergs C, Wissenbach U, Flockerzi V, Trost C (2001) Competitive regulation of CaT-like-mediated Ca^{2+} entry by protein kinase C and calmodulin. Proc Natl Acad Sci USA 98:3600–3605
Nijenhuis T, Hoenderop JG, Loffing J, van der Kemp AW, van Os CH, Bindels RJ (2003a) Thiazide-induced hypocalciuria is accompanied by a decreased expression of Ca^{2+} transport proteins in kidney. Kidney Int 64:555–564
Nijenhuis T, Hoenderop JG, van der Kemp AW, Bindels RJ (2003b) Localization and regulation of the epithelial Ca^{2+} channel TRPV6 in the kidney. J Am Soc Nephrol 14:2731–2740
Nijenhuis T, Hoenderop JG, Bindels RJ (2004) Downregulation of Ca^{2+} and Mg^{2+} transport proteins in the kidney explains tacrolimus (FK506)-induced hypercalciuria and hypomagnesemia. J Am Soc Nephrol 15:549–557
Nijenhuis T, Vallon V, van der Kemp AW, Loffing J, Hoenderop JG, Bindels RJ (2005) Enhanced passive Ca^{2+} reabsorption and reduced Mg^{2+} channel abundance explains thiazide-induced hypocalciuria and hypomagnesemia. J Clin Invest 115:1651–1658
Nijenhuis T, Renkema KY, Hoenderop JG, Bindels RJ (2006) Acid-base status determines the renal expression of Ca^{2+} and Mg^{2+} transport proteins. J Am Soc Nephrol 17:617–626
Nilius B, Mahieu F (2006) A road map for TR(I)Ps. Mol Cell 22:297–307
Nilius B, Vennekens R, Prenen J, Hoenderop JG, Bindels RJ, Droogmans G (2000) Whole-cell and single channel monovalent cation currents through the novel rabbit epithelial Ca^{2+} channel ECaC. J Physiol 527:239–248
Nilius B, Prenen J, Vennekens R, Hoenderop JG, Bindels RJ, Droogmans G (2001a) Modulation of the epithelial calcium channel, ECaC, by intracellular Ca^{2+}. Cell Calcium 29:417–428
Nilius B, Prenen J, Vennekens R, Hoenderop JGJ, Bindels RJM, Droogmans G (2001b) Pharmacological modulation of monovalent cation currents through the epithelial Ca^{2+} ECaC. Br J Pharmacol 134:453–462

Nilius B, Vennekens R, Prenen J, Hoenderop JG, Droogmans G, Bindels RJ (2001c) The single pore residue Asp542 determines Ca^{2+} permeation and Mg^{2+} block of the epithelial Ca^{2+} channel. J Biol Chem 276:1020–1025

Nilius B, Prenen J, Hoenderop JG, Vennekens R, Hoefs S, Weidema AF, Droogmans G, Bindels RJ (2002) Fast and slow inactivation kinetics of the Ca^{2+} channels ECaC1 and ECaC2 (TRPV5 and TRPV6). Role of the intracellular loop located between transmembrane segments 2 and 3. J Biol Chem 277:30852–30858

Nilius B, Weidema F, Prenen J, Hoenderop JG, Vennekens R, Hoefs S, Droogmans G, Bindels RJ (2003) The carboxyl terminus of the epithelial Ca^{2+} channel ECaC1 is involved in Ca^{2+}-dependent inactivation. Pflugers Arch 445:584–588

Nordin BE, Horsman A, Marshall DH, Simpson M, Waterhouse GM (1979) Calcium requirement and calcium therapy. Clin Orthop 140:216–239

Nordin BE, Need AG, Morris HA, Horowitz M, Robertson WG (1991) Evidence for a renal calcium leak in postmenopausal women. J Clin Endocrinol Metab 72:401–407

Nussey SS, Whitehead SA (2001) Endocrinology: an integrated approach. Taylor and Francis Group, London

O'Kelly I, Butler MH, Zilberberg N, Goldstein SA (2002) Forward transport. 14-3-3 binding overcomes retention in endoplasmic reticulum by dibasic signals. Cell 111:577–588

Obukhov AG, Nowycky MC (2004) TRPC5 activation kinetics are modulated by the scaffolding protein ezrin/radixin/moesin-binding phosphoprotein-50 (EBP50). J Cell Physiol 201:227–235

Odell AF, Scott JL, Van Helden DF (2005) Epidermal growth factor induces tyrosine phosphorylation, membrane insertion, and activation of transient receptor potential channel 4. J Biol Chem 280:37974–37987

Ogata N, Matsumura Y, Shiraki M, Kawano K, Koshizuka Y, Hosoi T, Nakamura K, Kuro OM, Kawaguchi H (2002) Association of klotho gene polymorphism with bone density and spondylosis of the lumbar spine in postmenopausal women. Bone 31:37–42

Okano T, Tsugawa N, Morishita A, Kato S (2004) Regulation of gene expression of epithelial calcium channels in intestine and kidney of mice by 1alpha,25-dihydroxyvitamin D3. J Steroid Biochem Mol Biol 89–90:335–338

Okazaki R, Chikatsu N, Nakatsu M, Takeuchi Y, Ajima M, Miki J, Fujita T, Arai M, Totsuka Y, Tanaka K, Fukumoto S (1999) A novel activating mutation in calcium-sensing receptor gene associated with a family of autosomal dominant hypocalcemia. J Clin Endocrinol Metab 84:363–366

Okunade GW, Miller ML, Pyne GJ, Sutliff RL, O'Connor KT, Neumann JC, Andringa A, Miller DA, Prasad V, Doetschman T, Paul RJ, Shull GE (2004) Targeted ablation of plasma membrane Ca^{2+}-ATPase (PMCA) 1 and 4 indicates a major housekeeping function for PMCA1 and a critical role in hyperactivated sperm motility and male fertility for PMCA4. J Biol Chem 279:33742–33750

Okuse K, Malik-Hall M, Baker MD, Poon WY, Kong H, Chao MV, Wood JN (2002) Annexin II light chain regulates sensory neuron-specific sodium channel expression. Nature 417:653–656

Ordaz B, Tang J, Xiao R, Salgado A, Sampieri A, Zhu MX, Vaca L (2005) Calmodulin and calcium interplay in the modulation of TRPC5 channel activity. Identification of a novel C-terminal domain for calcium/calmodulin-mediated facilitation. J Biol Chem 280:30788–30796

Palmada M, Poppendieck S, Embark HM, van de Graaf SF, Boehmer C, Bindels RJ, Lang F (2005) Requirement of PDZ domains for the stimulation of the epithelial Ca^{2+} channel TRPV5 by the NHE regulating factor NHERF2 and the serum and glucocorticoid inducible kinase SGK1. Cell Physiol Biochem 15:175–182

Pan TC, Liao BK, Huang CJ, Lin LY, Hwang PP (2005) Epithelial Ca^{2+} channel expression and Ca^{2+} uptake in developing zebrafish. Am J Physiol Regul Integr Comp Physiol 289:R1202–R1211

Panda DK, Miao D, Tremblay ML, Sirois J, Farookhi R, Hendy GN, Goltzman D (2001) Targeted ablation of the 25-hydroxyvitamin D 1alpha -hydroxylase enzyme: evidence for skeletal, reproductive, and immune dysfunction. Proc Natl Acad Sci USA 98:7498–7503

Paolisso G, Barbagallo M (1997) Hypertension, diabetes mellitus, and insulin resistance: the role of intracellular magnesium. Am J Hypertens 10:346–355

Pappenheimer JR (1987) Physiological regulation of transepithelial impedance in the intestinal mucosa of rats and hamsters. J Membr Biol 100:137–148

Parkinson DB, Thakker RV (1992) A donor splice site mutation in the parathyroid hormone gene is associated with autosomal recessive hypoparathyroidism. Nat Genet 1:149–152

Pearce SH, Trump D, Wooding C, Besser GM, Chew SL, Grant DB, Heath DA, Hughes IA, Paterson CR, Whyte MP, et al (1995) Calcium-sensing receptor mutations in familial benign hypercalcemia and neonatal hyperparathyroidism. J Clin Invest 96:2683–2692

Pearce SH, Williamson C, Kifor O, Bai M, Coulthard MG, Davies M, Lewis-Barned N, McCredie D, Powell H, Kendall-Taylor P, Brown EM, Thakker RV (1996) A familial syndrome of hypocalcemia with hypercalciuria due to mutations in the calcium-sensing receptor. N Engl J Med 335:1115–1122

Peces R, Drenth JP, Te Morsche RH, Gonzalez P, Peces C (2005) Autosomal dominant polycystic liver disease in a family without polycystic kidney disease associated with a novel missense protein kinase C substrate 80K-H mutation. World J Gastroenterol 11:7690–7693

Peng JB, Chen XZ, Berger UV, Vassilev PM, Tsukaguchi H, Brown EM, Hediger MA (1999) Molecular cloning and characterization of a channel-like transporter mediating intestinal calcium absorption. J Biol Chem 274:22739–22746

Peng JB, Chen XZ, Berger UV, Vassilev PM, Brown EM, Hediger MA (2000a) A rat kidney-specific calcium transporter in the distal nephron. J Biol Chem 275:28186–28194

Peng JB, Chen XZ, Berger UV, Weremowicz S, Morton CC, Vassilev PM, Brown EM, Hediger MA (2000b) Human calcium transport protein CaT1. Biochem Biophys Res Commun 278:326–332

Peng JB, Brown EM, Hediger MA (2001a) Structural conservation of the genes encoding CaT1, CaT2, and related cation channels. Genomics 76:99–109

Peng JB, Zhuang L, Berger UV, Adam RM, Williams BJ, Brown EM, Hediger MA, Freeman MR (2001b) CaT1 expression correlates with tumor grade in prostate cancer. Biochem Biophys Res Commun 282:729–734

Philipson KD, Nicoll DA (2000) Sodium-calcium exchange: a molecular perspective. Annu Rev Physiol 62:111–133

Phillips AM, Bull A, Kelly LE (1992) Identification of a Drosophila gene encoding a calmodulin-binding protein with homology to the trp phototransduction gene. Neuron 8:631–642

Picotto G, Massheimer V, Boland R (1997) Parathyroid hormone stimulates calcium influx and the cAMP messenger system in rat enterocytes. Am J Physiol 273:C1349–C1353

Pike JW, Dokoh S, Haussler MR, Liberman UA, Marx SJ, Eil C (1984) Vitamin D3—resistant fibroblasts have immunoassayable 1,25-dihydroxyvitamin D3 receptors. Science 224:879–881

Pitkin RM (1985) Calcium metabolism in pregnancy and the perinatal period: a review. Am J Obstet Gynecol 151:99–109

Plotkin MD, Kaplan MR, Verlander JW, Lee WS, Brown D, Poch E, Gullans SR, Hebert SC (1996) Localization of the thiazide sensitive Na-Cl cotransporter, rTSC1 in the rat kidney. Kidney Int 50:174–183

Pollak MR, Brown EM, Chou YH, Hebert SC, Marx SJ, Steinmann B, Levi T, Seidman CE, Seidman JG (1993) Mutations in the human Ca^{2+}-sensing receptor gene cause familial hypocalciuric hypercalcemia and neonatal severe hyperparathyroidism. Cell 75:1297–1303

Pollak MR, Brown EM, Estep HL, McLaine PN, Kifor O, Park J, Hebert SC, Seidman CE, Seidman JG (1994) Autosomal dominant hypocalcaemia caused by a Ca^{2+}-sensing receptor gene mutation. Nat Genet 8:303–307

Pollak MR, Delaney VB, Graham RM, Hebert SC (1996) Gitelman's syndrome (Bartter's variant) maps to the thiazide-sensitive cotransporter gene locus on chromosome 16q13 in a large kindred. J Am Soc Nephrol 7:2244–2248

Ponting C, Schultz J, Bork P (1997) SPRY domains in ryanodine receptors Ca^{2+}-release channels. Trends Biochem Sci 22:193–194

Poon WY, Malik-Hall M, Wood JN, Okuse K (2004) Identification of binding domains in the sodium channel NaV1.8 intracellular N-terminal region and annexin II light chain p11. FEBS Lett 558:114–118

Potts JT (2005) Parathyroid hormone: past and present. J Endocrinol 187:311–325

Prader A, Illig R, Heierli E (1961) An unusual form of primary vitamin D-resistant rickets with hypocalcemia and autosomal-dominant hereditary transmission: hereditary pseudo-deficiency rickets. Helv Paediatr Acta 16:452–468

Praga M, Vara J, Gonzalez-Parra E, Andres A, Alamo C, Araque A, Ortiz A, Rodicio JL (1995) Familial hypomagnesemia with hypercalciuria and nephrocalcinosis. Kidney Int 47:1419–1425

Prince RL, Dick I, Devine A, Price RI, Gutteridge DH, Kerr D, Criddle A, Garcia-Webb P, St John A (1995) The effects of menopause and age on calcitropic hormones: a cross-sectional study of 655 healthy women aged 35 to 90. J Bone Miner Res 10:835–842

Prod'hom B, Pietrobon D, Hess P (1989) Interactions of protons with single open L-type calcium channels. Location of protonation site and dependence of proton-induced current fluctuations on concentration and species of permeant ion. J Gen Physiol 94:23–42

Pronicka E, Gruszczynska B (1991) Familial hypomagnesaemia with secondary hypocalcaemia—autosomal or X-linked inheritance? J Inherit Metab Dis 14:397–399

Pu HX, Scanzano R, Blostein R (2002) Distinct regulatory effects of the Na,K-ATPase gamma subunit. J Biol Chem 277:20270–20276

Puliyanda DP, Ward DT, Baum MA, Hammond TG, Harris HW Jr (2003) Calpain-mediated AQP2 proteolysis in inner medullary collecting duct. Biochem Biophys Res Commun 303:52–58

Quamme GA (1997) Renal magnesium handling: new insights in understanding old problems. Kidney Int 52:1180–1195

Quamme GA, de Rouffignac C (2000) Epithelial magnesium transport and regulation by the kidney. Front Biosci 5:D694–711

Quamme GA, Dirks JH (1980) Intraluminal and contraluminal magnesium on magnesium and calcium transfer in the rat nephron. Am J Physiol 238:F187–F198

Quednau BD, Nicoll DA, Philipson KD (1997) Tissue specificity and alternative splicing of the Na^+/Ca^{2+} exchanger isoforms NCX1, NCX2, and NCX3 in rat. Am J Physiol 272:C1250–C1261

Raber G, Willems PH, Lang F, Nitschke R, van Os CH, Bindels RJ (1997) Coordinated control of apical calcium influx and basolateral calcium efflux in rabbit cortical collecting system. Cell Calcium 22:157–166

Ray WA, Griffin MR, Downey W, Melton LJ 3rd (1989) Long-term use of thiazide diuretics and risk of hip fracture. Lancet 1:687–690

Razzaque MS, Sitara D, Taguchi T, St-Arnaud R, Lanske B (2006) Premature aging-like phenotype in fibroblast growth factor 23 null mice is a vitamin D-mediated process. FASEB J 20:720–722

Reilly RF, Ellison DH (2000) Mammalian distal tubule: physiology, pathophysiology, and molecular anatomy. Physiol Rev 80:277–313

Reilly RF, Shugrue CA (1992) cDNA cloning of a renal Na^+-Ca^{2+} exchanger. Am J Physiol 262:F1105–F1109

Renigunta V, Yuan H, Zuzarte M, Rinne S, Koch A, Wischmeyer E, Schlichthorl G, Gao Y, Karschin A, Jacob R, Schwappach B, Daut J, Preisig-Muller R (2006) The retention factor p11 confers an endoplasmic reticulum-localization signal to the potassium channel TASK-1. Traffic 7:168–181

Renkema KY, Nijenhuis T, van der Eerden BC, van der Kemp AW, Weinans H, van Leeuwen JP, Bindels RJ, Hoenderop JG (2005) Hypervitaminosis D mediates compensatory Ca^{2+} hyperabsorption in TRPV5 knockout mice. J Am Soc Nephrol 16:3188–3195

Reuter H, Henderson SA, Han T, Ross RS, Goldhaber JI, Philipson KD (2002) The Na^+-Ca^{2+} exchanger is essential for the action of cardiac glycosides. Circ Res 90:305–308

Richard O, Freycon MT (1992) Congenital tubulopathy with magnesium loss. Pediatrie 47:557–563

Ritchie G, Kerstan D, Dai LJ, Kang HS, Canaff L, Hendy GN, Quamme GA (2001) 1,25(OH)2D3 stimulates Mg^{2+} uptake into MDCT cells: modulation by extracellular Ca^{2+} and Mg^{2+}. Am J Physiol Renal Physiol 280:F868–F878

Robben JH, Sze M, Knoers NV, Deen PM (2006) Rescue of vasopressin V2 receptor mutants by chemical chaperones: specificity and mechanism. Mol Biol Cell 17:379–386

Rocha AS, Kokko JP (1973) Membrane characteristics regulating potassium transport out of the isolated perfused descending limb of Henle. Kidney Int 4:326–330

Rocha AS, Magaldi JB, Kokko JP (1977) Calcium and phosphate transport in isolated segments of rabbit Henle's loop. J Clin Invest 59:975–983

Rodan GA (1992) Introduction to bone biology. Bone 13(Suppl 1):S3–6

Rodan GA, Martin TJ (2000) Therapeutic approaches to bone diseases. Science 289:1508–1514

Rohacs T, Lopes CM, Michailidis I, Logothetis DE (2005) PI(4,5)P2 regulates the activation and desensitization of TRPM8 channels through the TRP domain. Nat Neurosci 8:626–634

Romani AM, Maguire ME (2002) Hormonal regulation of Mg^{2+} transport and homeostasis in eukaryotic cells. Biometals 15:271–283

Rude RK, Gruber HE (2004) Magnesium deficiency and osteoporosis: animal and human observations. J Nutr Biochem 15:710–716

Rude RK, Oldham SB, Singer FR (1976) Functional hypoparathyroidism and parathyroid hormone end-organ resistance in human magnesium deficiency. Clin Endocrinol (Oxf) 5:209–224

Rude RK, Oldham SB, Sharp CF Jr, Singer FR (1978) Parathyroid hormone secretion in magnesium deficiency. J Clin Endocrinol Metab 47:800–806

Rude RK, Gruber HE, Norton HJ, Wei LY, Frausto A, Kilburn J (2006) Reduction of dietary magnesium by only 50% in the rat disrupts bone and mineral metabolism. Osteoporos Int 17:1022–1032

Runnels LW, Yue L, Clapham DE (2001) TRP-PLIK, a bifunctional protein with kinase and ion channel activities. Science 291:1043–1047

Runnels LW, Yue L, Clapham DE (2002) The TRPM7 channel is inactivated by PIP2 hydrolysis. Nat Cell Biol 4:329–336

Ryazanova LV, Dorovkov MV, Ansari A, Ryazanov AG (2004) Characterization of the protein kinase activity of TRPM7/ChaK1, a protein kinase fused to the transient receptor potential ion channel. J Biol Chem 279:3708–3716

Saitou M, Ando-Akatsuka Y, Itoh M, Furuse M, Inazawa J, Fujimoto K, Tsukita S (1997) Mammalian occludin in epithelial cells: its expression and subcellular distribution. Eur J Cell Biol 73:222–231

Saitou M, Fujimoto K, Doi Y, Itoh M, Fujimoto T, Furuse M, Takano H, Noda T, Tsukita S (1998) Occludin-deficient embryonic stem cells can differentiate into polarized epithelial cells bearing tight junctions. J Cell Biol 141:397–408

Sakai K, Hirai M, Minoshima S, Kudoh J, Fukuyama R, Shimizu N (1989) Isolation of cDNAs encoding a substrate for protein kinase C: nucleotide sequence and chromosomal mapping of the gene for a human 80 K protein. Genomics 5:309–315

Sands JM, Naruse M, Baum M, Jo I, Hebert SC, Brown EM, Harris HW (1997) Apical extracellular calcium/polyvalent cation-sensing receptor regulates vasopressin-elicited water permeability in rat kidney inner medullary collecting duct. J Clin Invest 99:1399–1405

Sato S, Ward CL, Krouse ME, Wine JJ, Kopito RR (1996) Glycerol reverses the misfolding phenotype of the most common cystic fibrosis mutation. J Biol Chem 271:635–638

Schenker T, Trueb B (2000) BSPRY, a novel protein of the Ro-Ret family. Biochim Biophys Acta 1493:255–258

Schlingmann KP, Weber S, Peters M, Niemann Nejsum L, Vitzthum H, Klingel K, Kratz M, Haddad E, Ristoff E, Dinour D, Syrrou M, Nielsen S, Sassen M, Waldegger S, Seyberth HW, Konrad M (2002) Hypomagnesemia with secondary hypocalcemia is caused by mutations in TRPM6, a new member of the TRPM gene family. Nat Genet 31:166–170

Schlingmann KP, Sassen MC, Weber S, Pechmann U, Kusch K, Pelken L, Lotan D, Syrrou M, Prebble JJ, Cole DE, Metzger DL, Rahman S, Tajima T, Shu SG, Waldegger S, Seyberth HW, Konrad M (2005) Novel TRPM6 mutations in 21 families with primary hypomagnesemia and secondary hypocalcemia. J Am Soc Nephrol 16:3061–3069

Schlumbohm C, Harmeyer J (2004) Dietary additions of lactose, casein and soy protein exerted only moderate effects on calcium homeostasis in calcitriol deficient piglets. J Steroid Biochem Mol Biol 89–90:605–609

Schmitz C, Perraud AL, Johnson CO, Inabe K, Smith MK, Penner R, Kurosaki T, Fleig A, Scharenberg AM (2003) Regulation of vertebrate cellular Mg^{2+} homeostasis by TRPM7. Cell 114:191–200

Schmitz C, Dorovkov MV, Zhao X, Davenport BJ, Ryazanov AG, Perraud AL (2005) The channel kinases TRPM6 and TRPM7 are functionally nonredundant. J Biol Chem 280:37763–37771

Schneeberger EE, Lynch RD (2004) The tight junction: a multifunctional complex. Am J Physiol Cell Physiol 286:C1213–C1228

Schuh K, Cartwright EJ, Jankevics E, Bundschu K, Liebermann J, Williams JC, Armesilla AL, Emerson M, Oceandy D, Knobeloch KP, Neyses L (2004) Plasma membrane Ca^{2+} ATPase 4 is required for sperm motility and male fertility. J Biol Chem 279:28220–28226

Schultheis PJ, Lorenz JN, Meneton P, Nieman ML, Riddle TM, Flagella M, Duffy JJ, Doetschman T, Miller ML, Shull GE (1998) Phenotype resembling Gitelman's syndrome in mice lacking the apical Na^+-Cl– cotransporter of the distal convoluted tubule. J Biol Chem 273:29150–29155

Schulze DH, Polumuri SK, Gille T, Ruknudin A (2002) Functional regulation of alternatively spliced Na^+/Ca^{2+} exchanger (NCX1) isoforms. Ann NY Acad Sci 976:187–196

Schweigel M, Martens H (2000) Magnesium transport in the gastrointestinal tract. Front Biosci 5:D666–D677

Schweigel M, Park HS, Etschmann B, Martens H (2006) Characterization of the Na^+-dependent Mg^{2+} transport in sheep ruminal epithelial cells. Am J Physiol Gastrointest Liver Physiol 290:G56–65

Scott K, Sun Y, Beckingham K, Zuker CS (1997) Calmodulin regulation of Drosophila light-activated channels and receptor function mediates termination of the light response in vivo. Cell 91:375–383

Scott RO, Thelin WR, Milgram SL (2002) A novel PDZ protein regulates the activity of guanylyl cyclase C, the heat-stable enterotoxin receptor. J Biol Chem 277:22934–22941

Seitz H, Jaworski ZF (1964) Effect of hydrochlorothiazide on serum and urinary calcium and urinary citrate. Can Med Assoc J 90:414–420

Servais L, Bearzatto B, Schwaller B, Dumont M, De Saedeleer C, Dan B, Barski JJ, Schiffmann SN, Cheron G (2005) Mono- and dual-frequency fast cerebellar oscillation in mice lacking parvalbumin and/or calbindin D28K. Eur J Neurosci 22:861–870

Shahsavarani A, McNeill B, Galvez F, Wood CM, Goss GG, Hwang PP, Perry SF (2006) Characterization of a branchial epithelial calcium channel (ECaC) in freshwater rainbow trout (Oncorhynchus mykiss). J Exp Biol 209:1928–1943

Shapiro RJ, Yong CK, Quamme GA (1987) Influence of chronic dietary acid on renal tubular handling of magnesium. Pflugers Arch 409:492–498

Shareghi GR, Agus ZS (1982) Magnesium transport in the cortical thick ascending limb of Henle's loop of the rabbit. J Clin Invest 69:759–769

Shenolikar S, Voltz JW, Minkoff CM, Wade JB, Weinman EJ (2002) Targeted disruption of the mouse NHERF-1 gene promotes internalization of proximal tubule sodium-phosphate cotransporter type IIa and renal phosphate wasting. Proc Natl Acad Sci USA 99:11470–11475

Shi J, Mori E, Mori Y, Mori M, Li J, Ito Y, Inoue R (2004) Multiple regulation by calcium of murine homologues of transient receptor potential proteins TRPC6 and TRPC7 expressed in HEK293 cells. J Physiol 561:415–432

Simon DB, Nelson-Williams C, Bia MJ, Ellison D, Karet FE, Molina AM, Vaara I, Iwata F, Cushner HM, Koolen M, Gainza FJ, Gitleman HJ, Lifton RP (1996) Gitelman's variant of Bartter's syndrome, inherited hypokalaemic alkalosis, is caused by mutations in the thiazide-sensitive Na-Cl cotransporter. Nat Genet 12:24–30

Simon DB, Lu Y, Choate KA, Velazquez H, Al-Sabban E, Praga M, Casari G, Bettinelli A, Colussi G, Rodriguez-Soriano J, McCredie D, Milford D, Sanjad S, Lifton RP (1999) Paracellin-1, a renal tight junction protein required for paracellular Mg^{2+} resorption. Science 285:103–106

Singh BB, Liu X, Tang J, Zhu MX, Ambudkar IS (2002) Calmodulin regulates Ca^{2+}-dependent feedback inhibition of store-operated Ca^{2+} influx by interaction with a site in the C terminus of TrpC1. Mol Cell 9:739–750

Skyberg D, Stromme JH, Nesbakken R, Harnaes K (1967) Congenital primary hypomagnesemia, an inborn error of metabolism? Acta Paediatr Scand Suppl 177:26–27

Smith EP, Boyd J, Frank GR, Takahashi H, Cohen RM, Specker B, Williams TC, Lubahn DB, Korach KS (1994) Estrogen resistance caused by a mutation in the estrogen-receptor gene in a man. N Engl J Med 331:1056–1061

Smythe E (2002) Direct interactions between rab GTPases and cargo. Mol Cell 9:205–206

Soleimani M (2002) Na^+:HCO3– cotransporters (NBC): expression and regulation in the kidney. J Nephrol 15(Suppl 5):S32–S40

Soleimani M, Aronson PS (1989) Effects of acetazolamide on Na^+-HCO3– cotransport in basolateral membrane vesicles isolated from rabbit renal cortex. J Clin Invest 83:945–951

Songyang Z, Fanning AS, Fu C, Xu J, Marfatia SM, Chishti AH, Crompton A, Chan AC, Anderson JM, Cantley LC (1997) Recognition of unique carboxyl-terminal motifs by distinct PDZ domains. Science 275:73–77

Sooy K, Schermerhorn T, Noda M, Surana M, Rhoten WB, Meyer M, Fleischer N, Sharp GW, Christakos S (1999) Calbindin-D28 k controls $[Ca^{2+}]_i$ and insulin release. Evidence obtained from calbindin-D28k knockout mice and beta cell lines. J Biol Chem 274:34343–34349

Sooy K, Kohut J, Christakos S (2000) The role of calbindin and 1,25dihydroxyvitamin D3 in the kidney. Curr Opin Nephrol Hypertens 9:341–347

Stauffer TP, Hilfiker H, Carafoli E, Strehler EE (1993) Quantitative analysis of alternative splicing options of human plasma membrane calcium pump genes. J Biol Chem 268:25993–26003

Straub M, Hautmann RE (2005) Developments in stone prevention. Curr Opin Urol 15:119–126

Street VA, McKee-Johnson JW, Fonseca RC, Tempel BL, Noben-Trauth K (1998) Mutations in a plasma membrane Ca^{2+}-ATPase gene cause deafness in deafwaddler mice. Nat Genet 19:390–394

Strehler EE, Zacharias DA (2001) Role of alternative splicing in generating isoform diversity among plasma membrane calcium pumps. Physiol Rev 81:21–50

Strewler GJ (2000) The physiology of parathyroid hormone-related protein. N Engl J Med 342:177–185

Stromme JH, Nesbakken R, Normann T, Skjorten F, Skyberg D, Johannessen B (1969) Familial hypomagnesemia. Biochemical, histological and hereditary aspects studied in two brothers. Acta Paediatr Scand 58:433–444

Suki WN (1979) Calcium transport in the nephron. Am J Physiol 237:F1–6

Suki WN, Rouse D (1981) Hormonal regulation of calcium transport in thick ascending limb renal tubules. Am J Physiol 241:F171–F174

Suki WN, Rouse D, Ng RC, Kokko JP (1980) Calcium transport in the thick ascending limb of Henle. Heterogeneity of function in the medullary and cortical segments. J Clin Invest 66:1004–1009

Sun F, Hug MJ, Lewarchik CM, Yun CH, Bradbury NA, Frizzell RA (2000) E3KARP mediates the association of ezrin and protein kinase A with the cystic fibrosis transmembrane conductance regulator in airway cells. J Biol Chem 275:29539–29546

Sutton RA, Dirks JH (1975) The renal excretion of calcium: a review of micropuncture data. Can J Physiol Pharmacol 53:979–988

Sutton RA, Domrongkitchaiporn S (1993) Abnormal renal magnesium handling. Miner Electrolyte Metab 19:232–240

Sutton RA, Walker VR (1980) Responses to hydrochlorothiazide and acetazolamide in patients with calcium stones. Evidence suggesting a defect in renal tubular function. N Engl J Med 302:709–713

Sutton RA, Wong NL, Dirks JH (1979) Effects of metabolic acidosis and alkalosis on sodium and calcium transport in the dog kidney. Kidney Int 15:520–533

Svenningsson P, Chergui K, Rachleff I, Flajolet M, Zhang X, El Yacoubi M, Vaugeois JM, Nomikos GG, Greengard P (2006) Alterations in 5-HT1B receptor function by p11 in depression-like states. Science 311:77–80

Takahashi K, Kitamura K (1999) A point mutation in a plasma membrane Ca^{2+}-ATPase gene causes deafness in Wriggle Mouse Sagami. Biochem Biophys Res Commun 261:773–778

Takeyama K, Kitanaka S, Sato T, Kobori M, Yanagisawa J, Kato S (1997) 25-Hydroxyvitamin D3 1alpha-hydroxylase and vitamin D synthesis. Science 277:1827–1830

Takezawa R, Schmitz C, Demeuse P, Scharenberg AM, Penner R, Fleig A (2004) Receptor-mediated regulation of the TRPM7 channel through its endogenous protein kinase domain. Proc Natl Acad Sci USA 101:6009–6014

Tamarappoo BK, Yang B, Verkman AS (1999) Misfolding of mutant aquaporin-2 water channels in nephrogenic diabetes insipidus. J Biol Chem 274:34825–34831

Tan CM, Nickols HH, Limbird LE (2003) Appropriate polarization following pharmacological rescue of V2 vasopressin receptors encoded by X-linked nephrogenic diabetes insipidus alleles involves a conformation of the receptor that also attains mature glycosylation. J Biol Chem 278:35678–35686

Tang J, Lin Y, Zhang Z, Tikunova S, Birnbaumer L, Zhu MX (2001) Identification of common binding sites for calmodulin and inositol 1,4,5-trisphosphate receptors on the carboxyl termini of trp channels. J Biol Chem 276:21303–21310

Tang VW, Goodenough DA (2003) Paracellular ion channel at the tight junction. Biophys J 84:1660–1673

Tang Y, Tang J, Chen Z, Trost C, Flockerzi V, Li M, Ramesh V, Zhu MX (2000) Association of mammalian trp4 and phospholipase C isozymes with a PDZ domain-containing protein, NHERF. J Biol Chem 275:37559–37564

Teitelbaum SL (2000) Bone resorption by osteoclasts. Science 289:1504–1508

Ten Bolscher M, Netelenbos JC, Barto R, Van Buuren LM, Van der vijgh WJ (1999) Estrogen regulation of intestinal calcium absorption in the intact and ovariectomized adult rat. J Bone Miner Res 14:1197–1202

Tohyama O, Imura A, Iwano A, Freund JN, Henrissat B, Fujimori T, Nabeshima Y (2004) Klotho is a novel beta-glucuronidase capable of hydrolyzing steroid beta-glucuronides. J Biol Chem 279:9777–9784

Torralbo A, Pina E, Portoles J, Sanchez-Fructuoso A, Barrientos A (1995) Renal magnesium wasting with hypercalciuria, nephrocalcinosis and ocular disorders. Nephron 69:472–475

Trombetta ES, Simons JF, Helenius A (1996) Endoplasmic reticulum glucosidase II is composed of a catalytic subunit, conserved from yeast to mammals, and a tightly bound noncatalytic HDEL-containing subunit. J Biol Chem 271:27509–27516

Trombetta ES, Fleming KG, Helenius A (2001) Quaternary and domain structure of glycoprotein processing glucosidase II. Biochemistry 40:10717–10722

Trost C, Bergs C, Himmerkus N, Flockerzi V (2001) The transient receptor potential, TRP4, cation channel is a novel member of the family of calmodulin binding proteins. Biochem J 355:663–670

Tsujikawa H, Kurotaki Y, Fujimori T, Fukuda K, Nabeshima Y (2003) Klotho, a gene related to a syndrome resembling human premature aging, functions in a negative regulatory circuit of vitamin D endocrine system. Mol Endocrinol 17:2393–2403

Tsukamoto Y, Saka S, Saitoh M (1992) Parathyroid hormone stimulates ATP-dependent calcium pump activity by a different mode in proximal and distal tubules of the rat. Biochim Biophys Acta 1103:163–171

Tsukita S, Furuse M (2000) Pores in the wall: claudins constitute tight junction strands containing aqueous pores. J Cell Biol 149:13–16

Tucker KL, Hannan MT, Chen H, Cupples LA, Wilson PW, Kiel DP (1999) Potassium, magnesium, and fruit and vegetable intakes are associated with greater bone mineral density in elderly men and women. Am J Clin Nutr 69:727–736

Ulloa-Aguirre A, Janovick JA, Brothers SP, Conn PM (2004) Pharmacologic rescue of conformationally-defective proteins: implications for the treatment of human disease. Traffic 5:821–837

Ullrich KJ, Schmidt-Nielson B, O'Dell R, Pehling G, Gottschalk CW, Lassiter WE, Mylle M (1963) Micropuncture study of composition of proximal and distal tubular fluid in rat kidney. Am J Physiol 204:527–531

Vainsel M, Vandevelde G, Smulders J, Vosters M, Hubain P, Loeb H (1970) Tetany due to hypomagnesaemia with secondary hypocalcaemia. Arch Dis Child 45:254–258

Van Abel M, Hoenderop JG, Dardenne O, St Arnaud R, Van Os CH, Van Leeuwen HJ, Bindels RJ (2002) 1,25-dihydroxyvitamin D3-independent stimulatory effect of estrogen on the expression of ECaC1 in the kidney. J Am Soc Nephrol 13:2102–2109

Van Abel M, Hoenderop JG, van der Kemp AW, van Leeuwen JP, Bindels RJ (2003) Regulation of the epithelial Ca^{2+} channels in small intestine as studied by quantitative mRNA detection. Am J Physiol 285:G78–85

Van Abel M, Hoenderop JG, van der Kemp AW, Friedlaender MM, van Leeuwen JP, Bindels RJ (2005) Coordinated control of renal Ca^{2+} transport proteins by parathyroid hormone. Kidney Int 68:1708–1721

Van Abel M, Huybers S, Hoenderop JG, van der Kemp AW, van Leeuwen JP, Bindels RJ (2006) Age-dependent alterations in Ca^{2+} homeostasis: role of TRPV5 and TRPV6. Am J Physiol Renal Physiol 291:F1177–F1183

Van Baal J, Raber G, de Slegte J, Pieters R, Bindels RJ, Willems PH (1996a) Vasopressin-stimulated Ca^{2+} reabsorption in rabbit cortical collecting system: effects on cAMP and cytosolic Ca^{2+}. Pflugers Arch 433:109–115

Van Baal J, Yu A, Hartog A, Fransen JA, Willems PH, Lytton J, Bindels RJ (1996b) Localization and regulation by vitamin D of calcium transport proteins in rabbit cortical collecting system. Am J Physiol 271:F985–F993

Van Baal J, Hoenderop JG, Groenendijk M, van Os CH, Bindels RJ, Willems PH (1999) Hormone-stimulated Ca^{2+} transport in rabbit kidney: multiple sites of inhibition by exogenous ATP. Am J Physiol 277:F899–906

Van Cromphaut SJ, Dewerchin M, Hoenderop JG, Stockmans I, Van Herck E, Kato S, Bindels RJ, Collen D, Carmeliet P, Bouillon R, Carmeliet G (2001) Duodenal calcium absorption in vitamin D receptor-knockout mice: functional and molecular aspects. Proc Natl Acad Sci USA 98:13324–13329

Van Cromphaut SJ, Rummens K, Stockmans I, Van Herck E, Dijcks FA, Ederveen AG, Carmeliet P, Verhaeghe J, Bouillon R, Carmeliet G (2003) Intestinal calcium transporter genes are upregulated by estrogens and the reproductive cycle through vitamin D receptor-independent mechanisms. J Bone Miner Res 18:1725–1736

Van de Graaf SF, Hoenderop JG, Gkika D, Lamers D, Prenen J, Rescher U, Gerke V, Staub O, Nilius B, Bindels RJ (2003) Functional expression of the epithelial Ca^{2+} channels (TRPV5 and TRPV6) requires association of the S100A10-annexin 2 complex. EMBO J 22:1478–1487

Van de Graaf SF, Bouillart I, Hoenderop JG, Bindels RJ (2004) Regulation of the epithelial Ca^{2+} channels TRPV5 and TRPV6 by 1,25-dihydroxyvitamin D3 and dietary Ca^{2+}. J Steroid Biochem Mol Biol 89–90:303–308

Van de Graaf SF, Chang Q, Mensenkamp AR, Hoenderop JG, Bindels RJ (2006a) Direct interaction with Rab11a targets the epithelial Ca^{2+} channels TRPV5 and TRPV6 towards the plasma membrane. Mol Cell Biol 26:303–312

Van de Graaf SF, Hoenderop JG, Bindels RJ (2006b) Regulation of TRPV5 and TRPV6 by associated proteins. Am J Physiol Renal Physiol 290:F1295–F1302

Van de Graaf SF, Hoenderop JG, van der Kemp AW, Gisler SM, Bindels RJ (2006c) Interaction of the epithelial Ca^{2+} channels TRPV5 and TRPV6 with the intestine- and kidney-enriched PDZ protein NHERF4. Pflugers Arch 452:407–417

Van de Graaf SF, van der Kemp AW, van den Berg D, van Oorschot M, Hoenderop JG, Bindels RJ (2006d) Identification of BSPRY as a novel auxiliary protein inhibiting TRPV5 activity. J Am Soc Nephrol 17:26–30

Van der Eerden BC, Hoenderop JG, de Vries TJ, Schoenmaker T, Buurman CJ, Uitterlinden AG, Pols HA, Bindels RJ, van Leeuwen JP (2005) The epithelial Ca^{2+} channel TRPV5 is essential for proper osteoclastic bone resorption. Proc Natl Acad Sci USA 102:17507–17512

van IJzendoorn SC, Tuvim MJ, Weimbs T, Dickey BF, Mostov KE (2002) Direct interaction between Rab3b and the polymeric immunoglobulin receptor controls ligand-stimulated transcytosis in epithelial cells. Dev Cell 2:219–228

Van Itallie C, Rahner C, Anderson JM (2001) Regulated expression of claudin-4 decreases paracellular conductance through a selective decrease in sodium permeability. J Clin Invest 107:1319–1327

Van Itallie CM, Fanning AS, Anderson JM (2003) Reversal of charge selectivity in cation or anion-selective epithelial lines by expression of different claudins. Am J Physiol Renal Physiol 285:F1078–F1084

Vasioukhin V, Bauer C, Yin M, Fuchs E (2000) Directed actin polymerization is the driving force for epithelial cell-cell adhesion. Cell 100:209–219

Vassilev PM, Peng JB, Hediger MA, Brown EM (2001) Single-channel activities of the human epithelial Ca^{2+} transport proteins CaT1 and CaT2. J Membr Biol 184:113–120

Vennekens R, Hoenderop JG, Prenen J, Stuiver M, Willems PH, Droogmans G, Nilius B, Bindels RJ (2000) Permeation and gating properties of the novel epithelial Ca^{2+} channel. J Biol Chem 275:3963–3969

Vennekens R, Droogmans G, Nilius B (2001a) Functional properties of the epithelial Ca^{2+} channel, ECaC. Gen Physiol Biophys 20:239–253

Vennekens R, Prenen J, Hoenderop JG, Bindels RJ, Droogmans G, Nilius B (2001b) Modulation of the epithelial Ca^{2+} channel ECaC by extracellular pH. Pflugers Arch 442:237–242

Vidal A, Sun Y, Bhattacharya SK, Ahokas RA, Gerling IC, Weber KT (2006) Calcium paradox of aldosteronism and the role of the parathyroid glands. Am J Physiol Heart Circ Physiol 290:H286–H294

Voets T, Janssens A, Prenen J, Droogmans D, Nilius G (2003) Mg^{2+}-dependent gating and strong inward rectification of the cation channel TRPV6. J Gen Physiol 121:245–260

Voets T, Janssens A, Droogmans G, Nilius B (2004a) Outer pore architecture of a Ca^{2+}-selective TRP channel. J Biol Chem 279:15223–15230

Voets T, Nilius B, van der Kemp AW, Droogmans G, Bindels RJ, Hoenderop JG (2004b) TRPM6 forms the Mg^{2+} influx channel involved in intestinal and renal Mg^{2+} absorption. J Biol Chem 279:19–25

Wabakken T, Rian E, Kveine M, Aasheim HC (2003) The human solute carrier SLC41A1 belongs to a novel eukaryotic subfamily with homology to prokaryotic MgtE Mg^{2+} transporters. Biochem Biophys Res Commun 306:718–724

Walder RY, Shalev H, Brennan TM, Carmi R, Elbedour K, Scott DA, Hanauer A, Mark AL, Patil S, Stone EM, Sheffield VC (1997) Familial hypomagnesemia maps to chromosome 9q, not to the X chromosome: genetic linkage mapping and analysis of a balanced translocation breakpoint. Hum Mol Genet 6:1491–1497

Walder RY, Landau D, Meyer P, Shalev H, Tsolia M, Borochowitz Z, Boettger MB, Beck GE, Englehardt RK, Carmi R, Sheffield VC (2002) Mutation of TRPM6 causes familial hypomagnesemia with secondary hypocalcemia. Nat Genet 31:171–174

Wallemacq PE, Reding R (1993) FK506 (tacrolimus), a novel immunosuppressant in organ transplantation: clinical, biomedical, and analytical aspects. Clin Chem 39:2219–2228

Walters JR, Balesaria S, Chavele KM, Taylor V, Berry JL, Khair U, Barley NF, van Heel DA, Field J, Hayat JO, Bhattacharjee A, Jeffery R, Poulsom R (2006) Calcium channel TRPV6 expression in human duodenum: different relationships to the vitamin D system and aging in men and women. J Bone Miner Res Jul 25 [Epub ahead of print]

Wang CL (1985) A note on Ca^{2+} binding to calmodulin. Biochem Biophys Res Commun 130:426–430

Wang CY, Shi JD, Yang P, Kumar PG, Li QZ, Run QG, Su YC, Scott HS, Kao KJ, She JX (2003) Molecular cloning and characterization of a novel gene family of four ancient conserved domain proteins (ACDP). Gene 306:37–44

Wang CY, Yang P, Shi JD, Purohit S, Guo D, An H, Gu JG, Ling J, Dong Z, She JX (2004a) Molecular cloning and characterization of the mouse Acdp gene family. BMC Genomics 5:7

Wang X, Kumar R, Navarre J, Casanova JE, Goldenring JR (2000) Regulation of vesicle trafficking in Madin-Darby canine kidney cells by Rab11a and Rab25. J Biol Chem 275:29138–29146

Wang Y, Zhang J, Yi XJ, Yu FS (2004b) Activation of ERK1/2 MAP kinase pathway induces tight junction disruption in human corneal epithelial cells. Exp Eye Res 78:125–136

Wasserman RH, Fullmer CS (1995) Vitamin D and intestinal calcium transport: facts, speculations and hypotheses. J Nutr 125:1971S–1979S

Weber K, Erben RG, Rump A, Adamski J (2001a) Gene structure and regulation of the murine epithelial calcium channels ECaC1 and 2. Biochem Biophys Res Commun 289:1287–1294

Weber S, Hoffmann K, Jeck N, Saar K, Boeswald M, Kuwertz-Broeking E, Meij II, Knoers NV, Cochat P, Sulakova T, Bonzel KE, Soergel M, Manz F, Schaerer K, Seyberth HW, Reis A, Konrad M (2000) Familial hypomagnesaemia with hypercalciuria and nephrocalcinosis maps to chromosome 3q27 and is associated with mutations in the PCLN-1 gene. Eur J Hum Genet 8:414–422

Weber S, Schneider L, Peters M, Misselwitz J, Ronnefarth G, Boswald M, Bonzel KE, Seeman T, Sulakova T, Kuwertz-Broking E, Gregoric A, Palcoux JB, Tasic V, Manz F, Scharer K, Seyberth HW, Konrad M (2001b) Novel paracellin-1 mutations in 25 families with familial hypomagnesemia with hypercalciuria and nephrocalcinosis. J Am Soc Nephrol 12:1872–1881

Weinman EJ, Steplock D, Wang Y, Shenolikar S (1995) Characterization of a protein cofactor that mediates protein kinase A regulation of the renal brush border membrane Na^+-H^+ exchanger. J Clin Invest 95:2143–2149

Weinman EJ, Hall RA, Friedman PA, Liu-Chen LY, Shenolikar S (2006a) The association of NHERF adaptor proteins with g protein-coupled receptors and receptor tyrosine kinases. Annu Rev Physiol 68:491–505

Weinman EJ, Mohanlal V, Stoycheff N, Wang F, Steplock D, Shenolikar S, Cunningham R (2006b) Longitudinal study of urinary excretion of phosphate, calcium, and uric acid in mutant NHERF-1 null mice. Am J Physiol Renal Physiol 290:F838–F843

White KE, Gesek FA, Reilly RF, Friedman PA (1998) NCX1 Na/Ca exchanger inhibition by antisense oligonucleotides in mouse distal convoluted tubule cells. Kidney Int 54:897–906

Wilcox ER, Burton QL, Naz S, Riazuddin S, Smith TN, Ploplis B, Belyantseva I, Ben-Yosef T, Liburd NA, Morell RJ, Kachar B, Wu DK, Griffith AJ, Friedman TB (2001) Mutations in the gene encoding tight junction claudin-14 cause autosomal recessive deafness DFNB29. Cell 104:165–172

Wilson FH, Disse-Nicodeme S, Choate KA, Ishikawa K, Nelson-Williams C, Desitter I, Gunel M, Milford DV, Lipkin GW, Achard JM, Feely MP, Dussol B, Berland Y, Unwin RJ, Mayan H, Simon DB, Farfel Z, Jeunemaitre X, Lifton RP (2001) Human hypertension caused by mutations in WNK kinases. Science 293:1107–1112

Wilson FH, Hariri A, Farhi A, Zhao H, Petersen KF, Toka HR, Nelson-Williams C, Raja KM, Kashgarian M, Shulman GI, Scheinman SJ, Lifton RP (2004) A cluster of metabolic defects caused by mutation in a mitochondrial tRNA. Science 306:1190–1194

Wilson IM, Freis ED (1959) Relationship between plasma and extracellular fluid volume depletion and the antihypertensive effect of chlorothiazide. Circulation 20:1028–1036

Wissenbach U, Niemeyer BA, Fixemer T, Schneidewind A, Trost C, Cavalie A, Reus K, Meese E, Bonkhoff H, Flockerzi V (2001) Expression of CaT-like, a novel calcium-selective channel, correlates with the malignancy of prostate cancer. J Biol Chem 276:19461–19468

Wittner M, Di Stefano A, Mandon B, Roinel N, de Rouffignac C (1991) Stimulation of NaCl reabsorption by antidiuretic hormone in the cortical thick ascending limb of Henle's loop of the mouse. Pflugers Arch 419:212–214

Wittner M, Mandon B, Roinel N, de Rouffignac C, Di Stefano A (1993) Hormonal stimulation of Ca^{2+} and Mg^{2+} transport in the cortical thick ascending limb of Henle's loop of the mouse: evidence for a change in the paracellular pathway permeability. Pflugers Arch 423:387–396

Wolf FI, Torsello A, Fasanella S, Cittadini A (2003) Cell physiology of magnesium. Mol Aspects Med 24:11–26

Wong NL, Quamme GA, Dirks JH (1986) Effects of acid-base disturbances on renal handling of magnesium in the dog. Clin Sci (Lond) 70:277–284

Wong V, Goodenough DA (1999) Paracellular channels! Science 285:62

Wood RJ, Tchack L, Taparia S (2001) 1,25-Dihydroxyvitamin D3 increases the expression of the CaT1 epithelial calcium channel in the Caco-2 human intestinal cell line. BMC Physiol 1:11

Woodhull AM (1973) Ionic blockage of sodium channels in nerve. J Gen Physiol 61:687–708

Xue Y, Karaplis AC, Hendy GN, Goltzman D, Miao D (2006) Exogenous 1,25-dihydroxyvitamin D3 exerts a skeletal anabolic effect and improves mineral ion homeostasis in mice which are homozygous for both the 1alpha-hydroxylase and parathyroid hormone null alleles. Endocrinology 147:4801–4810

Yaffe MB, Rittinger K, Volinia S, Caron PR, Aitken A, Leffers H, Gamblin SJ, Smerdon SJ, Cantley LC (1997) The structural basis for 14-3-3:phosphopeptide binding specificity. Cell 91:961–971

Yamauchi K, Rai T, Kobayashi K, Sohara E, Suzuki T, Itoh T, Suda S, Hayama A, Sasaki S, Uchida S (2004) Disease-causing mutant WNK4 increases paracellular chloride permeability and phosphorylates claudins. Proc Natl Acad Sci USA 101:4690–4694

Yang T, Hassan S, Huang YG, Smart AM, Briggs JP, Schnermann JB (1997) Expression of PTHrP, PTH/PTHrP receptor, and Ca^{2+}-sensing receptor mRNAs along the rat nephron. Am J Physiol 272:F751–F758

Yang W, Lee HW, Hellinga H, Yang JJ (2002) Structural analysis, identification, and design of calcium-binding sites in proteins. Proteins 47:344–356

Yeh BI, Sun TJ, Lee JZ, Chen HH, Huang CL (2003) Mechanism and molecular determinant for regulation of rabbit transient receptor potential type 5 (TRPV5) channel by extracellular pH. J Biol Chem 278:51044–51052

Yeh BI, Kim YK, Jabbar W, Huang CL (2005) Conformational changes of pore helix coupled to gating of TRPV5 by protons. EMBO J 24:3224–3234

Yildirim E, Dietrich A, Birnbaumer L (2003) The mouse C-type transient receptor potential 2 (TRPC2) channel: alternative splicing and calmodulin binding to its N terminus. Proc Natl Acad Sci USA 100:2220–2225

Yoshida T, Fujimori T, Nabeshima Y (2002) Mediation of unusually high concentrations of 1,25-dihydroxyvitamin D in homozygous klotho mutant mice by increased expression of renal 1alpha-hydroxylase gene. Endocrinology 143:683–689

Yoshizawa T, Handa Y, Uematsu Y, Takeda S, Sekine K, Yoshihara Y, Kawakami T, Arioka K, Sato H, Uchiyama Y, Masushige S, Fukamizu A, Matsumoto T, Kato S (1997) Mice lacking the vitamin D receptor exhibit impaired bone formation, uterine hypoplasia and growth retardation after weaning. Nat Genet 16:391–396

Young MM, Nordin BE (1967) Calcium metabolism and the menopause. Proc R Soc Med 60:1137–1138

Young MM, Jasani C, Smith DA, Nordin BE (1968) Some effects of ethinyl oestradiol on calcium and phosphorus metabolism in osteoporosis. Clin Sci 34:411–417

Yue L, Peng JB, Hediger MA, Clapham DE (2001) CaT1 manifests the pore properties of the calcium-release-activated calcium channel. Nature 410:705–709

Yun CC, Palmada M, Embark HM, Fedorenko O, Feng Y, Henke G, Setiawan I, Boehmer C, Weinman EJ, Sandrasagra S, Korbmacher C, Cohen P, Pearce D, Lang F (2002) The serum and glucocorticoid-inducible kinase SGK1 and the Na^+/H^+ exchange regulating factor NHERF2 synergize to stimulate the renal outer medullary K^+ channel ROMK1. J Am Soc Nephrol 13:2823–2830

Yun CH, Oh S, Zizak M, Steplock D, Tsao S, Tse CM, Weinman EJ, Donowitz M (1997) cAMP-mediated inhibition of the epithelial brush border Na^+/H^+ exchanger, NHE3, requires an associated regulatory protein. Proc Natl Acad Sci USA 94:3010–3015

Zerial M, McBride H (2001) Rab proteins as membrane organizers. Nat Rev Mol Cell Biol 2:107–117

Zhang Z, Tang J, Tikunova S, Johnson JD, Chen Z, Qin N, Dietrich A, Stefani E, Birnbaumer L, Zhu MX (2001) Activation of Trp3 by inositol 1,4,5-trisphosphate receptors through displacement of inhibitory calmodulin from a common binding domain. Proc Natl Acad Sci USA 98:3168–3173

Zheng W, Xie Y, Li G, Kong J, Feng JQ, Li YC (2004) Critical role of calbindin-D28k in calcium homeostasis revealed by mice lacking both vitamin D receptor and calbindin-D28k. J Biol Chem 279:52406–52413

Zhuang L, Peng JB, Tou L, Takanaga H, Adam RM, Hediger MA, Freeman MR (2002) Calcium-selective ion channel, CaT1, is apically localized in gastrointestinal tract epithelia and is aberrantly expressed in human malignancies. Lab Invest 82:1755–1764

Zizak M, Lamprecht G, Steplock D, Tariq N, Shenolikar S, Donowitz M, Yun CH, Weinman EJ (1999) cAMP-induced phosphorylation and inhibition of Na^+/H^+ exchanger 3 (NHE3) are dependent on the presence but not the phosphorylation of NHE regulatory factor. J Biol Chem 274:24753–24758

Zolotnitskaya A, Satlin LM (1999) Developmental expression of ROMK in rat kidney. Am J Physiol 276:F825–F836

Zsurka G, Gregan J, Schweyen RJ (2001) The human mitochondrial Mrs2 protein functionally substitutes for its yeast homologue, a candidate magnesium transporter. Genomics 72:158–168

MIX
Papier aus verantwortungsvollen Quellen
Paper from responsible sources
FSC® C105338

If you have any concerns about our products,
you can contact us on
ProductSafety@springernature.com

In case Publisher is established outside the EU,
the EU authorized representative is:
Springer Nature Customer Service Center GmbH
Europaplatz 3, 69115 Heidelberg, Germany

Printed by Libri Plureos GmbH
in Hamburg, Germany